CARNIVORES
OF THE WORLD

CARNIVORES
OF THE WORLD

Luke Hunter
Illustrated by Priscilla Barrett

PANTHERA
LEADERS IN WILD CAT CONSERVATION

PRINCETON UNIVERSITY PRESS
PRINCETON AND OXFORD

For Sophie

First published in the United States and Canada
by Princeton University Press,
41 William Street, Princeton, New Jersey 08540
nathist.press.princeton.edu

First published in the United Kingdom in 2011 by
New Holland Publishers (UK) Ltd
www.newhollandpublishers.com
Garfield House, 86–88 Edgware Road, London W2 2EA, United Kingdom
80 McKenzie Street, Cape Town 8001, South Africa
Unit 1, 66 Gibbes Street, Chatswood, New South Wales, Australia 2067
218 Lake Road, Northcote, Auckland, New Zealand

Colour plate illustrations (pages 17–187) and sketches
(pages 7 and 10) by Priscilla Barrett
Skull and footprint illustrations (pages 188–233)
by Sally McClarty

Library of Congress Control Number: 2011926335
Cloth ISBN: 978-0-691-15227-1
Paperback ISBN: 978-0-691-15228-8

Senior Editor: Krystyna Mayer
Designer: Alan Marshall
Index: Krystyna Mayer
Production: Melanie Dowland
Publisher: Simon Papps

This book has been composed in Adobe Garamond and Avenir
Reproduction by Modern Age Repro Co. Ltd, Hong Kong
Printed and bound in Singapore by Tien Wah Press (Pte) Ltd

1 3 5 7 9 10 8 6 4 2

CONTENTS

INTRODUCTION

This book describes all of the world's terrestrial carnivores, 245 species that are united in a shared ancestry of subsisting mainly on meat. Many other species, humans included, eat meat, but this does not make them carnivores in scientific nomenclature. That label belongs exclusively to the members of the order Carnivora which, despite remarkable variation in size and shape, all descend from a small, civet-like carnivorous ancestor that lived over 60 million years ago. Some modern carnivores eat little meat or, as in the case of the Giant Panda, none at all, but all members of the Carnivora trace their ancestry back to the same predatory origins, and retain many of the physical, behavioural and ecological adaptations common to their truly carnivorous relatives.

The Carnivora is the fifth largest mammalian order (of 29 extant orders), occurs on every large landmass including Antarctica, and inhabits every major habitat on Earth, from the hyper-arid interior of the Sahara Desert to Arctic ice sheets. The world's smallest carnivore, the tiny Least Weasel (page 182), can squeeze through a wedding ring and weighs ten thousand times less than the largest terrestrial species, the Polar Bear

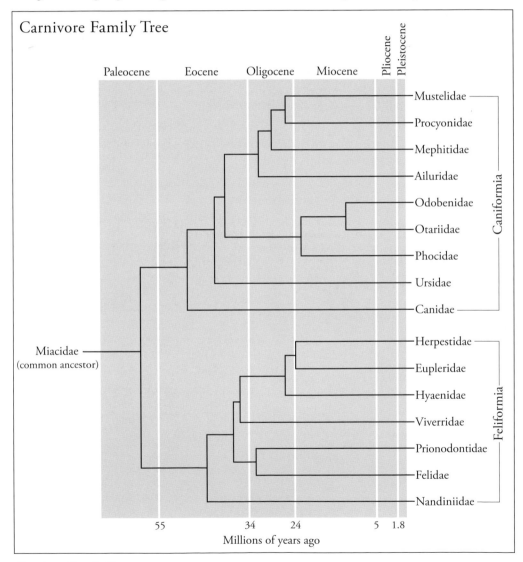

Carnivore Family Tree

Carnivore Family Tree

Phylogeny of the Carnivora, showing family level divisions and the approximate dates at which related families shared a common ancestor. All modern carnivores descend from miacids, which were small, civet-like carnivores that lived approximately 60 million years ago.

(page 132). The Carnivora includes some of the world's most iconic, magnificent and admired species – and, regrettably, some of the most endangered.

This book covers the world's 13 terrestrial carnivore families. It does not include three marine carnivore families, sea-lions (family Otariidae, 16 species), seals (family Phocidae, 19 species) and the Walrus (family Odobenidae). At times, these three families have been classified as a separate order, the Pinnipedia, though the consensus today is that they belong within the Carnivora. Although they do not appear here, the pinnipeds are covered in many excellent field guides to marine mammals.

Carnivores are separated into two major suborders, reflecting a divergence early in the order's evolution an estimated 45–50 million years ago (though given the paucity of fossil remains, possibly considerably earlier). Suborder Feliformia contains the 'cat-like' families Felidae, Hyaenidae, Herpestidae, Eupleridae, Prionodontidae, Viverridae and Nandiniidae. Suborder Caniformia comprises the 'dog-like' families Canidae, Ursidae, Procyonidae, Ailuridae, Mephitidae and Mustelidae, as well as the three pinniped families.

CARNIVORE FAMILIES: THE KEY FEATURES

Suborder FELIFORMIA, 7 families, 120 species

FAMILY FELIDAE CATS, 37 species
Size range Rusty-spotted Cat and Black-footed Cat (1–2.5kg) to Tiger (75–325kg)

The cat family arose approximately 30 million years ago in what is now Eurasia, and today occurs globally except in the Arctic, Antarctica and Australasia. Cats are divided into two major subfamilies:

- Big cats, subfamily Pantherinae (7 species), comprising the Panthera genus plus Clouded Leopards.
- Small cats, subfamily Felinae (30 species), which includes the Cheetah and Puma; despite their size, both these species are essentially outsized small cats that evolved larger bodies to fill similar niches to the large Panthera cats.

Thirty-seven cat species are currently recognized, though molecular analyses may produce further revisions. Preliminary genetic data suggest that the Chinese Mountain Cat (page 16) may be a subspecies of the Wildcat rather than a separate species, while genetics only recently revealed that Clouded Leopards on Borneo and Sumatra were sufficiently distinct from mainland animals to be reclassified as a separate species (page 40).

Cats are hypercarnivores that subsist almost entirely on animal prey, which is killed by a suffocating bite to the throat in the case of large prey, or by crushing the skull of small prey. Most cats are solitary, territorial and nocturno-crepuscular. The Lion (page 44) is the only cat that lives in large and complex social groups, though male Cheetahs (page 38) form small, enduring coalitions, and free-living domestic cats in colonies sometimes form small and stable social groups.

Domestic Cat
The domestic cat, Felis silvestris catus, descends from the Wildcat (page 16), with which it is still able to interbreed. They are considered the same species, but some authorities argue that the genetic differences selected by human-mediated breeding warrant the domestic cat being considered a separate full species, Felis catus. *Worldwide, the number of domestic cats, including feral and semi-wild populations, may now exceed a billion.*

FAMILY HYAENIDAE HYAENAS, 4 species
Size range Aardwolf (7.7–14kg) to Spotted Hyaena (49–86kg)

The hyaena family arose at least 23 million years ago in Eurasia and reached an evolutionary peak 6–12 million years ago, when as many as 24 different species existed. Despite their dog-like appearance, hyaenas belong in the Feliformia and thus are more closely related to cats and their allies, than to dogs. There are four extant species, all of which evolved in Africa, which remains the stronghold of modern hyaena distribution, with one species, the Striped Hyaena (page 50), also found in the Middle East through to India. The family is divided into two main subfamilies:

- Protelinae, a clade of relatively gracile, dog-like hyaenas with a single extant member, the Aardwolf, which diverged from the rest of the family around 10.6 million years ago.
- Hyaeninae, the bone-cracking hyaenas, which contains the other three modern species.

Hyaenas are hypercarnivores with a prodigious digestive ability that probably arose early in the family's evolution. In the Aardwolf (page 50), this capacity evolved to deal with noxious defensive terpenes secreted by termites, which are its primary prey. The bone-cracking hyaenas are capable of digesting all parts of animal prey except the hooves, the hair and the keratin sheaths of ungulate horns, and they tolerate the extremely high bacterial loads present in rotting carrion.

All hyaena species live in enduring social groups that take their simplest form in the Aardwolf's monogamous, cooperatively breeding pairs. Striped Hyaenas also live in monogamous pairs, but small groups comprising one female and multiple males have recently been documented; the full range of their sociality is still poorly known. Brown Hyaenas (page 52) form small family groups that share a territory, but otherwise spend most of their time alone. The most complex social patterns are displayed by the Spotted Hyaena (page 52), which lives in large clans with a unique matrilineal social structure. Females are larger than and dominant to males, and usually live their entire lives in the same clan; female cubs inherit their mother's rank and outrank immigrant males, even adults. This is unknown from any other carnivore, and in fact it most closely resembles the societies of primates such as baboons.

FAMILY HERPESTIDAE MONGOOSES, 34 species
Size range Common Dwarf Mongoose (210–340g) to White-tailed Mongoose (1.8–5.2kg)

Mongooses were formerly classified within the Viverridae, but are now recognized in their own family, the Herpestidae. Within the Feliformia, the family is most closely related to the Eupleridae (which emerged as a Herpestidae offshoot) and the Hyaenidae. The mongoose family is subdivided into two large subfamilies:

- Herpestinae (solitary mongooses, 23 species).
- Mungotinae (social mongooses, 11 species).

This major division is thought to reflect a divergence early in mongoose evolution, in which the opening up of forested habitats favoured sociality in an ancestral group-living species that ultimately gave rise to the modern social species.

Mongooses occur in Africa, the Middle East and South Asia, with one species, the Egyptian Mongoose (page 58), present in Portugal and Spain. Members of this family are primarily carnivorous and eat mainly small vertebrates and invertebrates; fruit and vegetable matter is eaten to a limited degree by some species. Reflecting the subfamily classification, mongooses are either largely solitary (though some Herpestinae have semi-social tendencies, for example denning together) or live in complex social groups. Social patterns are best understood in dwarf mongooses (page 66), Banded Mongoose (page 68) and Meerkat (page 64); the rest of the Mungotinae are believed to be similarly social, though they are not nearly as well studied.

FAMILY EUPLERIDAE EUPLERIDS, 9 species
Size range Narrow-striped Boky (450–740g) to Fosa (6.2–8.6kg)

Members of the Eupleridae have historically been classified as cats, mongooses or civets, but it is now known that the family arose from a single mongoose-like ancestor that colonized Madagascar from mainland Africa an estimated 16.5–24 million years ago. Subsequent rapid adaptive radiation on Madagascar led to the nine living species, now grouped in two subfamilies:

- Euplerinae ('civet-like' species, Fanaloka, Fosa and Falanouc).
- Galidiinae ('mongoose-like' species, vontsiras and Boky).

It is possible that more species remain undiscovered. Durrell's Vontsira *Salanoia durrelli* was first formally described in late 2010 as this book was going to press. Genetic analysis shows it to be very closely related to the Brown-tailed Vontsira (page 74) but distinct in having a more robust skull and dentition, pale reddish-brown fur, and broad, well-developed pads on the feet with a fringe of stiff hair along the outer margins (possibly to assist with living on floating vegetation mats in marshland, the only habitat from which it is currently known). Also in 2010, the Falanouc (page 72) was tentatively divided into two species on the basis of differences in external and craniodental morphology: the Eastern Falanouc *Eupleres goudotii* and the larger Western or Giant Falanouc *E. major*. With only five known specimens of Giant Falanouc and no genetic analysis, the classification remains provisional.

Euplerids eat mostly animal prey; their feeding habits range from almost exclusive insectivory in the Falanouc to the mammal-dominated diet of the Fosa (page 72). Most species are thought to be chiefly solitary, though both temporary and enduring sociality has been observed in five species. The majority of euplerids remain largely unstudied.

FAMILY PRIONODONTIDAE LINSANGS, 2 species
Size range Banded Linsang (590–800g) to Spotted Linsang (550g–1.2kg)

The Prionodontidae is an ancient carnivoran family originally classified among the Viverridae and once thought to be most closely related to African oyans (page 94; formerly also called linsangs), which are morphologically and ecologically very similar. In fact, recent molecular analysis reveals that linsangs represent an early sister group to the Felidae, with a shared common ancestor around 42 million years ago. Linsangs are only distantly related to oyans (family Viverridae), a remarkable case of evolutionary convergence.

Linsangs are restricted to South-east Asia, where they inhabit evergreen and moist forested habitats. They are highly arboreal and hypercarnivorous nocturnal hunters of small prey. They are solitary, but little detail is known of their social and spatial organization.

FAMILY VIVERRIDAE GENETS, OYANS AND CIVETS, 33 species
Size range Leighton's Oyan (500–700g) to Binturong (9–20kg)

The Viverridae is an ancient lineage of the Feliformia thought to have arisen at least 34 million years ago in Eurasia, followed by later colonization of Africa. It is subdivided into four subfamilies:

- Viverinnae (large terrestrial civets, 6 species).
- Genettinae (genets and oyans, 16 species).
- Paradoxurinae (palm civets and Binturong, 7 species).
- Hemigalinae (Otter Civet and allies, 4 species).

The species limits within the Viverridae are mostly well defined, although the Critically Endangered Malabar Civet (page 84) is possibly the same species as the Large-spotted Civet (page 84), and the classification of genets is controversial, with as many as 17 species proposed in the genus *Genetta* (14 are recognized in this book).

Viverrids are restricted to Africa and South Asia; one African species, the Small-spotted Genet (page 90), also occurs in Europe, though this is possibly as a result of human introduction. They are largely solitary and nocturnal. Many species are semi to highly arboreal and have protractile claws, as in felids. Viverrids are primarily carnivorous, with a diet dominated by small vertebrates and invertebrates, or in the case of the Paradoxurinae, largely frugivorous.

FAMILY NANDINIIDAE AFRICAN PALM CIVET, 1 species
Size African Palm Civet (1.2–3kg)

The African Palm Civet (page 94) is a primitive species that retains some unique ancestral features (mainly in the structure of the skull and carnassials) that no longer occur in any other modern carnivore. It was formerly classified in the Viverridae as an African member of the Paradoxurinae, giving rise to its erroneous common name. Molecular data confirm that it represents an ancient sister species to all other feliform carnivores; it shared a common ancestor with all other Feliformia an estimated 36–54 million years ago.

Morphologically and ecologically, the African Palm Civet probably closely resembles the earliest feliform carnivores. Endemic to equatorial Africa, it inhabits forest and woodland savannahs. It is mainly frugivorous, and also takes vertebrate and invertebrate prey. It is arboreal, nocturnal and mainly solitary, with defined and defended territories.

Suborder CANIFORMIA, 9 families, 161 species (including pinnipeds, 36 species)

FAMILY CANIDAE DOGS, 35 species
Size range Fennec Fox and Blanford's Fox (0.8–1.9kg) to Grey Wolf (18–79.4kg)

The Canidae is thought to be the most ancient living caniform family, whose origins began over 40 million years ago in North America. This remained the centre of canid evolution until around six million years ago, when the formation of the Beringian land bridge connected Asia to North America, allowing canids to flood into Eurasia. Canids similarly colonized South America with the emergence of the Isthmus of Panama three million years ago. All modern canids are considered members of the subfamily Caninae (there are two extinct subfamilies), which is divided further into two distinct lineages that diverged 5–9 million years ago:

- Large wolf-like canids (wolves, jackals, the Coyote, Dhole and African Wild Dog, 9 species).
- Small fox-like canids (all other species, 26 species).

Canid species are largely well defined, though there is debate over whether the Dingo (page 98) represents a Grey Wolf (page 100) subspecies or an entirely separate species, and some authorities consider eastern North American populations of the Grey Wolf to be a distinct species, *Canis lycaon*. In early 2011, genetic analysis showed that certain Golden Jackal (page 104) populations from Egypt to Ethiopia actually represent an ancient Grey Wolf subspecies that should be reclassified as the African Wolf. At the other extreme, the very similar Kit Fox and Swift Fox (page 108) are sometimes classified as the same species.

The Canidae is the most widespread family within the Carnivora, with at least one species found on every continent except Antarctica (colonization of Australia by the Dingo was assisted by humans 3500–4000 years ago). It is also the most social family. All canids form enduring social relationships centred around a monogamous male-female pair that cooperates to raise pups; in some cases, yearling offspring remain with their parents to act as 'helpers' in raising subsequent litters. In most foxes and jackals, the mated pair remains the basic social unit, while it forms the nucleus for larger, more complicated social groups in many other canids such as Grey Wolves, African Wild Dogs (page 96) and Dholes (page 102). Within this range, canid sociality is extremely flexible, with some species shifting across a continuum from monogamous pairs to pack living, depending on the availability of resources. Canids are obligate carnivores that eat mainly animal prey; fruits and vegetables are additionally consumed by some species.

Domestic Dog
The domestic dog descends from the Grey Wolf, with which it is still able to interbreed, and is classified as Canis lupus familiaris, *a subspecies of the Grey Wolf. The number of domestic dogs worldwide, including feral and stray populations, is thought to number in the mid-hundreds of millions. There are more than 115 million pet dogs just in the United States, Brazil and China combined.*

FAMILY URSIDAE BEARS, 8 species
Size range Sun Bear (25–80kg) to Polar Bear (150–800kg)

The bear family arose early in carnivoran evolution and, together with the Canidae, is thought to be one of the most ancient families within the Caniformia. The earliest putative species are approximately 33–37 million years old, though their similarity to early canids (with which bears share an ancient common ancestor) obscures precise dating of the family's origins. Today's eight species of bear are divided into the following three subfamilies:

- Ailuropodinae (Giant Panda).
- Tremarctinae (Andean Bear).
- Ursinae ('typical bears'; 6 species).

The Giant Panda (page 122) and Andean Bear (page 124) are the most ancient and distinctive species. The inter-relationships of the other six species (the 'typical bears') are poorly understood, except it is clear that Polar Bears evolved recently and rapidly from a population of Brown Bears isolated during the mid-Pleistocene perhaps only 200,000 years ago.

Bears occur mainly in Eurasia and North America, with one species, the Andean Bear, found in northern South America. Most species are omnivorous, shifting their diet seasonally depending on food availability to focus on the energetically richest diet; the family's extremes are represented by the completely herbivorous Giant Panda, and the Polar Bear, which subsists largely on seals.

Bears are distinctive among carnivores in weathering severe winters by going into hibernation, essentially a strategy to survive the leanest period of the year. During hibernation, bears do not eat, drink, urinate or defecate, but they use as much as 4000 calories daily by burning fat reserves. Hibernating American Black Bears (page 128) reduce oxygen consumption and metabolic rate by half, breathe only once every 45 seconds and reduce their heart rate to as low as eight beats per minute.

In concert with hibernation, breeding females undergo delayed implantation, or 'embryonic diapause', in which development of the embryo is postponed shortly after conception, which typically occurs in the northern spring–summer. This allows recently impregnated females to gain sufficient fat for winter hibernation without having to nourish developing embryos. It is also likely to reduce the period in which hibernating females nourish embryos, though birth still occurs about midway through the fast while in the den. Both of these features probably arose early in ursid evolution. All modern bears display some degree of delayed implantation, though only species (or populations) that experience harsh winters hibernate. Populations that enter hibernation undergo a period of hyperphagia before denning, in which individuals spend up to 20 hours each day foraging for high-quality food (especially hard mast) to lay down fat reserves.

Bears are largely cathemeral and solitary, and occupy stable ranges generally without strict territorial defence. The range size of the Polar Bear is the largest recorded among carnivores, and among the largest recorded in mammals.

FAMILY PROCYONIDAE RACCOONS, COATIS AND ALLIES, 13 species
Size range Ringtail (0.8–1.1kg) to Northern Raccoon (1.7–28kg)

The Procyonidae arose approximately 27–30 million years ago as an offshoot from the lineage that gave rise to the Mustelidae, hence these two families are considered the other's closest relative within the Caniformia. The earliest procyonids evolved in Europe, from which they colonized Asia and North America; the family died out in Eurasia and today occurs only in North and South America. Divisions within the family, including the exact number of species, remain controversial. The olingo genus *Bassaricyon* (page 138) is undergoing revision at the time this book is going to press (the expected classification of four species appears here), and the currently described Mountain Coati (page 136) may prove to be two distinct species.

Procyonids are among the least carnivorous of carnivores, with most species having broadly omnivorous diets. The Northern Raccoon (page 134) is one of the most omnivorous species of mammal on Earth, while the Kinkajou (page 140) eats fruit almost exclusively. Procyonids are largely nocturnal, and vary between being solitary to highly social, and from being terrestrial to almost exclusively arboreal.

FAMILY AILURIDAE RED PANDA, 1 species
Size Red Panda (3–3.6kg)

The Red Panda (page 140) is the only member of a unique family, the Ailuridae, and has an uncertain phylogenetic position. It belongs in the Caniformia, where it is thought to be most closely related to procyonids, but there is also evidence for grouping it closely to mustelids, mephitids and ursids. It was formerly grouped with the Giant Panda in a separate family due mainly to a similar diet of bamboo and associated adaptations (the Red Panda was actually discovered and named first, making a misnomer of the Giant Panda's common name). Although this classification is no longer accepted and the Giant Panda is unequivocally a bear, the Red Panda's closest relatives remain uncertain.

Red Pandas are restricted to forests of south-east China and bordering countries. Their diet is almost exclusively bamboo. They are cathemeral, solitary and occupy stable ranges.

FAMILY MEPHITIDAE SKUNKS AND STINK BADGERS, 12 species
Size range Pygmy Spotted Skunk (130–230g) to Striped Skunk (0.6–5.5kg)

Skunks and stink badgers were formerly classified in the mustelid family, but are now recognized in their own family as an early offshoot of a branch of carnivore evolution that also gave rise to the mustelids, procyonids and Red Panda. They are further subdivided into two subfamilies:

- Mephitinae, comprising 10 species of true skunk.
- Myadinae, containing the two extant stink badger species.

Many skunk species are poorly defined and genetic analysis is likely to identify more 'hidden' species, or subsume species into one, particularly among the spotted skunks (genus *Spilogale*) and hog-nosed skunks (genus *Conepatus*).

True skunks are restricted to North and South America, and stink badgers are endemic to insular South-east Asia. All members of the family have enlarged muscular anal scent glands that spray potent fluid in self-defence, and all have associated black-and-white aposematic colouration. Mephitids are omnivorous, with invertebrates and small vertebrates dominating the diet. They are largely solitary with little evidence of territoriality; many species can reach high densities, congregate at food patches and have extensively overlapping ranges.

FAMILY MUSTELIDAE WEASELS, MARTENS, BADGERS AND OTTERS, 56 species
Size range Least Weasel (25–300g) to Sea Otter (14.5–45 kg)

The Mustelidae is the largest family of the Carnivora, and arose in Eurasia at least 24 million years ago. Subdivision within this large family is complicated and undergoes constant revisions as new molecular and fossil discoveries are made. The clearest partition occurs between the otters (subfamily Lutrinae) and the rest of the family, which is further subdivided into as many as seven subfamilies. Even the number of species is not static. After decades of classification as a single species, hog badgers (page 162) were recognized as three distinct species only in 2008. Similar scientific scrutiny with modern molecular techniques for other branches of the family may yet throw out more species, for example among the grisons (page 168).

From their Eurasian origins, mustelids underwent repeated colonizations into the Americas and Africa. Today the family occurs globally, with species on every continent except Antarctica and Australasia (though Least Weasels and Stoats, page 182, have been introduced to New Zealand by humans). Most modern species are variations on the family's earliest evolutionary form, a solitary long-bodied, terrestrial hunter of small mammals. However, as befits such a large and diverse family, mustelids have evolved to adopt a wide variety of lifestyles, from aquatic and social in the case of otters to semi-arboreal in the martens.

Similar to their close relatives, the Mephitidae, most mustelids have anal glands that produce strongly smelling secretions. These glands are best developed in the Striped Weasel, Libyan Weasel, Zorilla (page 170) and Marbled Polecat (page 172), which are able to spray secretions defensively.

The Mustelidae is remarkable for the prevalence of delayed implantation, in which development of the embryo in the womb is temporarily postponed, in some cases for as long as 11 months. This adaptation allows both mating and birth to occur during optimal summer–spring periods when finding mates and raising young is most benign. Thus, a breeding female American Badger typically conceives in the summer and gives birth the following spring. Approximately a third of mustelids are thought to display some degree of delayed implantation; this compares to less than 0.05 per cent of mammals overall (it is also prevalent among bears).

CONSERVATION OF CARNIVORES

Carnivores are rare. Their positions at the tops of intricate food pyramids dictate that they are naturally far less common than the species on which they prey. Every Tiger (page 42) needs to kill about 50 medium or large ungulates a year to survive. In naturally functioning ecosystems, this represents about 10 per cent of available prey; that is, a population of 500 prey animals is required to sustain a single Tiger for a year. Therefore, a tiny population of 10 Tigers requires 5000 prey animals (not accounting for the needs of coexisting carnivores such as Leopards, page 46, and Dholes); in turn, these require large expanses of habitat to survive. The outcome of these calculations has been called the Large Carnivore Problem; predators require large tracts of suitable habitat with abundant prey populations. The problem is most acute for large and medium-sized carnivores such as big cats, hyaenas, wolves, African Wild Dogs, bears, Fosas and Wolverines (page 166), but *natural* vulnerability to extinction is an inherent feature of the entire order.

The primary threat to most carnivores is the combined loss of habitat and prey. More than two-thirds of Earth's terrestrial land area is now devoted to supporting humans, with the remaining natural habitat disappearing at an estimated rate of 1 per cent per year. Where people replace forests, woodlands and grasslands with cities, agriculture and livestock, most carnivores decline or disappear. Even maintaining habitat is valueless if there is no food for carnivores. Tracts of relatively intact but 'empty forest' across Asia, Latin America and central Africa are worthless to carnivores because people have hunted out their prey.

Compounding the relentless depletion of resources on which carnivores depend – *indirect* threats in conservation nomenclature – are the reasons why humans kill them directly. People have hunted carnivores for millennia and for many reasons, but the two most critical contemporary motives contributing to carnivore declines are the killing of carnivores as a perceived or real threat to livestock (and, less so, human life), and the hunting of carnivores because their body parts are considered valuable. The former affects large carnivores wherever they encounter people and their herds; subsistence yak shepherds in Central Asia trap Snow Leopards (page 40) for essentially the same reasons that commercial cattle ranchers in the western United States clash with reintroduced Grey Wolves.

The killing of carnivores for their parts occurs globally, but is particularly problematic in Asia, where the consumption and use of wildlife for 'traditional medicine' has a history of thousands of years. The primary threat to the Tiger today is intense poaching pressure to feed this trade, which is growing exponentially as a burgeoning Chinese middle class covets the parts of Tigers (and many other species). Not surprisingly, these two drivers are often interleaved; Mongolian herders make extra money by selling the furs of sheep-killing wolves (or any they can shoot, for that matter), and the claws, fat and other sought-after parts of Lions poisoned by African pastoralists are often sold or traded.

Although not nearly as widespread, ancillary anthropogenic threats to carnivores can be locally damaging to populations. Recreational hunting, whether by big-game hunters for trophies or by trappers for fur-bearing carnivores, can provoke declines where poorly regulated or in concert with other factors, for example natural fluctuations in prey numbers.

Infectious disease is a natural part of wildlife populations worldwide, but it can be particularly problematic to carnivores when introduced by humans and their domestic animals. Wild canids are especially vulnerable to rabies and canine distemper transmitted by domestic dogs, and outbreaks have devastated populations of Ethiopian Wolves (page 98) and African Wild Dogs. Finally, a handful of species is threatened by hybridization with domestic carnivores. The Wildcat and Dingo are unlikely to remain genetically distinct due to interbreeding with feral domestic cats and dogs respectively.

Today, four modern carnivore species are extinct, all as a direct result of human impacts, primarily hunting (the last record is given in parenthesis):

- Falkland Island Wolf *Dusicyon australis* (1876).
- Sea Mink *Neovison macrodon* (1894).
- Japanese Sea Lion *Zalophus japonicus* (1951).
- Caribbean Monk Seal *Monachus tropicalis* (1952).

Eight carnivores (including two pinnipeds) are Critically Endangered with an extremely high risk of extinction in the wild. Twenty-four (including four pinnipeds) are Endangered with a high risk of extinction.

The key to saving most carnivores relies on the existence of large expanses of wilderness relatively free from human influences. The foundation of any meaningful effort to conserve carnivores lies in setting aside vast protected areas and ensuring that they truly protect; this does not necessarily entail excluding people, but it does mean vigorously limiting their worst impacts such as clearing habitat, hunting wildlife and introducing disease. Worldwide, there are hundreds of globally significant parks that protect carnivores. However, as human populations continue to grow, the pressure for their land and resources intensifies, while the opportunity for expanding or creating more protected areas dwindles. Accordingly, parks alone will not be sufficient to guarantee the survival of many carnivore species.

Equally as important, attention must be devoted to the human-modified landscapes that now dominate the globe and which, historically, have been omitted from conservation planning. Despite their demanding ecological requirements, many carnivores are able to survive in modified habitats – a habitat need not necessarily be pristine for carnivores to maintain a presence. Even the most difficult to conserve carnivores, such as wolves, bears and big cats, can inhabit landscapes where people and their livestock dominate, provided they are tolerated. The key is fostering mechanisms for coexistence, typically through reducing the problems that carnivores create (for example by improving livestock husbandry to reduce its vulnerability to predators), or by giving carnivores a dollar value, for example through tourism or hunting. Many communities that live with carnivores today are experimenting with a combination of both.

ABOUT PANTHERA

Panthera was created in 2006 to focus solely on conserving the world's wild cats. Today, the organization is the largest single entity devoted to cat conservation (though numerous other conservation organizations also work on conserving cats).

Rather than focusing on individual sites, Panthera's approach relies on producing rigorous science to understand how best to address the challenges of conserving cats across their range. Panthera's biologists design strategies that aim to conserve cats 'range-wide' by protecting core populations and maintaining the biological and genetic corridors that connect them. Implementing the resulting blueprints relies on developing partnerships with governments, universities, non-government organizations and individuals to protect critical habitats, mitigate the conflicts between people and cats, and reduce the illegal or unsustainable killing of cats and their prey.

Panthera focuses most of its efforts on big cats that act as 'umbrella species'. Successfully conserving a significant population of large cats entails protecting the entire ecosystem on which it relies, and which sustains hundreds of thousands or millions of other species. To date, Panthera has developed range-wide conservation approaches for Tigers, Lions, Jaguars and Snow Leopards. The organization plans to build similar strategies for Leopards, Cheetahs, Clouded Leopards and Pumas. For more information, see www.panthera.org

STRUCTURE OF THE SPECIES ACCOUNTS

Every species account in the following section is written in a standard format, starting with the most widely accepted common name, scientific name and other common names in usage. Standard measurements as explained below are provided under the names.

Each species is introduced in a brief description noting the main features useful for identification, including regional and seasonal variation. The accompanying plates depict every species, with a range of forms shown for more variable species. Variants are labelled where they represent discrete regional or morphological types, for example melanistic Jaguar (page 48) and the various forms of Arctic Fox (page 106). Labels do not appear where the depicted forms represent a cross-section of the variation present in a species regardless of geography, season or population, such as in the Bobcat (page 34), African Wild Dog (page 96) and Sun Bear (page 126).

Due to constraints of space and the very wide geographic coverage of the book, maps are not included here; newly produced maps for every species are available at www.panthera.org/carnivoreguide

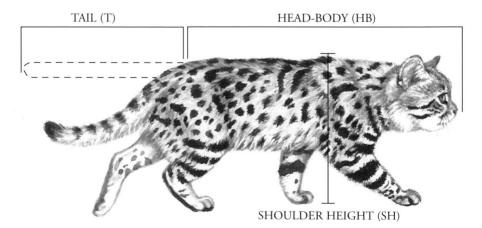

TAIL (T)   HEAD-BODY (HB)

SHOULDER HEIGHT (SH)

Measurements
Animals are usually measured laid out on their side with the tail extended in a straight line behind the body. Values are provided for head-body length (HB), *tail length (T), shoulder height (SH) and weight (W). Where possible, measurements are given for both sexes, especially for species with marked sexual dimorphism.*

The species accounts summarize current knowledge in four major categories as described below. Much of this information is lacking for many carnivores, illustrating the inequity between the few well-studied species and the many that are poorly studied or virtually unknown. Observations or features indicative of a specific region or population are noted with the site name in parenthesis; where multiple figures are available (for example for territory size), a range is provided from the typical habitats or conditions in which the species occurs. Certain site names are abbreviated, for example in the case of protected area types; these include NP (National Park), WS (Wildlife Sanctuary) and GR (Game Reserve). The four major categories of information in the species accounts are:

FEEDING ECOLOGY
Diet, including primary prey species, other food items and whether humans, livestock or crops are eaten; hunting strategies and behaviour, including when foraging occurs and whether it is social or solitary; estimates of hunting success; other notable features of feeding ecology, including whether the species scavenges or caches food.

SOCIAL AND SPATIAL BEHAVIOUR
Degree of sociality, monogamy and territoriality; features of dispersal behaviour; estimates of range size and density.

REPRODUCTION AND DEMOGRAPHY
Degree of seasonality; length of gestation; litter size; breeding patterns; inter-litter interval; development of young, including age of weaning and dispersal; age at sexual maturity or first breeding; mortality rates (for cubs and adults where known) and main natural causes of death; lifespan for wild individuals (where known) and/or in captivity.

STATUS AND THREATS
Summary of the species' status and the main threats. Where applicable, CITES and IUCN Red List categories (below) are provided.

CITES (Convention on International Trade in Endangered Species) is an agreement between governments (currently 176) to control international trade in wild animals and plants. It covers the importing and exporting of live wildlife and its parts, including furs, hunting trophies and souvenirs. The species covered by CITES are listed in three Appendices, according to the degree of protection they need. Many carnivores are listed in Appendices I or II. Carnivores not listed are not considered threatened, or international trade is not considered a possible threat. CITES assessments are usually made for a species across its range, but locally endangered populations or subspecies are often listed separately.

- Appendix I covers species threatened with extinction. Trade in these species is permitted only in exceptional circumstances.

- Appendix II includes species not necessarily threatened with extinction, but in which trade must be controlled in order to avoid utilization that may threaten their survival.

- Appendix III contains species that are protected in at least one country, which has asked other CITES signatories for assistance in controlling trade. For more information, see www.cites.org

IUCN Red List IUCN (International Union for Conservation of Nature) is the largest professional global conservation organization. It produces the Red List of Threatened Species, a comprehensive and expert-driven process assessing the status of wildlife species and classifying them according to the degree of threat and likelihood of extinction. Assessment is a complex process based on multiple criteria, including population size, number of sub-populations, number of breeding individuals, degree of various threats, and so on.

Each Red List category has a precise definition and criteria (for details, see www.iucnredlist.org). From most threatened to least threatened, the categories are: Extinct (EX), Extinct in the Wild (EW), Critically Endangered (CR), Endangered (EN), Vulnerable (VU), Near Threatened (NT) and Least Concern (LC). Where the data are not available for a full assessment, species are classified as Data Deficient (DD). Those species not yet assessed by the Red List are Not Evaluated (NE). Red List assessments are usually made at the species level, but more endangered populations or subspecies are often evaluated separately.

Chinese Desert Cat, Chinese Steppe Cat

HB 68.5–84cm; T 32–35cm; W 6.5–9kg

Light yellow-grey in winter, darkening to tawny or grey-brown in summer, with very faint markings or none at all on the body except for a dark dorsal line. Tail bushy, conspicuously banded and has a dark tip. Ears are tufted. Recent genetic data indicate this is a subspecies of Wildcat, but classification is still disputed. **Distribution and Habitat** Known only from Qinghai, Sichuan and Gansu Provinces, C China. Records from elsewhere in China and Tibet are equivocal. Inhabits alpine grassland, meadows, shrubland and forest edges at 2500–5000m. May also occur in true montane forest, semi-desert and cold desert, but this is unconfirmed. **Feeding Ecology** Virtually unstudied in the wild. A single study on diet showed that small rodents such as voles, mole rats and pikas make up 90% of the diet. Also eats birds, including a record of a pheasant. **Social and Spatial Behaviour** Unknown. Probably solitary and dens in rock outcrops, burrows under tree roots and dense thickets. **Reproduction and Demography** A handful of records suggest seasonal breeding, which is likely given the harsh winters of its range. Male-female pairs are mostly observed January–March, which is the likely breeding season, with kittens born around May. MORTALITY and LIFESPAN Unknown. **Status and Threats** Very restricted range, and thought to be naturally rare. Killed for its fur, which is mostly for local use, but hunting is widespread and pelts are common in fur markets. Large-scale government-mandated poisoning of rodents and lagomorphs prevalent in C China is likely to constitute a serious threat both by reducing prey populations and secondary poisoning. CITES Appendix II, Red List VU.

WILDCAT *Felis silvestris*

HB ♀ 40.6–64cm; ♂ 44–75cm; T 21.5–37.5cm;
W ♀ 2–5.8kg, ♂ 2–7.7kg

Same species as domestic cat, thought to have been domesticated in the Fertile Crescent over 9000 years ago. Appearance is very similar to that of domestic cat, though wild individuals are genetically distinct and generally larger, have longer legs and are more robust. Wild populations intergrade, but cluster into three major subspecies (sometimes controversially considered separate species): Asiatic Wildcat *F. s. ornata*, typically distinctly spotted on an isabelline background; African Wildcat *F. s. lybica*, sandy-grey with banded legs and red-backed ears; and European Wildcat *F. s. silvestris*, which looks like a heavily built striped tabby with a bushy tail and white chin and throat. Recent genetic data suggests that African Wildcat should be further divided into two subspecies in sub-Saharan Africa and N Africa/SW Asia. Piebald, ginger and black variants are usually the result of hybridization with domestic cats. **Distribution and Habitat** Western Europe including Scotland, Africa (except C Africa and the Sahara), the Middle East and much of W and S Asia to C China. Very broad habitat tolerance. Except for dense forest and open desert interiors, occurs in virtually all habitats with cover to 3000m. Avoids very open and high montane habitats, and deep snow. Readily occupies agricultural lands, fields and plantations, but avoids intensively farmed habitat with little cover. **Feeding Ecology** Through almost its entire range, the diet is dominated by small rodents such as mice, rats, voles, jirds, gerbils and jerboas; a notable exception is Scotland, where hares and rabbits comprise up to 70% of prey. Other important prey includes small birds, especially ground foragers such as doves, pigeons, partridges, sandgrouse, guineafowl, quails and weavers. Reptiles including large venomous snakes, e.g. puff adders and cobras, amphibians and arthropods are also eaten. Readily kills poultry and (rarely) very young domestic goats and lambs. Drinks daily when water is available, but often occurs far from water in the Kalahari, Namib and Sahara, suggesting it is water independent. Hunts mostly on the ground, and is chiefly nocturno-crepuscular; when protected, may be active during the day, especially in cold winters. Scavenges, and sometimes caches food by covering it with debris, soil and leaf litter. **Social and Spatial Behaviour** Solitary and territorial. Displays typically feline territorial behaviours such as marking with urine and faeces, but extent of territorial defence probably varies widely between different habitats. Range size also varies widely, typically (but not universally) with larger male ranges overlapping multiple female ranges. Range estimates include 1.7–2.75km² (♀s) to 13.7km² (1 ♂) in Portugal; 1.75km² (average for both sexes) in E Scotland with abundant lagomorphs; 3.5km² (♀s, average) to 7.7km² (♂s, average) in S Kalahari; 8–10km² (both sexes) in W Scotland with scarce prey; 11.7km² (average for both sexes) in Saudi Arabia and 51.2km² (1 ♀) in UAE. Density estimates include 0.7–10 cats/km². **Reproduction and Demography** Breeds seasonally in areas with extreme seasonality such as the Sahara and most of Europe, mating winter–early spring and giving birth spring–early summer. Elsewhere, kittens may be born year-round, though birth peaks often coincide with prey flushes during or after the rainy season, e.g. E and S Africa. Gestation 56–68 days. Litter size typically 2–4, rarely to 8. Weaning at 3–4 months, independent at 5–10 months. Sexually mature at 9–12 months, but wild individuals probably first breed at 18–22 months. MORTALITY Rates poorly documented; most mortality in studied populations is due to human factors. Known predators include large cats, Golden Eagle, Honey Badger (of kittens) and domestic dogs. Starvation of kittens and subadults contributes to low survival in harsh winters. LIFESPAN 11 years in the wild, 15 in captivity. **Status and Threats** Widely distributed, adaptable and tolerant of human activity such as agriculture and forestry, which often elevates rodent populations. Not endangered in any traditional sense, but hybridizes with domestic cats (producing fertile hybrids) throughout its range, especially in W Europe; in Scotland, up to 88% of wild-living cats may be hybrids. Only the most remote populations are thought to lack hybrids, and the species may soon cease to exist as a genetically distinct form. Other important threats include domestic cat diseases, persecution for killing poultry and for perceived killing of game species (Europe) and small stock, hunting for fur (mainly C Asia) and roadkills. CITES Appendix II, Red List LC (European Wildcat VU).

CHINESE
MOUNTAIN CAT

WILDCAT

Asiatic
subspecies

African
subspecies

European
subspecies

Small-spotted Cat

HB ♀ 35.3–41.5cm ♂ 36.7–52cm; T 12–20cm;
W ♀ 1–1.6kg, ♂ 1.5–2.45kg
One of the smallest cats. Light tawny to buff-grey with ginger to black markings. Northern individuals are generally paler, with the darkest, most strongly marked animals found in the E Cape, S Africa, but there is wide variation within populations. **Distribution and Habitat** Endemic to southern Africa and restricted to open short-grass habitat, dry savannah, Karoo scrub and semi-desert. Independent of drinking water, but does not occur in open hyper-arid desert interiors. **Feeding Ecology** Specializes in rodents <100g and ground-roosting birds to the size of Black Bustard. Heaviest prey is Cape Hare; attacks on newborn Springbok lambs are unsuccessful. Reptiles, amphibians, eggs and invertebrates are occasionally eaten. Nocturno-crepuscular, hunting up to 70% of the night with three strategies: rapid bounding through cover to flush birds; painstaking weaving around cover searching for prey; and waiting in ambush at rodent burrows. Succeeds in 60% of hunts (Benfontein, S. Africa) with 10–14 rodents or birds caught per night, averaging a kill every 50 minutes. Surplus food is cached in shallow diggings or hollow termitaria; scavenges. **Social and Spatial Behaviour** Solitary and territorial, with very frequent urine-marking and vocalizing. Male ranges encompass up to four female ranges. Males guard oestrous females and fight intruding males. Ranges average 8.6km² (♀s) and 16.1km² (♂s). **Reproduction and Demography** Seasonal (S Africa). Birth peaks coincide with spring–summer rains and prey flushes. Oestrus 36 hours; gestation 63–68 days. Litter size 1–4, exceptionally to 6 (captivity). Kittens weaned by 2 months and independent at 3–4 months. Sexual maturity at 7 months (♀s) and 9 months (♂s). MORTALITY Poorly known; predators include Black-backed Jackal, Caracal, domestic dogs and large owls. LIFESPAN 6 years in the wild, 16 in captivity. **Status and Threats** Probably a naturally uncommon species, and threatened by agricultural expansion into semi-arid areas bringing overgrazing, burning and insecticides, which impact rodent and insect populations. Hundreds of Black-footed Cats are killed annually during control activities intended for jackals in S Africa. CITES Appendix I, Red List VU.

SAND CAT *Felis margarita*

HB ♀ 39–52cm, ♂ 42–57cm; T 23.2–31cm;
W ♀ 1.35–3.1kg, ♂ 2–3.4kg
Very small, strikingly pale cat with indistinct markings on the body, resolving to dark stripes on the legs and tail. Flat broad head topped by oversized ears. Feet densely covered in fur, probably for traction and insulation on loose hot sand. **Distribution and Habitat** N Africa, the Middle East and C Asia. A desert specialist, capable of occupying true desert with rainfall of 20mm/year. Inhabits a variety of sandy and stony desert habitats with cover, and arid shrub-covered steppes. Absent from heavily vegetated valleys in these habitats. **Feeding Ecology** Eats mainly small rodents including gerbils, spiny mice, jirds and jerboas, plus occasional kills of young hares, small birds, reptiles (including poisonous snakes) and invertebrates. Independent of drinking water. Foraging mainly nocturnal. Capable of very rapid digging to excavate burrowing prey, and sometimes caches food with a covering of sand. **Social and Spatial Behaviour** Solitary. It is unknown if it is territorial. Based on a few observations, ranges of males in Israel overlap and one male used 16km². Nightly movements average 5.4km. **Reproduction and Demography** Breeding appears seasonal in the Sahara. Mating November–February; births January–April. Oestrus 5–6 days; gestation 59–67 days. Litter size 2–8, typically 3–4. Weaning at around 5 weeks, and independence from 4 months. Sexually mature at 9–14 months. LIFESPAN 14 years in captivity. MORTALITY Unknown. **Status and Threats** Probably naturally rare, but its habitat is so remote that it is somewhat insulated from human activities (at least in Africa). Threats include expansion of cultivation, and feral domestic cats and dogs, which result in predation, competition and possible disease transmission. Caught around human settlements in traps for jackals and foxes. CITES Appendix II, Red List NT.

JUNGLE CAT *Felis chaus*

Swamp Cat, Reed Cat

HB ♀ 56–85cm, ♂ 65–94cm; T 20–31cm;
W ♀ 2.6–9kg, ♂ 5–12.2kg
Leggy cat, uniformly coloured with indistinct body markings that become more obvious on the limbs, and a short tail. Temperate animals tend to be darker and more richly marked than tropical ones. Melanism reported from India and Pakistan. **Distribution and Habitat** Temperate and tropical S Asia, extending into Egypt along the Nile Valley. Prefers dense reed-beds, long-grass and scrub habitats in swamps, wetland and coastal areas, but also inhabits dry and evergreen forests. Tolerates cultivated marshy landscapes including sugar-cane fields and rice paddies. **Feeding Ecology** Small mammals (<1kg) comprise primary prey, mainly small rodents, Muskrat and squirrels, but hares, Coypu (5–9kg), and gazelle and Chital neonates are recorded. Kills birds, especially waterfowl, francolins, pheasants, peafowl and jungle fowl. Also eats small reptiles and amphibians, and has been observed diving into shallow water for fish. Tajikistan individuals are recorded eating large quantities of Russian Silverberry fruits during winter. Readily preys on domestic poultry, and sometimes scavenges from the kills of larger carnivores. **Social and Spatial Behaviour** Poorly known. Solitary. Characteristically feline scent-marking and vocalization suggest maintenance of exclusive core areas, and larger male home ranges overlap several smaller female ranges in Israel. There are no density estimates, but it is frequently the most common felid where it is found. **Reproduction and Demography** Thought to be weakly seasonal. Most observed mating November–February. Kittens born December–June. Gestation 63–66 days. Litters average 2–3, exceptionally to 6. Kittens independent at 8–9 months. Sexually maturity at 11 months (♀s) and 12–18 months (♂s). LIFESPAN 20 years in captivity. MORTALITY Unknown. **Status and Threats** Common in many parts of its range and tolerates agricultural landscapes with cover. Destruction and development of wetland habitats are a particular threat in arid areas, e.g. Egypt. Persecuted for poultry raiding, trapped heavily around Copyu fur farms (former USSR republics) and hunted for fur in its northern range. CITES Appendix II, Red List LC.

BLACK-FOOTED CAT

Dark form

Pale form

SAND CAT

JUNGLE CAT

Dark form

Pale form

HB ♀ 38.8–65.5cm, ♂ 43–75cm; T 17.2–31.5cm;
W ♀ 0.55–4.5kg, ♂ 0.74–7.1kg

Varies very widely in size and appearance. Individuals in tropical Asia are boldly spotted and may be a third the weight of cats in temperate Russia and NE China, which are pale ginger-grey with faint markings ('Amur Leopard Cat'). Small dark Iriomote Leopard Cat occurs on Iriomote Island off Japan; individuals on Tsushima Island near the Korean peninsula are closely related. Melanism not recorded. Amur, Iriomote, Tsushima and some other island populations are sometimes treated as separate species, though molecular data indicate that all comprise a single species. **Distribution and Habitat** Most widespread of all small Asian felids, found in tropical and temperate Asia from the Russian Far East to S India, Indonesia, Borneo and the Philippines. Inhabits all forest types at sea level to 3000m, from lowland tropical rainforest to cold temperate forest with winter snowfall (Russia), as well as dense shrubland, marshland and coastal scrub. Occurs in human-modified habitats with cover, e.g palm-oil plantations, secondary forest and farmland. Avoids open grassland and steppes. **Feeding Ecology** Hunts a wide variety of small prey, chiefly small rodents (especially mice, rats and squirrels), birds, snakes and lizards. Surprisingly aquatic and sometimes forages in shallow water for freshwater crabs, amphibians and invertebrates. Harmless to hoofstock, but raids domestic poultry and is easily baited with chickens. Hunting activity variable, ranging from being strictly nocturnal at some sites to cathemeral. **Social and Spatial Behaviour** Solitary and weakly territorial. Male ranges generally overlap numerous smaller female ranges (though range size differs little between sexes in some populations, e.g. Phu Khieo WS, Thailand). Overlap between same-sex adults is considerable at range edges, and usually minimal in exclusive core areas (which overlap significantly in Phu Khieo). Range size 1.4–37.1km² (♀s) and 2.8–28.9km² (♂s). Density estimates include 34 cats/100km² (Iriomote) to 37.5/100km² (Tabin WR, Sabah). **Reproduction and Demography** Breeding varies from highly seasonal in Russia to aseasonal in the tropics. Captive individuals are able to have two litters a year, though a single litter is probably typical in the wild. Gestation 60–70 days. Litter size 1–4. Sexual maturity at 8–12 months (captivity). MORTALITY Estimated annual adult mortality varies from 8% (remote sanctuary, Phu Khieo) to 47% (accessible protected area, Khai Yai NP, Thailand). Leopard is a known natural predator; human hunting and roadkills are factors in accessible areas. LIFESPAN 13 years in captivity. **Status and Threats** Common, adaptable and lives near humans when it is tolerated. Reaches high densities in suitable habitat, and is the most abundant felid in most of its range. However, it is very heavily hunted for fur in its temperate range, where densities are naturally lower than in the tropics. At least 150,000 are killed annually in China for the fur trade. Impacts of such high harvests are unknown. CITES Appendix I (Bangladesh, India and Thailand), Appendix II elsewhere, Red List LC.

PALLAS'S CAT *Otocolobus manul*

Manul, Steppe Cat
HB ♀ 46–53cm, ♂ 54–57cm; T 23–29cm;
W ♀ 2.5–5kg, ♂ 3.3–5.3kg

Stocky, heavily furred small cat, silvery-grey to rufous-grey with faint striping on the body. Long winter coat has a pale, frosted appearance; spring–summer coat is darker with more obvious stripes and often a reddish tinge. Face has dark cheek stripes, and the crown is distinctively marked with small spots. Bushy tail is banded with narrow stripes and ends in a dark tip. Colouration provides excellent camouflage in open rocky habitat: Pallas's Cat is poorly adapted for running, and when threatened freezes and flattens itself to the ground, conferring very effective concealment. **Distribution and Habitat** Central Asia, from the Caspian Sea through N Iran, Afghanistan, Pakistan and N India to C China, Mongolia and S Russia. Lives to 4800m in cold arid habitats with cover, especially dry grassland steppes with stone outcrops and stony semi-desert. Prefers valleys and rocky areas, and avoids completely open habitat. Though well adapted for extreme cold, usually avoids areas with deep snow. **Feeding Ecology** Hunts mainly small rodents and lagomorphs. Pikas are especially important prey, typically comprising over 50% of the diet; voles, mice, hamsters, gerbils and ground squirrels are also important. Occasional prey includes hares, hedgehogs, small birds, lizards and invertebrates. Hunts by three distinct techniques: 'stalking' by very deliberate creeping around cover; 'moving and flushing', used mainly in long summer undergrowth; and waiting in ambush at rodent burrows. Not known to kill any livestock or poultry. Recorded scavenging from carcasses. **Social and Spatial Behaviour** Solitary. Both sexes maintain enduring home ranges, with large overlapping male ranges encompassing multiple smaller female ranges that overlap little with each other. Likely to be territorial, at least in the breeding season; breeding males often have injuries consistent with fighting. Female territories 7.4–125.2km², averaging 23.1km², compared with male territories of 21–207km², averaging 98.8km² (Hustain Nuruu NP, Mongolia). Density estimates include 4–8 cats/100km². **Reproduction and Demography** Highly seasonal. Mating December–March; births late March–May. Oestrus very short at 24–48 hours; gestation 66–75 days. Litters average 3–4, exceptionally to 8 (captivity). Kittens independent at 4–5 months. Sexual maturity at 9–10 months for both sexes. MORTALITY 68% of kittens do not survive to disperse (Hustain Nuruu NP). Adult mortality estimated at 50%. Most mortality occurs in winter in October–April. Known predators include large eagles, Red Fox and domestic dogs. LIFESPAN 11.5 years in captivity. **Status and Threats** Lives in remote areas but at low densities. Poorly adapted to avoid predators, and depends on fairly specific habitats, making it naturally vulnerable to threats. Hunted for fur in much of its range, and domestic dogs often constitute a key predator; human factors (including dogs) account for 56% of deaths in C Mongolia. State-sanctioned rodent-poisoning campaigns in China, Mongolia and Russia are a serious threat to Pallas's Cat prey. CITES Appendix II, Red List NT.

LEOPARD CAT

Amur
form

Iriomote
form

Spotted
form

Defensive
hiding

PALLAS'S CAT

HB ♀ 44.6–52.1cm, ♂ 41–61cm; T 12.8–16.9cm; W ♀ 1.5–1.9kg, ♂ 1.5–2.2kg Very small, unusual cat with a short tubular body, stubby tail, compact foreshortened face with a flattened forehead, large, closely set eyes and small ears. Feet partially webbed and claws protrude partially from claw sheaths. Dark brown becoming rusty-brown on the head, with contrasting facial stripes. Body largely unmarked except for light dappling and banding on the underparts. **Distribution and Habitat** Peninsular Malaysia (possibly barely across the border into S Thailand), Borneo and Sumatra. Inhabits moist forested habitats; all records are from near rivers and water sources in primary and secondary forests, swamp forest, mangrove and coastal scrub-forest. Sometimes recorded from rubber and palm plantations, suggesting tolerance of some habitat modification. **Feeding Ecology** Poorly known but unique morphology, behaviour and habitat preferences suggest it is adapted for aquatic prey. Captive animals are attracted to water, readily submerging themselves and feeling for food in pools with spread paws, similar to raccoons. Stomach contents of dead wild individuals contained fish and crustaceans; likely to also take small mammals, and sometimes killed in traps set at poultry coops. **Social and Spatial Behaviour** Unknown. **Reproduction and Demography** Poorly known. Gestation 56 days (captivity). Litters 1–2 (based on only three captive births). MORTALITY Unknown. LIFESPAN 14 years in captivity. **Status and Threats** Known only from a handful of physical records and sightings, but it is unclear if the species is rare or simply elusive. Its relatively restricted distribution and close association with moist forested habitats is alarming given the heavy pressure on these areas for draining and development in SE Asia. CITES Appendix I, Red List EN.

RUSTY-SPOTTED CAT *Prionailurus rubiginosus*

HB 35–48 cm; T 15–29.8cm; W ♀ 1–1.1kg, ♂ 1.5–1.6kg One of the smallest cats, the size of a slight, very small domestic cat. Rust-tinged brown or grey-brown with rows of reddish spots that sometimes form complete stripes on the nape, shoulders and upper flanks. Tail solid rust-brown, sometimes with faint bands. **Distribution and Habitat** Endemic to India and Sri Lanka. Formerly regarded as a moist forest specialist, but recent records show that it also occurs in dry forest, bamboo forest, wooded grassland, arid shrubland, scrubland and rocky hill-slopes. Occurs near settlements in modified habitats including cropland, tea plantations and abandoned dwellings in villages. **Feeding Ecology** Poorly known. Has a reputation for being especially fierce and taking large prey, but its diet is likely to be mainly small rodents, birds, reptiles, frogs and invertebrates. Killed fairly frequently while raiding domestic poultry. Most sightings of foraging cats are on the ground, but it is a highly agile climber and possibly hunts both on the ground and arboreally. Chiefly nocturnal. **Social and Spatial Behaviour** Unknown. **Reproduction and Demography** Unknown from the wild. In captivity, reproduction is aseasonal; two wild litters, one each from India and Sri Lanka, both found February. Gestation 66–79 days. Litter size 1–2. MORTALITY Unknown. LIFESPAN 12 years in captivity. **Status and Threats** Regarded as rare, but recent observations suggest it is more widespread and common than previously thought. Found fairly often in association with human settlements; given its tiny size and potential utility in controlling rodents, it can prosper provided it is tolerated by people. That said, it is often killed for skins and meat, by domestic dogs and for killing poultry; mistakenly persecuted as Leopard cubs in Sri Lanka. CITES Appendix I (India), Appendix II (Sri Lanka), Red List VU.

FISHING CAT *Prionailurus viverrinus*

HB ♀ 57–74.3cm, ♂ 66–115cm; T 24–40cm; W ♀ 5.1–6.8kg, ♂ 8.5–16kg Robust cat with a blocky powerful head and short thick tail. Fur olive-grey covered in dark spots that often coalesce into stripes on the nape, shoulders and back. Feet partially webbed, with large claws that protrude partially from the claw sheaths. **Distribution and Habitat** SE Asia from NE Bangladesh and NE India to S Thailand, and isolated populations in Nepal, extreme SE Pakistan (possibly extinct), SW India, Sri Lanka and Java; records from Sumatra are equivocal. Closely associated with wetland habitats including marshes, riverine woodland, dense Terai grassland (Nepal) and mangroves. Sometimes found in degraded habitats around aquaculture ponds and rice paddies, but generally intolerant of wetland modification. **Feeding Ecology** Paws and claws are adapted for aquatic foraging and a diet dominated by fish. Capable swimmer that submerges itself in pursuit of prey, and hunts along the water's edge or in shallows, where it scoops up fish with its paws. Despite its aquatic adaptations, its dentition is more generalized, indicating a broader diet. Also kills small mammals, birds (including ducks and coots hunted in the water), reptiles, amphibians and invertebrates. Occasionally kills prey to the size of Chital fawns. Sometimes kills livestock, mainly very young goats and poultry, and there are confirmed records of kills of very young calves. Reports of it killing children are unsubstantiated. Scavenges. **Social and Spatial Behaviour** Poorly known. Solitary. Only documented range sizes are 4–6km² (2 ♀s) and 22km² (1 ♂) from Terai grassland, Chitwan NP, Nepal. **Reproduction and Demography** Unknown from the wild. In captivity, gestation 63–70 days, litter size 1–4. Sexual maturity in one captive female occurred at 15 months. MORTALITY Unknown. LIFESPAN 12 years in captivity. **Status and Threats** Until recently considered widespread and relatively common, but accelerated development of wetlands and floodplains throughout Asia, combined with illegal hunting, have prompted a rapid decline. Now rare in Java, Laos and Vietnam, and has lost significant range in India and Thailand. May no longer occur in Pakistan. Main strongholds appear to be S Thailand, Sri Lanka and isolated areas of Bangladesh, NE India and Nepal. CITES Appendix II, Red List EN.

FLAT-HEADED CAT

RUSTY-SPOTTED CAT

Fishing

FISHING CAT

MARBLED CAT *Pardofelis marmorata*

Plate 5

HB 45–62cm; T 35.6–53.5cm; W 2.5–5kg
Resembles a small Clouded Leopard with thick grey-buff to red-brown fur, patterned with large, dark-bordered blotches that become small dabs on the limbs. Tubular bushy tail proportionally very long, sometimes exceeding the head-body length and distinctive in the field. When walking relaxed, the tail is held horizontally in a continuous straight line from the body. **Distribution and Habitat** SE Asia, south of the Himalayas in Nepal and Bhutan to SW China, and through Indochina, Borneo and Sumatra. Restricted to forested habitats, chiefly undisturbed evergreen, deciduous and tropical forests. Can occupy secondary and logged forests, though it is unknown whether modified habitat is suboptimal. **Feeding Ecology** Except for one radio-collared female tracked for a month in Thailand, the species has never been studied in the wild. Diet is likely to be dominated by small vertebrates. Highly agile climber and has been observed hunting in trees,

perhaps for arboreal mammals such as squirrels, as well as birds. From limited camera trapping in protected areas, thought to be mostly diurnal. **Social and Spatial Behaviour** Virtually unknown. Occasional sightings of adult pairs have fostered speculation that it forms long-term pair bonds, but it is more likely to be solitary. A collared Thai female used a range of 5.3km² in 1 month. Rare in camera-trap surveys and Asian wildlife markets, possibly reflecting naturally low densities. **Reproduction and Demography** Very poorly known. Gestation 66–82 days (captivity). Litters average 2 kittens (based on only two captive births). Females sexually mature at 21–22 months (captivity). MORTALITY Unknown. LIFESPAN 12 years in captivity. **Status and Threats** Appears to be naturally rare and forest dependent, suggesting particular vulnerability to habitat loss and hunting, which is very prevalent throughout its range. CITES Appendix I, Red List VU.

BAY CAT *Pardofelis badia*

HB 53.3–67cm; T 32–39.1cm; W (emaciated ♀) 2kg
Resembles a small and slender Asiatic Golden Cat with a proportionally smaller, rounded head and stubby rounded ears. Occurs in two morphs: rich rusty-red, and grey with variable red undertones especially along the transition from the upper body colour to the paler underparts. Unmarked except for stripes on the forehead and cheeks, and faint spotting along the transition between the upper body colour and pale underparts. Bright white underside to the tail with a dark dorsal stip that is distinctive in the field. **Distribution and Habitat** Endemic to Borneo. Closely associated with dense forested habitats, with most historical records in primary, riverine, swamp and mangrove forests. Tolerates moist plantation forests with dense understorey, and has been camera trapped from recently logged secondary forest, suggesting some tolerance for habitat modification. **Feeding Ecology**

Unknown, but presumably small vertebrates make up major food items. Two Bay Cats were trapped in 2003 when they entered an animal dealer's pheasant aviaries, suggesting that it may attack domestic poultry. **Social and Spatial Behaviour** Unknown. Rarely photographed during camera-trapping surveys, suggesting that it occurs at very low densities; e.g. Bay Cats were photographed 25 times at 4 sites in Eastern Sabah over 4 years, compared with 259 images of Clouded Leopards and more than 1000 images of Leopard Cats. **Reproduction and Demography** Unknown. **Status and Threats** Apparent extreme rarity and forest dependence raises concern for the species' conservation prospects. Forest conversion, especially to palm-oil plantations, is regarded as a serious threat. Rarity and value are known to animal dealers, elevating illegal trapping pressure. CITES Appendix II, Red List EN.

ASIATIC GOLDEN CAT *Pardofelis temminckii*

Temminck's Golden Cat
HB ♀ 66–94cm, ♂ 75–105cm; T 42.5–58cm;
W ♀ 8.5kg, ♂ 12–15.8kg
Usually rich russet-brown, but varying from pale tawny to dark greyish-brown. Largely unmarked except for the face and faint spotting on the chest and belly. A richly spotted 'ocelot' morph is recorded from Bhutan, China and Myanmar. Melanism occurs. Except in black individuals, the underside of the tail is always conspicuously bright white with a dark upper tip. **Distribution and Habitat** Sub-Himalayan Nepal, NE India and Bhutan to S China, SE Asia and Sumatra. Found in a variety of moist and dry forests, usually under 3000m, but to 3738m in open shrub-grassland mosaic in Bhutan. Has been sighted or killed near human settlements, including in open agricultural areas, and appears to be more tolerant of open habitat than Clouded Leopard, Bay Cat and Marbled Cat; even so, it is never far from cover. **Feeding Ecology** Poorly known. Confirmed prey includes mice, rats, Berdmore's Ground Squirrel, mouse deer, Dusky Leaf Monkey, snakes, lizards and birds. Powerfully built and reputed to

kill medium-sized ungulates, including muntjacs and livestock to the size of very young cattle and buffalo calves. Confirmed records of livestock kills are mostly from hunters shooting it over depredated carcasses, in which it may have been the predator or possibly only a scavenger. Sometimes raids poultry. Nocturno-crepuscular, but diurnal activity is recorded under protection. **Social and Spatial Behaviour** Poorly known. Solitary. Only range sizes known are 32.6km² (1 ♀) and 47.7km² (1 ♂) from Phu Khieo Wildlife Sanctuary, Thailand. **Reproduction and Demography** Unknown from the wild. In captivity, reproduction aseasonal, gestation 78–80 days and litter size 1–3 (typically 1). Sexual maturity in captive animals at 18–24 months. MORTALITY Unknown. LIFESPAN 17 years in captivity. **Status and Threats** Threatened by forest loss and illegal hunting, which are widespread throughout its range, but status and degree of threat are poorly known. Skins of Asiatic Golden Cats are traded heavily in China and Myanmar, where hunting pressure is regarded as high. CITES Appendix I, Red List NT.

MARBLED
CAT

BAY
CAT

Red form

Grey form

'Ocelot' form

Typical form

ASIATIC
GOLDEN CAT

HB ♀ 63–82cm, ♂ 59–92cm; T 20–38cm;
W ♀ 6–12.5kg, ♂ 7.9–18kg
Normally dappled with bold spots, coalescing into blotches on the extremities and nape. A buff-coloured morph with a faint freckled appearance known as 'servaline' occurs in W and C Africa. Melanism is recorded from the E African highlands. **Distribution and Habitat** Africa. Inhabits mainly woodland savannah, grassland, forest and alpine moorland to 3850m, usually in association with rivers, marshes and floodplains. Absent from rainforest and desert. Tolerates agricultural areas with cover. **Feeding Ecology** Specializes in hunting small mammals in long-grass or shubby habitats, with rodents and shrews comprising 80–93.5% of the diet. Small grassland birds are the next most important prey, and it sometimes hunts large birds including flamingos and storks. Occasionally takes genets, mongooses, hares, juvenile antelopes, small reptiles and arthropods. Livestock depredation is unusual, but it occasionally kills poultry and untended young goats. Hunting mostly crepuscular, becoming nocturnal near humans. Rarely scavenges. **Social and Spatial Behaviour** Solitary and territorial, though same-sex adults appear relatively tolerant. Home ranges overlap considerably and aggressive confrontations are rarely observed. Range size 15.8–19.8km² (♀s) and 31.5km² (1 ♂) in KwaZulu-Natal, S Africa. Density estimated at 41 Servals/100km² in exceptionally optimum habitat (Ngorongoro Crater, Tanzania). **Reproduction and Demography** Breeding appears weakly seasonal; births peak November–March (southern Africa) and August–November (Ngorongoro Crater). Gestation 65–75 days. Litters average 2–3, exceptionally to 6. Kittens independent at 6–8 months. MORTALITY Known predators include Lion, Leopard, Nile Crocodile and domestic dogs. Predation by Martial Eagle on kittens recorded. LIFESPAN 11 years (♀s) in the wild, 20 in captivity. **Status and Threats** Tolerates agricultural areas with sufficient cover, water and enlightened management, but conversion of wetland and grassland by draining, burning and over-grazing by livestock is a significant threat. Popular in local fur trade in NE Africa and W African Sahel belt, and for fetish and traditional use in S Africa. Sport hunted with few restrictions in Tanzania and southern Africa. CITES Appendix I, Red List LC.

CARACAL *Caracal caracal*

HB ♀ 61–103cm, ♂ 62.1–108cm; T 18–34cm;
W ♀ 6.2–15.9kg, ♂ 7.2–26kg
Uniformly coloured, ranging from pale sandy-brown to brick-red, unmarked except for faint spotting on the underside. Ears have conspicuous silvery-black backs and long black tufts. Melanistic cases are actually dark chocolate-brown rather than true black. Formerly classified as a lynx, but they are not closed related. **Distribution and Habitat** Most of Africa (except for true desert and rainforest), the Middle East and SW Asia. Prefers dry woodland savannah, dry forest, grassland, coastal scrub, semi-desert and arid mountainous habitat. Sometimes inhabits evergreen and montane forests, exceptionally to 3300m. Tolerates pastoral and agricultural landscapes with cover. **Feeding Ecology** Formidable hunter, recorded killing adult Bushbuck, Springbok and female Impalas, but most prey weighs <5kg; small rodents, hyraxes, hares, Springhare and birds are the most important. Reptiles comprise 12–17% of the diet (West Coast NP, S Africa); amphibians, fish and invertebrates are occasionally consumed. Readily kills small untended livestock, which comprises to 55% of the diet in farming areas in southern Africa. Occasionally hoists kills into trees, and scavenges. **Social and Spatial Behaviour** Solitary. Adults maintain enduring home ranges with exclusive core areas and overlap at the edges. Female territories 3.9–26.7km² (S. Africa) to an average of 57km² (Israel), compared with male territories of 5.1–65km² to an average of 220km² (Israel). A Kalahari adult male used 308km² with a core area of 93km². **Reproduction and Demography** Weakly seasonal. Breeds year-round, but births peak October–February (S Africa), and November–May (E Africa). Gestation 68–81 days. Litters average 2–3, exceptionally to 6. Kittens independent at 9–10 months. MORTALITY Most known mortality is anthropogenic; occasionally killed by larger carnivores, including domestic dogs in rural areas. LIFESPAN 19 years in captivity. **Status and Threats** Habitat degradation, loss of prey and human hunting are significant threats in C, W and N Africa, and Asia, where Caracals are rare. Persecuted intensely on livestock land in E and southern Africa, but resilient and difficult to extirpate there. Sport hunted with few restrictions in E and southern Africa. CITES Appendix I (Asia), Appendix II (Africa). Red List LC.

AFRICAN GOLDEN CAT *Profelis aurata*

HB 61.6–101cm; T 16.3–37cm; W ♀ 5.3–8.2kg,
♂ 8–16kg
Two distinct colour phases, red-brown and grey, ranging from heavily spotted to plain in either phase. There is some intergradation between morphs, and melanism occurs. Does not change colour as is often claimed. **Distribution and Habitat** Endemic to equatorial Africa. Strongly associated with undisturbed moist forests to 3600m, including alpine bamboo forest, dense coastal forest and riverine forest strips in woodland savannah. Occurs in banana plantations inside forest, and in abandoned logged areas with secondary undergrowth. Avoids open and dry habitats. **Feeding Ecology** Eats a wide variety of small prey typically weighing 1.5–3.6kg; birds, shrews, rodents and small forest duikers are the most important. Actively hunts forest monkeys, but hunting is mainly terrestrial; scavenging of eagle kills on the forest floor may account for most primate occurrences in the diet. Reported to raid poultry in villages and to scavenge from wire snares. **Social and Spatial Behaviour** Unstudied and very poorly known. Solitary. Adults urine-mark and leave faeces exposed on trails, suggesting territorial behaviour. **Reproduction and Demography** Gestation 75 days. Litter size 1–2. Kittens weaned at around 6 weeks (captivity). Sexual maturity in captive animals 11 months (♀s) and 18 months (♂s). MORTALITY Poorly known; Leopard is a confirmed predator. LIFESPAN 12 years in captivity. **Status and Threats** Thought to be naturally rare. Forest dependent and most threatened by deforestation; many W and E African moist forests are now heavily degraded and converted to savannah. Bushmeat hunting in W and C Africa heavily impacts prey species, and African Golden Cats are killed frequently in some areas for bushmeat and fetish markets. CITES Appendix II, Red List NT.

SERVAL

Typical form

Servaline form

CARACAL

Red
form

**AFRICAN
GOLDEN CAT**

Grey form

HB ♀ 43–74cm; ♂ 44–88cm; T 23–40cm;
W ♀ 2.6–4.9kg, ♂ 3.2–7.8kg
Yellow-brown to silver-grey with small, solid dab-like spots; southern, temperate animals are typically paler and larger. Melanism is common in Uruguay, SE Brazil and E Argentina, rare elsewhere. Hydridizes with Oncillas in S Brazil. **Distribution and Habitat** C Bolivia to Uruguay–S Brazil, and most of Argentina. Inhabits subtropical and temperate brushland, forest, semi-arid scrub, pampas grassland and marshland from sea level to 3300m. **Feeding Ecology** Eats small rodents such as grass mice, rice rats and cavies, as well as birds and herptiles. Occasionally takes Six-banded Armadillo, tree porcupines and small opossums. Introduced Brown Hare dominates the diet in Patagonia. Mostly nocturno-crepsucular and terrestrial, but swims readily taking Copyu, Marsh Rat, frogs and fish at the water's edge. Raids domestic poultry; sheep records probably scavenged. **Social and Spatial Behaviour** Solitary. Deposits faeces in trees, creating arboreal middens. Range sizes average 1.5–5.1km² (♀s) and 2.2–9.2km² (♂s). Density estimates include 0.4 cats/10km² (Argentine pampas during a prey shortage) to 13.9/10km² (protected Argentine scrubland). **Reproduction and Demography** Seasonal in temperate southern range (possibly elsewhere), with most births in summer. Gestation 62–78 days. Litter size 1–3. MORTALITY Starvation and high parasite loads elevate mortality during droughts. Puma is a known predator. LIFESPAN 14 years in captivity. **Status and Threats** Widespread and abundant in good habitat, but often killed by vehicles, domestic dogs and for killing poultry. CITES Appendix I, Red List NT.

ONCILLA *Leopardus tigrinus*

Little-spotted Cat, Tiger Cat
HB ♀ 43–51.4cm, ♂ 38–59.1cm; T 20.4–42cm;
W ♀ 1.5–3.2kg, ♂ 1.8–3.5kg
Very small slender cat with thick soft fur marked with small rosettes or blotches. Melanism occurs. Hydridizes with Geoffroy's and Pampas Cats in Brazil. **Distribution and Habitat** N Venezuela to N Argentina, and a disjunct population in the central Cordillera of Costa Rica–N Panama. Inhabits wooded savannah, dense scrubland and forest to 3200m (exceptionally to 4800m in Colombia). Tolerates modified habitats near humans provided there is cover. **Feeding Ecology** Prey typically weighs <1kg, mainly rodents, shrews, small opossums, birds, eggs, reptiles and invertebrates. Foraging mainly nocturno-crepuscular, but shifts depending on prey activity, e.g. to diurnalism in Caatinga. **Social and Spatial Behaviour** Solitary. Range size 0.9–25km² (♀s) and 4.8–17.1km² (♂s). Density estimates include 0.01 Oncillas/100km² (lowland Amazon forest) to 1–5/100km². **Reproduction and Demography** Unknown from the wild. Gestation 62–76 days. Litter size 1–2. **Longevity** 17 years in captivity. **Status and Threats** Appears to be naturally rare, elevating its vulnerability to threats, but status is poorly known. Formerly heavily hunted for fur trade. CITES Appendix I, Red List VU.

MARGAY *Leopardus wiedii*

Tree Ocelot
HB ♀ 47.7–62cm, ♂ 49–79.2cm; T 30–52cm;
W ♀ 2.3–3.5kg, ♂ 2.3–4.9kg
Lightly built, resembling a small lean Ocelot, but with a proportionally much longer tail, rounded head and distinctive large eyes. **Distribution and Habitat** N Mexico to Uruguay. One specimen from S Texas around 1850 is the only US record. Closely associated with dense lowland forest habitats, usually to 1500m. Avoids converted landscapes except for dense plantations, e.g. of coffee, cocoa, Eucalyptus and pine. **Feeding Ecology** Mostly eats mice, rats, squirrels, cavies, opossums, agoutis, pacas and Brazilian Rabbit, as well as birds, herptiles, invertebrates and small amounts of fruits. Forages terrestrially and arboreally: a spectacularly acrobatic climber able to hunt the most agile prey, including small primates. Occasionally kills domestic poultry. **Social and Spatial Behaviour** Solitary. Range sizes average 0.9–20km² (♀s) and 4–15.9km² (♂s). Based on camera trapping, reaches lower densities than Ocelot. **Reproduction and Demography** Aseasonal (captivity). Gestation 76–84 days. Litter size usually 1, rarely 2. MORTALITY Poorly known. LIFESPAN 20 years in captivity. **Status and Threats** Strongly forest dependent and responds poorly to forest conversion, the main threat. Localized illegal hunting and persecution for killing poultry also occurs. CITES Appendix I, Red List NT.

OCELOT *Leopardus pardalis*

HB ♀ 69–90.9cm, ♂ 67.5–101.5cm; T 25.5–44.5cm;
W ♀ 6.6–11.3kg, ♂ 7–18.6kg
Latin America's third largest cat, powerfully built with a relatively short tail. Very richly marked with open blotches, streaks and rosettes with russet-brown centres. **Distribution and Habitat** N Mexico to N Argentina, with two relict populations in extreme S Texas, USA, numbering 60–100. Inhabits arid scrubland to lowland rainforest from sea level to 1200m, strongly preferring dense habitats. **Feeding Ecology** Kills howler monkeys, coatis, sloths, tamanduas, and juvenile peccaries and deer, but most prey weighs 1.5–8kg, e.g. agoutis, pacas, opossums and armadillos. Also eats birds, reptiles, frogs, fish, land crabs and insects. Sometimes raids poultry coops. Foraging chiefly nocturnal and terrestrial. Scavenges and sometimes caches large carcasses by covering with debris. **Social and Spatial** **Behaviour** Solitary and territorial. Male ranges (average 4–90.5km²) overlap multiple female ranges (average 1.3–75km²). Ranges are smallest in the Brazilian Pantanal and largest in cerrado savannah (Emas NP, Brazil). Density estimates include 2.3–3.8 Ocelots/100km² (pine forest, Belize) to 11–12/100km² (rainforest, Belize). **Reproduction and Demography** Aeasonal. Gestation 79–82 days. Litter size 1–2, exceptionally 3. Kittens independent at 17–22 months. MORTALITY Annual mortality (Texas) is 8% (resident adults) to 47% (dispersers). Predators include large cats, Coyote and domestic dogs. LIFESPAN 20 years in captivity. **Status and Threats** Widespread but reliant on dense habitat and has low reproductive potential. Vulnerable to habitat loss, illegal hunting and persecution for depredation. Texan Ocelots die mainly from anthropogenic factors, especially roadkills. CITES Appendix I, Red List LC.

GEOFFROY'S CAT

ONCILLA

MARGAY

OCELOT

Guigna, Kodkod

HB ♀ 37.4–51cm, ♂ 41.8–49cm; T 19.5–25cm; W ♀ 1.3–2.1kg, ♂ 1.7–3kg

Tiny cat, grey-brown to russet-brown with small, dark dab-like spots coalescing into irregular lines on the back and nape. Face distinctively marked with dark stripes under the eyes bordering the muzzle, resembling that of a Puma kitten. Melanism is common, sometimes with rich brown (rather than black) extremities on which the markings are obvious. **Distribution and Habitat** C and S Chile including Chiloé Island and marginally in adjacent Argentina. Strongly associated with dense temperate habitats, especially evergreen forest, montane forest, thicket and scrubland. Avoids open land and plantations with little understorey, but uses secondary forest, forested ravines and coastal forest strips in cleared habitat. **Feeding Ecology** Hunts small rodents and marsupials, e.g. Monito del Monte, ground-foraging birds such as tapaculos, ovenbirds, thrushes and lapwings, and small reptiles and insects. Regularly kills domestic poultry in fragmented human-dominated landscapes, e.g. Chiloé Is, and considered a pest. Reports of goat killing and hunting in groups are implausible. Cathemeral. **Social and Spatial Behaviour** Solitary. Ranges overlap considerably, including in core areas, suggesting limited territoriality. Range size 0.6–2.5km² (♀s) and 1.6–4.4km² (♂s). Density estimates from radio-telemetry, counting adults and subadults, 1–3.3 Guiñas/km². **Reproduction and Demography** Unknown from the wild; cold winters possibly drive seasonal breeding. Gestation 72–78 days. Litter size 1–3. MORTALITY Poorly known: where studied, humans are the main cause of death. LIFESPAN 11 years in captivity. **Status and Threats** Very restricted distribution and dependent on dense habitat. Forest loss for agriculture and pine plantations has reduced its range to many small fragmented populations, which are further threatened by illegal killing, mainly over poultry depredation. CITES Appendix I, Red List VU.

COLOCOLO *Leopardus colocolo*

Pampas Cat

HB 42.3–79cm; T 23–33cm; W 1.7–3.7kg

Highly variable, ranging from smoky-grey to dark rusty-brown, with little or no spotting to rich russet-coloured blotches on the body. Based on morphology, the species clusters into three major groups: 'Colocolo' in Chile, W side of the Andes; 'Pampas Cat' in Colombia to S Chile, E side of the Andes; and 'Pantanal Cat' in Brazil, Paraguay and Uruguay. They are sometimes considered separate species, but all forms intergrade, and genetic data indicate only moderate differences. Melanism is recorded from Brazil. **Distribution and Habitat** S Colombia to S Chile, extending into C Brazil, Paraguay and Uruguay. Occupies more habitat types than any other Latin American felid, including pampas and cerrado grassland, woodland savannah, marshland, open forest, cloud forest, semi-arid desert and Andean steppes to 5000m. Does not occur in rainforest. Tolerates plantations and agricultural habitat with cover, e.g. maize cropland. **Feeding Ecology** Focuses mainly on small mammals, especially Tuco-tuco, Mountain Viscacha, chinchilla rats, leaf-eared mice, rats and introduced European Hare. Other notable prey includes flamingos (probably chicks), tinamous, Magellanic Penguin chicks, small reptiles, eggs and beetles. Kills domestic poultry. Foraging generally terrestrial and nocturnal, but varies with the region; almost entirely diurnal in the Brazilian cerrado, perhaps due to the presence of nocturnal large cats. Scavenges from carcasses, including those of livestock, Vicuña and Guanaco. **Social and Spatial Behaviour** Unstudied in the wild. Solitary. Radio-telemetry studies are underway in the Brazilian cerrado and Bolivian Andes. **Reproduction and Demography** Unknown from the wild. Gestation 80–85 days. Litter size 1–3. MORTALITY Poorly known. Apparently frequently killed by domestic dogs in some locations, e.g. NW Argentina. LIFESPAN 16.5 years in captivity. **Status and Threats** Widely distributed with a broad habitat tolerance, and often the most common felid present (in camera-trap surveys). Killed for raiding poultry and vulnerable to shepherds' dogs, especially in open Andean habitats. In rural areas, it is killed for religious ceremonial uses in which the skin or a stuffed cat is believed to confer fertility and productivity on domestic livestock and crops. CITES Appendix I, Red List NT.

ANDEAN CAT *Leopardus jacobita*

Andean Mountain Cat

HB 57.7–85cm; T 41–48cm; W (single ♂) 4kg

Silver-grey marked with russet blotches on the body that darken to rich grey-brown on the face, limbs and tail. Tail very thick and bushy, with distinctive thick banding that becomes paired brown or russet rings, often with mid-brown centres, towards the tip. **Distribution and Habitat** S Peru to NW Argentina; restricted to High Andean habitats mostly at 3000–5100m. Has been recorded from the Patagonian steppe at 1800m (Mendoza, Argentina), and 600–700m (Neuquén, Argentina). Occurs only in semi-arid to arid treeless habitats with rocky slopes and cliffs, and associated shrubland and grassland. **Feeding Ecology** Rodents are the primary prey, especially Mountain Viscacha, in which it seems to specialize. Mice, chinchilla rats, cavies, European Hare and tinamous are also recorded. Mostly nocturnal, with crepuscular activity peaks in high-altitude areas reflecting activity patterns of viscachas. Sometimes scavenges from the carcasses of dead ungulates. **Social and Spatial Behaviour** Solitary. Only one animal, a female in the Bolivian altiplano, has ever been radio-collared; home range estimated at 47.1km² for 4 months of study. **Reproduction and Demography** Poorly known. Kittens have been observed in October–April, suggesting seasonal breeding with spring/summer births. Litter size 1–2. MORTALITY and LIFESPAN Unknown. **Status and Threats** Status virtually unknown. During surveys, evidence of Andean Cat is found far less frequently than that of Colocolo and other carnivores, suggesting that the species is naturally rare. It has a very restricted distribution and narrow habitat preference, which is vulnerable to livestock grazing and agriculture. This may also impact prey numbers, especially combined with hunting of prey, particularly viscachas, which is considered a serious threat. Killed for religious beliefs (see Colocolo) and for suspected poultry and livestock killing; killed by goat-herders and their dogs in Neuquén Province, Argentina, even though it does not prey on stock. CITES Appendix I, Red List EN.

GUIÑA

Colocolo
form

COLOCOLO

Pantanal
form

Pampas
form

**ANDEAN
CAT**

HB ♀ 85–130cm, ♂ 76–148cm; T 12–24cm;
W ♀ 13–21kg, ♂ 11.7 29kg
Largest lynx. Colour varies from silver-grey and tawny to red-brown, with highly variable spotting, including unspotted, small coin-like spots, large dab-like spots and elongated brown rosettes. Colouration varies considerably between and within populations. **Distribution and Habitat** Fenno-Scandinavia, Russia (75% of range), China and temperate C Asia, with scattered populations in E and W Europe. Inhabits mainly forest, montane areas with cover, cold semi-desert, tundra, open woodland and scrub. Occurs to 4700m in the Himalayas, exceptionally to 5500m. **Feeding Ecology** In contrast to other lynx, hunts mainly small to medium-sized ungulates. The most important species across much of its range is Roe Deer, followed by Chamois, Musk Deer and juveniles of Red Deer, Sika, Moose, ibex and Wild Boar. Ungulates are especially important prey in winter, when snow elevates vulnerability to predation: exceptional kills of adult Red Deer occur in deep snow. Hares, small rodents, squirrels, marmots and birds increase in importance during spring and summer. In far northern forests, where ungulates are less common, Mountain and Brown Hares are the most important prey year-round. Kills livestock and poultry, including frequent predation on semi-domestic Reindeer in Finland, Norway and Sweden. Mainly nocturno-crepuscular, but may be cathemeral, especially in winter and during the breeding season. Scavenges, and sometimes caches large carcasses with a covering of ground debris. **Social and Spatial Behaviour** Solitary. Male ranges are larger than female ranges, and overlap more extensively.

Both sexes demarcate territorial boundaries with urine, scent and faecal marks, but ranges are generally too large to permit exclusivity, except among females with small kittens. Range sizes increase from south to north, reflecting prey availability; 98–1850km² (♀ s) and 180–3000km² (♂ s). Average range size in W/C Europe 106–264km² (both sexes combined), compared with 307–1515km² in Scandinavia. Largest recorded ranges are in Norway. Density estimates include 0.25 lynx/100km² (S Norway) to 1.9–3.2/100km² (Poland). **Reproduction and Demography** Seasonal. Mates February–mid-April; births May–June. Oestrus 3–5 days; gestation 67–74 days. Litter size 1–5, typically 2–3. Independent at 9–11 months, with dispersal occurring before the mother's next litter. Females first breed at 22–24 months, males at 3 years. MORTALITY Kitten mortality usually at least 50%; 59–60% of Swiss kittens die before independence. Natural adult mortality low, only 2% annually in Scandinavia, but anthropogenic factors increase that by a factor of eight; 44–60% of subadults die during dispersal (Switzerland). Predation occurs occasionally by Grey Wolf, Tiger and Wolverine (on young animals). LIFESPAN 17 years in the wild, 25 in captivity. **Status and Threats** Considered secure, with large areas of its massive range still intact, especially in Russia (where the population is estimated at 30,000–35,000), Mongolia and China. Extirpated from most of W and C Europe, where remaining populations are small and isolated. No longer legally hunted for fur, but illegal trade continues in its Russo-Asian range. Sport hunting is legal in much of Europe and Russia. CITES Appendix II, Red List LC.

IBERIAN LYNX *Lynx pardinus*

Spanish Lynx, Pardel Lynx
HB ♀ 68.2–75.4cm, ♂ 68.2–82cm; T 12.5–16cm;
W ♀ 8.7–10kg, ♂ 7–15.9kg
Tawny-grey to reddish-brown with solid spotting or blotches, sometimes breaking up into freckling. Both sexes have a prominent white facial mane with black streaks, and the ears have long black tufts. Tail ends in a black tip. **Distribution and Habitat** Restricted to two populations in Doñana and Sierra Morena, S Spain. Occurs in dense mosaics of forest, thicket, brushland and Mediterranean scrub, with open pastures and edges favoured for hunting. Shuns agricultural land and exotic plantations, but uses pine plantations for dispersal. **Feeding Ecology** Highly reliant on European Rabbit, which comprises 75–93% of the diet, depending on the location and season. Cannot live in areas without abundant rabbits, and requires an estimated 277 a year for a female without kittens, to 379 for a male. Incidental prey includes small rodents, hares and birds, including ducks, geese, partridges, magpies and pigeons. Juvenile Red and Fallow Deer are sometimes killed during autumn and winter. Kills other carnivores, including Red Fox, Egyptian Mongoose, Common Genet and feral domestic cats, but these are rarely eaten and probably killed as competitors for rabbit prey. Kills rabbits by biting the skull; larger prey such as deer are killed by suffocation. Does not kill livestock or poultry, due largely to very limited opportunity in its remaining range. Covers larger kills with leaf litter and debris to consume over a number of days.

Social and Spatial Behaviour Solitary and territorial. Each male range overlaps one and sometimes two female ranges, with intra-sexual overlap at the edges and exclusive core areas. Territorial fights are occasionally fatal. Range size 8.5–24.6km², averaging 12.6km² (♀ s), and 8.5–25km², averaging 16.9km² (♂ s). Density 0.1–0.2 lynx/km² in poor habitat to 0.72–0.88/km² in optimal habitat. **Reproduction and Demography** Seasonal. Usually mates January–February, with births peaking March and occasional births April–June. Litter size 2–4, averaging 3. Kittens independent at 7–8 months, remaining in their natal range until dispersal at 18–20 months. Females can breed at 2 years, but usually first give birth in the wild at 3 years, and breed until age 9. MORTALITY Kitten mortality around 33%; two kittens usually survive from most litters of three. Adult mortality around 10% for resident animals insulated from human factors, but anthropogenic mortality is now the major reason why Iberian Lynxes die. LIFESPAN 10 years in the wild, 14 in captivity. **Status and Threats** The world's most endangered felid, estimated to number 84–143 adults in two isolated populations in Spain. Not detected unequivocally in Portugal since the 1990s. Decline has been driven by massive habitat conversion of forest to exotic plantations, combined with disease epidemics in rabbits and direct human killing of lynxes. Anthropogenic killing remains a serious threat, responsible for 75% of lynx deaths mostly by illegal trapping, shooting and roadkills. CITES Appendix I, Red List CE.

EURASIAN LYNX

IBERIAN
LYNX

BOBCAT *Lynx rufus*

Plate 10

Bay Lynx, Red Lynx
HB ♀ 50.8–95.2cm, ♂ 60.3–105cm; T 9–19.8cm;
W ♀ 3.6–15.7kg, ♂ 4.5–18.3kg
Various shades of grey to rusty-brown, with markings ranging
from very minimal spotting to large, Ocelot-like blotches.
Bob-tail has 3–6 dark half-stripes and a vivid white underside
and tip, distinguishing the species from Canada Lynx.
Melanistic individuals occasionally occur, mostly recorded
from the SE USA. **Distribution and Habitat** Southern
Canada, the USA and N Mexico. Very wide habitat
tolerance, including all forest types, brushland, scrub,
prairies, semi-desert and mountainous terrain. Tolerant of
farmland, agricultural land and peri-urban landscapes
provided there is cover. **Feeding Ecology** Recorded killing
adult deer weighing up to 68kg, but typical prey is hare-sized
or smaller. Lagomorphs are key prey throughout its range,
especially Snowshoe Hare, Cottontail, jackrabbits and
Marsh Rabbit. White-tailed Deer, Mule Deer and Pronghorn
are taken primarily as fawns, but northern Bobcats kill more
adults, especially in winter. Other prey includes rodents to
the size of porcupines, smaller carnivores, opossums, birds,
herptiles, fish, arthropods and eggs. Kills sheep, goats and
poultry, though problems are usually localized. Sometimes
kills small pets in peri-urban areas. Forages mostly on the
ground. Generally nocturnal with crepuscular activity peaks.
Sometimes caches carcasses with a covering of dirt or snow to
consume over time, e.g. up to 14 days for an adult deer kill.
Eats carrion: road- and winter-killed deer are important food
sources, especially in northern winters. **Social and Spatial
Behaviour** Solitary and territorial. Range size and overlap
decrease with increasing prey availability. Territorial fights are
occasionally fatal. Average range size 1–2km² (Alabama,
California, Louisiana, Oregon) to 86km² (Adirondacks, NY)
for females, and 2–11km² (Alabama, California, Louisiana,
Oregon) to 325km² (Adirondacks, NY) for males. Ranges
contract during prey peaks, especially of hares and rabbits.
Nightly movements are as large as 20km. Density estimates
include 4–6 Bobcats/100km² (Minnesota), 20–28/100km²
(Arizona, Nevada), exceptionally to >100/100km² (e.g.
coastal California, when protected from hunting). **Reproduction and Demography** Weakly seasonal. Births occur
year-round, but peak spring–summer, strongly so in northern
areas. Gestation 62–70 days. Litter size averages 2–3,
exceptionally to 6. Weaning at 2–3 months. Kittens
independent at 8–10 months. Females can breed at 9–12
months, but usually first give birth after 24 months.
MORTALITY Kitten mortality fluctuates extensively, depending
mainly on prey numbers, e.g. 29–82% mortality in
Wyoming in different years. Adult mortality from 20–33%
for unharvested populations to 33–81% for hunted
populations. Mostly killed by humans, as well as winter
starvation, predation by Puma, Coyote and domestic dogs,
and episodic disease outbreaks in dense populations.
LIFESPAN 16 years in the wild, 32.2 in captivity. **Status and
Threats** Widespread, resilient to human pressures and secure
in most of its range. Nonetheless, some populations are
exposed to intense hunting pressure and vulnerable to over-
harvesting. Around 40,000–50,000 Bobcats are legally killed
in the USA and Canada annually. Persecuted for supposed
livestock depredation, e.g. Mexico, and 2000–2500 are killed
annually in legal control in the USA. CITES Appendix II,
Red List LC.

CANADA LYNX *Lynx canadensis*

Canadian Lynx
HB ♀ 76.2–96.5cm, ♂ 73.7–107cm; T 5–12.7cm;
W ♀ 5–11.8kg, ♂ 6.3–17.3kg
Uniformly coloured, typically buff-grey with silver or bluish
frosting in winter, and brownish in summer, unspotted or
lightly spotted on the limbs. Tail shorter than Bobcat's, with
a completely black tip. Lynx-Bobcat hybrids occasionally
occur where where the two species overlap in Maine,
Minnesota and New Brunswick. **Distribution and Habitat**
Most of Canada south of the treeline and some US border
states, south to Utah. Successfully reintroduced into S
Colorado and unsuccessfully into New York. Closely tied to
dense boreal and coniferous forests; rarely uses open
habitat. **Feeding Ecology** Strongly dependent on
Snowshoe Hare, which comprises up to 97% of the diet
locally and seasonally. Northern hare populations cycle
every 8–11 years, sometimes spectacularly, from 2300
hares/km² to 12/km². Lynx numbers are closely linked,
lagging 1–2 years behind. Lynx switch prey during declines
and during summer. Southern populations (which
experience weak or non-existent hare cycles) have more
diverse diets year-round, though Snowshoe Hare remains
the primary prey. Other important prey includes Red
Squirrel, other rodents, small birds, and game birds like
grouse and Ptarmigan. Ungulate lambs, especially of
Caribou, are sometimes killed, but most ungulates are
scavenged. Rarely kills livestock or poultry. Hunting mainly
crepuscular and nocturnal. Occasionally caches prey by
covering it with snow or leaves. **Social and Spatial
Behaviour** Solitary and probably territorial, but spatial
behaviour varies extensively depending on hare availability.
Southern populations with stable but low densities of hares
tend to maintain large enduring home ranges with high
overlap between neighbours. Northern populations
maintain smaller and possibly more exclusive ranges during
hare peaks, but ranges expand during declines, sometimes
leading to nomadism. Average range size estimates include:
39–133km² (♀s) and 69–277km² (♂s) in southern
populations; 13–18km² (♀s) and 14–44km² (♂s) in
northern populations with high hare numbers; to
63–506km² (♀s) and 44–266km² (♂s) in northern
populations with few hares. Density estimates include 2–4
lynx/100km² during low hare density to 10–45/100km²
during peaks. **Reproduction and Demography** Seasonal.
Mates March–May; births May–early July. Gestation
63–70 days. Litter size 1–8, with more females breeding
and larger litters produced during hare peaks. Independent
at 10–17 months. Females breed as early as 10 months
during high hare years, but usually first breed at 22–23
months. MORTALITY Kitten survival linked closely to hare
numbers, reaching 60–95% mortality in poor years.
Estimates of adult mortality include 11–27% for
unharvested populations to 45–88% for harvested
populations. Starvation and trapping by humans are
responsible for most deaths. LIFESPAN 16 years in the wild,
26.9 in captivity. **Status and Threats** Generally widespread
and secure, but extirpated from most of its US distribution
and its range appears to be retreating north with forestry
and possibly climatic warming. On average, at least 11,000
lynx are legally harvested annually, most in Canada and
Alaska; they are vulnerable to over-harvesting during hare
declines. CITES Appendix II, Red List LC.

BOBCAT

CANADA LYNX

Eyra

HB ♀ 53–73.5cm, ♂ 48.8–83.2cm; T 27.5–59cm; W ♀ 3.5–7kg, ♂ 3–7.6kg

Uniformly coloured with two distinct phases, red-brown (often with a bright white muzzle and chin), and dark grey. Litters can include kittens of both colours. Body and head elongated and lean with a long slender tail, giving the impression of a mustelid in the field. **Distribution and Habitat** C Argentina to N Mexico, and formerly in the Rio Grande Valley, Texas, USA, but not confirmed there since 1986. Occurs from sea level to 2000m (occasionally to 3200m) in all types of forest, scrub, chaparral, brush, dense grassland and pasture. Does well in human-modified or recovering habitat with cover and high rodent densities, e.g. pasture grassland, old fields and secondary forest. **Feeding Ecology** Makes kills to the size of an armadillo, but most prey weighs less than 1kg. Principal prey includes rats, mice and birds, especially ground-dwellers like tinamous and quail. Also eats small primates (rarely), cavies, rabbits, opossums, reptiles, fish and arthropods. Brocket deer are recorded in scats, but probably from scavenging. Occasionally raids poultry coops. Appears to be primarily crepuscular/diurnal; a capable climber, but most hunting takes place on the ground. **Social and Spatial Behaviour** Solitary. Exhibits typical felid marking behaviour and is possibly territorial, though radio-tracked individuals overlap extensively. Mean home ranges in Mexico similar for sexes, 8.9km² (♀ s) and 9.3km² (♂ s); different in Brazil, 1.4–18km² (♀ s) and 8.5–25.3km² (♂ s); unusually large in Belize, 20km² (1 ♀) to 88–100km² (2 ♂ s). **Reproduction and Demography** Poorly known; possibly weakly seasonal, though reports are contradictory. In captivity, gestation 72–75 days, and litters number 1–4 (average 1.8–2.3). Weaning begins at around 5–6 weeks. Sexual maturity at 17–26 months. MORTALITY Poorly known. Recorded predators include Puma, and domestic dogs near villages. LIFESPAN 10.5 years in captivity. **Status and Threats** Tolerant of human activity and seldom hunted for its unicolour fur. Persecuted for killing poultry, and common roadkill in some areas, but most populations are secure. Widespread and relatively common in S America. Endangered in C America; considered Endangered in the USA, but possibly extinct there. CITES Appendix I (C and N America), otherwise Appendix II, Red List LC.

PUMA *Puma concolor*

Cougar, Mountain Lion, Panther (Florida)

HB ♀ 95–141cm, ♂ 107–168cm; T 57–92cm; W ♀ 22.7–57kg, ♂ 39–80kg (exceptionally to 120kg)

Uniformly coloured, ranging from light grey through tawny-brown to brick-red, with creamy white underparts. Temperate Pumas tend to be larger with paler, greyish colouration, while tropical individuals are smaller with richer reddish tones. Tail-tip and backs of the ears are dark brown to black, and the white muzzle is bordered by black. Long tubular tail is distinctive in the field. Cubs have rich dark brown spots that usually fade within the first year. **Distribution and Habitat** Relatively widely distributed in SW Canada, the W USA and S America; more restricted in Mexico and C America, and extirpated from the E USA except for 100–120 in S Florida ('Florida Panther'). Very wide habitat tolerance provided there is vegetation or rocky terrain, including temperate and tropical forests, woodland, coastal and desert scrublands, and rocky desert from sea level to >4000m. Mostly shuns open areas, but readily passes though marginal habitat. Lives close to humans provided cover and prey are available. **Feeding Ecology** Broad diet encompassing arthropods to adult male Elk. Large kills are more common in temperate populations, where deer, Moose, Elk and wild sheep form the principal prey; Guanaco is a key prey species in Chile. Tropical Pumas focus more on smaller prey such as brocket deer, Pudu, peccaries, Capybara, Paca, agoutis and armadillos. Locally, the diet may be dominated by feral livestock, e.g. pigs (Florida) and wild horses (Nevada). Kills reptiles to the size of adult caimans and alligators, and birds such as Wild Turkey, various geese and Rhea. Readily kills livestock, and sometimes takes domestic pets in peri-urban areas. Humans are rarely killed, e.g. 20 recorded fatalities in N America in 1890–2004. Hunting mainly nocturno-crepuscular. Typically discards entrails of large kills before covering them in dirt or leaf litter, consuming the carcasses over a period of 3 days to 4 weeks (winter). Scavenges, though this usually represents a small amount of its intake. **Social and Spatial Behaviour** Solitary and territorial, but often with considerable overlap in ranges. Territorial fights are sometimes fatal. Home range varies from 25–1500km², averaging 33km² (Venezuela) to 685km² (Utah) for females, and 60km² (Venezuela) to 826km² (Utah) for males. Florida females disperse 6–32km, males 24–208km: a Wyoming male dispersed 1037km and was killed by a train in Oklahoma. Density varies from 0.3 Pumas/100km² (e.g. Utah, Texas) to 1–3/100km² (Alberta, California, Utah, Wyoming), exceptionally to 7/100km² (Vancouver Island). **Reproduction and Demography** Weakly seasonal. Births occur year-round, but typically peak summer (Yellowstone NP and Canada) or spring (Florida). Oestrus lasts 1–16 days; gestation averages 92 days (range 82–98 days). Litter size averages 2–3 kittens, exceptionally to 6. Weaning at around 4–5 months. Cubs independent at around 18 months (range 10–24 months). Females first give birth at 18 months (typically >24 months) and males first breed at around 3 years. MORTALITY 36–58% of cubs die in their first year. Estimates of adult mortality include 9% (♂ s) to 18% (♀ s) in New Mexico, and 3% (♀ s) to 61% (♂ s) in NW Montana. Mostly killed by humans (legal hunting, roadkills, illegal killing); principal natural factors are other Pumas and, less so, disease and hunting accidents. LIFESPAN 16 years in the wild, 20 in captivity. **Status and Threats** Tolerant of human activity, but extirpated from around 40% of its Latin American range and most of its eastern N American range. Habitat loss combined with intense persecution in livestock areas are the major threats, especially in Latin America. Legal sport hunting kills 2500–3500 Pumas per year in the USA, triggering population declines in some states. Road accidents are a major threat to Florida Panther, e.g. 17 killed in 2009. CITES Appendix I (Florida, Nicaragua through Panama), elsewhere Appendix II; legally hunted in Argentina, Canada, Mexico, Peru and the USA. Red List LC.

Red form

JAGUARUNDI

Grey form

Temperate
form

PUMA

Tropical
form

Cubs

HB ♀ 105–140cm, ♂ 108–152cm; T 60–89cm; SH ♀ 67–89cm; W ♀ 21–51kg, ♂ 29–64kg
Yellow-blond fading to white underparts, with black coin-like spots and unique facial tear streaks. Saharan Cheetahs have very short pale fur, ranging from brown-spotted beige to near-white with faint cinnamon spots. King Cheetahs are a recessive colour phase and may be born to normally spotted parents. Cubs have a fluffy smoky-grey mantle that dwindles by 4–5 months to a short mane on the shoulders, which is inconspicuous in most adults but often obvious in Asiatic Cheetahs. Mantle's function is unclear, but it probably assists with camouflage and thermoregulation; mimicry of Honey Badger to deter predators, as is often claimed, is doubtful. Claws dog-like and lack fleshy claw-sheaths present in other cats, but are partially protractile. Claws appear in tracks, except for the sharp, strongly curved dewclaw, which is used for prey capture.

Distribution and Habitat
Relatively widely distributed in southern and E Africa, rare in W Africa, and extinct in N Africa except southern Algeria and perhaps Egypt. Extinct in Asia except for 70–110 in Iran. Favours woodland-savannah mosaics and open grassland, becoming sparser in dense humid woodland, e.g. Zambian miombo, and absent from rainforest. Tolerates arid habitats including the deserts of C Iran (which experience winter snowfall), Namib and Sahara, but transient in the driest areas. Recorded to 3500m (Mt Kenya, Kenya).

Feeding Ecology
Usually hunts ungulates weighing 20–60kg, particularly gazelles and Impala, occasionally up to Nyala (62–108kg), and juveniles of larger ungulates including wildebeest, zebra and (rarely) buffalo and Giraffe. Male coalitions take large prey, e.g. adult hartebeest, oryx and wildebeest. Iranian Cheetahs eat mainly Urial Sheep and Persian Ibex following the widespread extirpation of gazelles. Hares are important prey to recently independent young adults and in marginal habitats, e.g. Iran. Kills small untended livestock, but is easily deterred by people or dogs, and is not recorded killing humans. Hunting mostly diurnal to maximize visibility and avoid nocturnal competitors. Following a short and careful stalk, pursues prey for up to 500m at a maximum recorded speed of 105km/h. Prey is bowled over or pulled off balance using the dewclaw, and usually killed by suffocation. Around 30% of hunts are successful, more so for small prey; 87–93% for hares, 86–100% for juvenile gazelles (Serengeti NP, Tanzania). Rarely defends kills, losing up to 13% (Serengeti NP), chiefly to Spotted Hyaena and Lion. Very rarely scavenges and seldom returns to previously fed-upon carcasses.

Social and Spatial Behaviour
Unusual social system in which females are solitary and non-territorial, while male sociality and territoriality vary widely. Females occupy large home ranges that are not defended. Where prey is scarce or migratory, female ranges reach 1500km² (average 833km²; Serengeti NP) to 6353.7km² (average 1400–1836km²; Namibia). Areas with resident or abundant prey produce smaller ranges (e.g. 185–246km²; Kruger NP), though females are non-territorial regardless. Dispersing females usually settle near their natal range, so that neighbouring females are often related. Males typically disperse further to avoid breeding with female relatives, and live in permanent coalitions of 2–4, usually littermates; around a third of Serengeti coalitions include an unrelated member. Coalitions defend territories where profitable, repelling rival males in fights that may be fatal. Serengeti coalitions target small areas with high female overlap averaging 37km². Males in woodland habitats with resident females establish medium-sized territories, e.g. Kruger NP 126km² (3 ♂ coalition) to 195km² (single ♂). Coalitions defend territories more successfully than loners, and single males are often nomadic 'floaters' with much larger home ranges, averaging 777km² (Serengeti) to 1829km² (Namibia). An pair of Iranian males (likely nomadic) used 1737km² in 5 months of radio tracking. Coalitions sometimes also float on large ranges (average 1608.4km²; Namibia) as an alternative to territorial defence. Cheetahs naturally occur at low densities: 0.16/100km² (Iran), 0.25–2/100km² (Namibian farmland), 0.5–2.30/100km² (Kruger NP) to 9/100km² (Serengeti NP). Seasonal densities exceptionally reach 40/100km² (Serengeti NP).

Reproduction and Demography
Cheetahs have a reputation for poor reproduction arising from the difficulties of breeding them in captivity, but wild Cheetahs are prolific. Aseasonal, though Iranian Cheetahs apparently mate in winter (January–February) and give birth in spring (April–May). Gestation 90–98 days. Litter size typically 3–6, rarely to 8; litter of 9 from Kenya may have included an adoption. Inter-litter interval averages 20.1 months (Serengeti). Weaning begins at 6–8 weeks and is complete by 4–5 months. Cubs independent at 12–20 months (average 17–18 months) and disperse as a sib-group; females leave the group before sexual maturity, while males remain together. Females can conceive at 21–24 months; first give birth around 29 months (Serengeti) and can reproduce to 12 years. Males are sexually mature at 12 months, though they rarely breed before 3 years. MORTALITY 95% of Serengeti cubs die before independence, most killed by Lions. Losses in the denning period are the highest, and only 20% of Serengeti cubs that emerge at >8 weeks survive. Mortality lower in denser habitat and/or with fewer Lions; post-emergence estimates include 38% (Phinda, S Africa), 43% (Nairobi NP, Kenya) and 50% (Kruger NP). Adults occasionally killed by Lion, Leopard and Spotted Hyaena; males killed in territorial fights, and some adults killed in hunting accidents. Despite high genetic homogeneity, wild Cheetahs suffer little disease. LIFESPAN Maximum 13.5 (average 6.2) years for Serengeti females, 9.3 (average 5.3) for males; 21 in captivity.

Status and Threats
Reduced to approximately 23% of its historic African range and a single population in Asia. Widely persecuted by livestock farmers despite causing relatively minor damage, and profoundly affected by reduction of prey in pastoral areas. Human hunting of prey is critical in Sahel, N Africa, and Iran, where Cheetahs are naturally very rare. Limited hunting for skins occurs, e.g. Sahel and NE Africa, where there is also a significant trade in live cubs and adults, mainly to the Arabian Peninsula. High genetic homogeneity has had little impact on wild populations. CITES Appendix I permitting trade of approximately 200 hunting trophies (Botswana, Namibia and Zimbabwe) and live animals. Red List VU (Global), CE (Asia).

CHEETAH

Saharan
form

Typical
form

Cub

King Cheetah

Ounce

HB ♀ 86–117cm, ♂ 104–125cm; T 78–105cm; SH to 60cm; W ♀ 21–53kg, ♂ 25–55kg

Dark cream to smoky-grey with large dark grey or black open blotches, and smaller solid markings on the head and legs. Muscular tail proportionally the longest of any felid, used for balance during hunts and wrapped around the body to insulate it against extreme cold. The species is unmistakable, but central Asian Leopard (*Panthera pardus*) in pale long winter coat is sometimes mistaken for it (chiefly in fur markets). **Distribution and Habitat** Found in 12 countries in C Asia throughout the world's highest mountain ranges, including the Altai, Himalayas, Karakoram, Hindu Kush, Pamirs and Tien Shan. Uniquely adapted to high and steep rugged terrain, and copes well with deep snow. Also uses meadows, steppes, wide valleys and open montane desert, but mainly to move between rocky habitat. Occurs at 900–1500m, e.g. in Mongolia's Gobi, to 5500m. **Feeding Ecology** Dependent on mountain ungulates, especially Asiatic Ibex and Blue Sheep, which are the main prey in much of its range. Other prey includes Argali, Markhor, Urial, Himalayan Tahr, Musk Deer, and occasionally gazelles, and juveniles of Eurasian Wild Boar, Wild Yak and Asiatic Wild Ass. Large kills are supplemented by small prey, especially during the summer, when herbivores disperse to higher elevations. Marmots are the most important of smaller prey; also opportunistically takes rabbits, hares, pikas and game birds such as Tibetan Snowcock and partridges. Kills domestic animals, especially sheep and goats, but also cattle, yaks, camels, horses and dogs; depredation can reach significant levels locally and seasonally when wild prey is scarce. No recorded fatal attacks on humans. Stalks in typical felid fashion before rushing prey at close range; superbly agile over extraordinarily steep and rugged terrain. Large prey is usually killed by a suffocating throat bite. Eats carrion and has been observed displacing Dholes from carcasses. **Social and Spatial Behaviour** Solitary. Maintains stable ranges that are regularly marked, but extent of actual territorial defence is unknown. Ranges are very large given naturally low prey densities of habitat; published estimates from ground-based radio-telemetry are probably gross under-estimates, e.g. a Mongolian female's calculated range went from 58km^2 to at least 1590km^2 (and possibly >4500km^2) when fitted with a satellite collar. Male ranges 181–1628km^2 (averaging 540km^2; Gobi, Mongolia). Daily movements may be as much as 28km. Species is extremely difficult to count, but estimates from camera trapping vary from 0.15 leopards/100km^2 (SaryChat, Kyrgyzstan) to 4.5/100km^2 in prey-rich habitat (Hemis NP, India). **Reproduction and Demography** Poorly known from the wild; likely to be a seasonal breeder given that it experiences extreme winters. Calling and scent-marking peaks January–March, which would coincide with spring births; captive births peak May. Gestation 90–105 days. Litter size averages 2–3 cubs, exceptionally to 5. Weaning at 2–3 months. Age at independence unknown, but probably occurs at around 18 months. MORTALITY Unknown in the wild. LIFESPAN 20 years in captivity. **Status and Threats** Somewhat insulated from human activities given that it lives in such remote inhospitable areas, but it is naturally rare, and human populations and their livestock are increasing in its habitat. Main threats relate to prey availability; widely killed for livestock depredation, often compounded by prey loss from human hunting and competition with livestock. Furs and especially body parts have commercial value, chiefly in China; regular confiscations indicate widespread hunting. CITES Appendix I, Red List EN.

INDOCHINESE CLOUDED LEOPARD *Neofelis nebulosa*
DIARDI'S CLOUDED LEOPARD *Neofelis diardi*

HB ♀ 68.6–94cm, ♂ 81.3–108cm; T 60–92cm; W ♀ 10–11.5kg, ♂ 17.7–25kg

Recent molecular and morphological analyses have shown that island (Sumatra and Borneo) Clouded Leopards comprise a separate species, *N. diardi*, from the mainland species, *N. nebulosa*. Diardi's (or Sunda) Clouded Leopards are generally darker, with smaller cloud markings each often enclosing small spots. Clouded Leopards display an exceptionally large gape and elongated canine teeth, similar to those of the extinct sabertooth cats (to which they are not closely related), the reasons for which are unclear. **Distribution and Habitat** Mainland Indochina (Indochinese Clouded Leopard), and Borneo and Sumatra (Diardi's Clouded Leopard). Closely associated with dense evergreen tropical forest, but also occurs in secondary and logged forests, swamp forest, dry woodland and mangroves. Radio-collared individuals used grassland patches in open-forest mosaics for hunting. **Feeding Ecology** Ecology poorly known, but confirmed prey includes primates (ranging from Slow Loris to adult male Proboscis Monkey), Hog Deer, muntjacs, Bearded Pig, Malayan Pangolin, Asiatic Bush-tailed Porcupine, small rodents, civets and birds. Anecdotes of predation on Orang-utans have not been confirmed. Unambiguous kills of deer and pigs revealed a deep killing bite to the back of the neck, an unusual technique for large prey and perhaps related to the unique dentition. Recorded occasionally killing small livestock and poultry. Probably hunts mostly on the ground, but is highly agile and arboreal; there are four published observations of Clouded Leopards attacking Proboscis Monkeys in trees. **Social and Spatial Behaviour** Only eight animals have ever been radio-collared, in Borneo, Nepal and Thailand. Probably follows felid pattern of overlapping territories with exclusive core areas. Limited data suggest sexes have similar territory sizes: 34–40km^2 (♀s) and 35.5–43.5km^2 (♂s). **Reproduction and Demography** Probably aseasonal. Gestation 85–95 days (rarely to 109 days). Litters average 2–3 (range 1–5). Weaning begins at around 7–10 weeks. Sexual maturity at 20–30 months. MORTALITY Humans are responsible for most deaths in studied areas; otherwise mortality is unknown. LIFESPAN 17 years in captivity. **Status and Threats** More resilient than larger felids, but status is uncertain. Threatened by widespread habitat loss and conversion, especially to rubber and palm plantations or agriculture. Skins, bones and meat have commercial value, and Clouded Leopards are illegally hunted in much of their range, including in protected areas, e.g. at last seven killed in Kerinci Seblat NP (Sumatra), 2000–2001. CITES Appendix I, Red List VU.

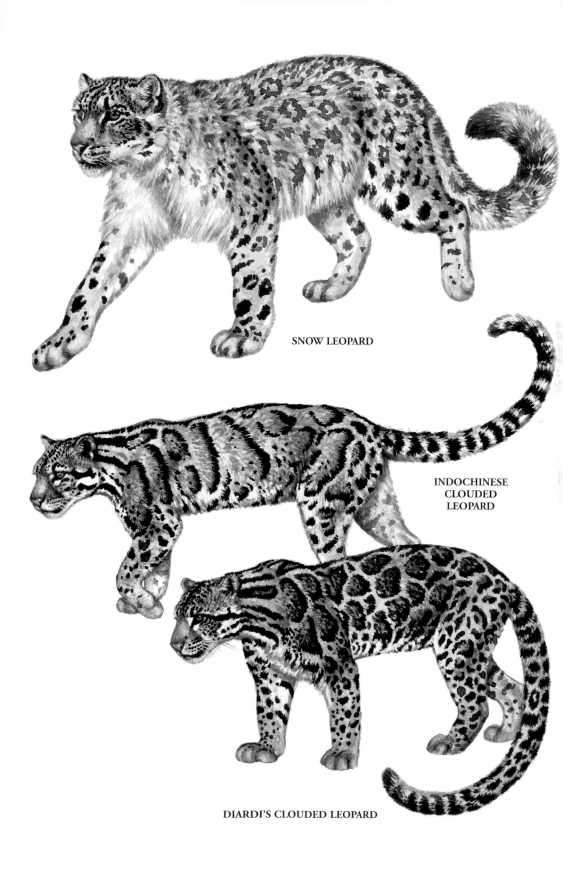

SNOW LEOPARD

INDOCHINESE
CLOUDED
LEOPARD

DIARDI'S CLOUDED LEOPARD

HB ♀ 146–177cm, ♂ 189–300cm; T 72–109cm; SH 80–110cm; W ♀ 75–177kg, ♂ 100–261kg (exceptionally to 325kg)

Background colour varies from pale yellow to rich red with white or cream underparts. Generally darker and more richly striped in tropical S Asia, and paler and more lightly striped in temperate areas. White Tigers are not albino, and arise from a recessive mutation producing blue eyes and chocolate-coloured stripes on a white background. There is only one record from the wild (Madhya Pradesh, India) since 1951, a male cub from which all captive white Tigers are descended (and, hence, are extremely inbred). An intermediate form called 'golden tabby' or 'strawberry' is known only from captivity. True melanism is unknown, though individuals occasionally appear with extensive coalesced striping, producing an almost entirely black appearance. Size varies very widely: Sumatran Tigers are the smallest with males up to 140kg, while individuals from the Indian subcontinent and Russia are the largest. Largest wild male on record, from Nepal, weighed 261kg (up to 325kg recorded from captivity). Classified into nine subspecies, three of them extinct. Sumatran Tiger is clearly sufficiently genetically isolated to be accorded its own subspecies; however, continental populations were probably continuous in historical times, so current subspecies classification may reflect only modest differences between populations.

Distribution and Habitat
Restricted to 7% of its historic range, from W India to the Himalayas, through Indochina to Malaysia, Sumatra, the Russian Far East and extreme NE China. Extinct in Bali (1940s), C Asia (1968) and Java (1980s), and likely extinct in SC China. Occurs mainly in various temperate and tropical forests, forest-grassland mosaics and associated dense cover such as Terai (dense floodplain grassland), thickets, scrub, marshes, mangroves and reedbeds. Reaches the highest densities on the Indian subcontinent in dry and mesic forests, and Terai. Typically avoids human-modified habitats such as agricultural land, palm plantations and monocultures. Occurs exceptionally up to 4110m (Himalayas, Bhutan), but typically under 2000m.

Feeding Ecology
Adults can kill almost anything they encounter, with the exception of adult rhinos and elephants, but the diet is dominated by various deer species and wild pigs. Typically focuses on ≤5 species of locally common prey such as Sambar, Red Deer, Chital, Hog Deer, Sika, muntjacs and Wild Boar. Capable of killing adult Gaur and Asiatic Water Buffalo weighing up to 1000kg, but most kills of these species are subadults and juveniles. Smaller prey taken relatively often includes primates, porcupines, small carnivores and hares. Readily kills other predators, though they are not always eaten; Leopard, Dhole, Brown Bear and Asiatic Black Bear are recorded prey. Preys on livestock, mainly when untended in forest. Amur Tigers regularly kill domestic dogs, particularly when these are accompanying hunters in forest, and during severe winters that force Tigers into villages. Probably kills more people than any other large carnivore (mainly in India), in part because human populations in Asia are so dense and often utilize Tiger habitat intensively; true 'man-eaters' that focus on humans as prey are rare. Hunting mainly nocturno-crepuscular. Scavenges and appropriates kills of other carnivores.

Social and Spatial Behaviour
Solitary and territorial. Adults establish exclusive territories where possible, but total exclusivity is rare. Territorial overlap is least in high-density populations with abundant prey and small ranges, e.g. Nepal, India. Territorial fights are rare, but sometimes fatal when they occur, more frequently in males than females. Recorded territory size varies from 10km² (♀s; Chitwan NP, Nepal) to 2058km² (♂s; Russia). Range size for Terai and productive forest in Nepal or India 10–51km² (♀s) and 24–243km² (♂s), compared with 181–761km² (♀s) and 434–2058km² (♂s) in Russia. Females mostly settle near their natal range, while males disperse more widely. Dispersal distances at Chitwan average 9.7km for females (maximum 33km) and 33km for males (maximum 65km), compared with 14km for females (maximum 72km) and 103km for males (maximum 195km) in Russia. Densities of populations, even in high-quality habitat, are often depressed due to human hunting of prey (even if Tigers are not hunted). Density estimates vary from 0.2–2.6 Tigers/100km² in lowland tropical forest where poaching is prevalent (Laos, Malaysia, Myanmar, Sumatra), 0.5–1.4/100km² (temperate forest, Russia) to 11.5–19/100km² in well-protected forest and Terai (India).

Reproduction and Demography
Generally aseasonal, but over 50% of Amur Tiger cubs are born late summer, August–October, and winter births are rare. Oestrus 2–5 days; gestation 95–107 days, averaging 103–105. Litter size 2–5, averaging 2–3. Weaning at around 3–5 months. Inter-litter interval 20–24 months, averaging 21 months. Cubs independent at 17–24 months. Females often inherit part of their mother's range, while males disperse more widely. Sexually maturity 2.5–3 years (both sexes): earliest breeding 3 years for females (average 3.4 years in Chitwan) and 3.4 years for males (average 4.8 years, Chitwan). Reproduction by females is possible until at least age 15.5. **MORTALITY** 34% (Chitwan) to 41–47% (Russia) of cubs die in their first year, most related to anthropogenic causes and infanticide. Estimates of adult mortality include 23% (both sexes combined) in Nagarahole, India, and 19% (♀s) to 37% (♂s) in Russia. Humans are the main cause of death for most populations, and adults also die in territorial fights. Accidents (e.g. a male fell through a frozen river in Russia) and disease occur, but are uncommon. **LIFESPAN** To 16 years for females and 12 for males in the wild; 26 in captivity.

Status and Threats
The most endangered large cat, having suffered a calamitous decline in the 20th century that continues today. An estimated 70% of the world's Tigers now live in around 0.5% of their historic range. Combined with loss of habitat to forestry, commercial palm plantations and agriculture, Tigers are particularly threatened by intense illegal hunting to supply the traditional Asian medicinal trade. This is compounded by widespread hunting of their prey to feed a massive demand for bushmeat, especially in SE Asia. Given a respite from human hunting, Tiger populations recover quickly; unfortunately, there are now few areas where this is taking place. CITES Appendix I, Red List EN (Globally), CE (China, Russia, Sumatra).

TIGER

White
form

Bengal
subspecies

Sumatran
subspecies

Amur subspecies

HB ♀ 158–192cm, ♂ 172–250cm; T 60–100cm; SH 100–128cm; W ♀ 110–168kg, ♂ 150–272kg
Typically tawny or sandy with pale underparts, but varying from ash-grey or cream to ginger and (rarely) dark brown. Backs of the ears and tail-tip are a contrasting black or dark brown. White Lions from Kruger NP region, S. Africa, are leucistic (not albino), arising from a recessive gene for coat colour, and have pigmented eyes, nose and pads; they can be born to normally coloured parents. Mane colour ranges from platinum to black with highly variable length. 'Maneless' males occur most often where it is extremely hot, e.g. Tsavo NP, Kenya, and in most Sahelian populations. Cubs are born with dark brown rosettes that fade with age, retained as vestigial spotting in some adults. Mane growth begins at 6–8 months. Asiatic Lions are 10–20% smaller than African Lions, usually have a distinct belly fold (occasionally present in African Lions), and males have somewhat reduced manes.

Distribution and Habitat
Patchy distribution south of the Sahara, chiefly in and around protected areas, with the largest populations in E and southern Africa. Outside Africa, there is only one population of 300–350 in Gujurat, India. Naturally absent only from true desert and equatorial rainforest. Optimum habitat is mesic open woodland and grassland savannah. Inhabits dry deciduous teak forest in India. Recorded to 4200–4300m (Bale Mountains, Ethiopia; Mt Kilimanjaro, Tanzania).

Feeding Ecology
Opportunistic, killing virtually everything it encounters, but prefers large herbivores weighing 60–550kg; cannot persist without large prey. In any given population, the diet is dominated by 3–5 ungulates such as wildebeest, zebra, buffalo, Giraffe, Gemsbok, Impala, Nyala, Kob, Thomson's Gazelle, Chital, Sambar and Warthog. Smallest preferred prey species documented is Springbok (27–48kg) in Etosha NP, Namibia. Smaller prey, especially Warthog, often dominates during the lean dry season where large prey is migratory, e.g. Chobe NP, Botswana and Serengeti NP, Tanzania. Only healthy mature elephants are invulnerable to Lion predation; large prides kill young or unwell adults, as well as adult rhinos and hippos. Kills untended livestock and occasionally preys on humans; isolated pockets of persistent man-eating still occur, e.g. SE Tanzania and N Mozambique. Hunting mainly nocturno-crepuscular. Though females make most kills, males are capable hunters that increase hunting success of very large prey, e.g. buffalo and Giraffe. Success estimates include 15% (Etosha NP), 23% (Serengeti NP) and 38.5% (Kalahari). Readily scavenges, comprising 5.5% of intake in Etosha NP to almost 40% in Serengeti NP. Frequently appropriates kills from other carnivores.

Social and Spatial Behaviour
The only communally living cat. Prides comprise 1–20 (usually 4–11) related lionesses, their offspring and 1–9 (usually 2–4) immigrant males unrelated to breeding females. Pride size is correlated with prey biomass, smaller in arid areas and exceptionally reaching 45–50 under ideal conditions. Female membership of the pride is mostly stable, but the entire pride is together rarely: small subgroups come and go within pride range in a continuous 'fission-fusion' pattern. Females defend their range against other prides and strange males. Coalition males are usually related to each other, but smaller coalitions (2–3) often include unrelated members. Females generally stay with the pride for life, but occasionally disperse following a male takeover or to avoid mating with male relatives. Young males are evicted or leave at 25–48 months, entering a nomadic phase lasting 2–3 years before attempting to acquire their own pride. Coalition members are highly cooperative and remain together for life, defending their territory and females from male intruders. Following a takeover, new males usually kill or evict all unrelated cubs younger than 12–18 months to hasten the lionesses' return to oestrus. Coalition tenure is generally 2–4 years. Serengeti pride territories average 65km² (woodland) to 184km² (grassland), reaching a maximum of 500km². Arid areas produce much larger ranges: to 2800km² (Kalahari), 1055–1745km² (Kaudom GR, Namibia) and 2721–6542km² (Kunene, NW Namibia). Two Kunene male coalitions (possibly nomads) had ranges of 13,365–17,221km². Density varies from 0.05–0.62 Lions/100km² (Kunene), 1.5–2/100km² (Kalahari), 6–12/100km² (Kruger NP), 12–14/100km² (Gir) to 38/100km² (Lake Manyara NP, Tanzania).

Reproduction and Demography
Aseasonal. Oestrus averages 4–5 days; gestation 98–115 days (mean 110). Litter size typically 2–4, up to 7. Lionesses often give birth synchronously and communally care for cubs; females suckle all cubs, but only carry their own. Weaning begins at around 6–8 weeks, but suckling may continue to 8 months. Cubs can hunt independently at around 18 months, but rarely disperse before 2 years. Inter-litter interval 20–24 months. Lionesses can conceive at 30–36 months, but typically first give birth at around 42–48 months and cease reproducing after age 15. Males are sexually mature at 26–28 months, but rarely breed before 4 years. MORTALITY Up to 50–70% of cubs die in their first year, mostly from infanticide, predation and starvation; mortality in the second year drops below 20%. Apart from human-caused mortality, adult Lions usually die in fights with other Lions (especially males), from injuries while hunting large prey and from starvation when old or debilitated. Disease is uncommon, but episodes are occasionally severe; over 1000 Lions (40% of the population) died during the 1993–1994 canine distemper outbreak, Serengeti NP. LIFESPAN 19 years for females, 16 (but rarely over 12) for males in the wild; 30 in captivity.

Status and Threats
Conservation dependent. Now reduced to approximately 20% of its original African range and one tiny Asiatic population, due mainly to eradication by humans, coupled with combined loss of habitat and prey from agriculture and livestock herding. Persecuted intensely by herders, and suffers from widespread official killing of 'problem' animals. Habit of scavenging makes it highly vulnerable to poisoned baits and snares set for bushmeat. Generally declining outside protected areas, and may ultimately persist only in reserves. Genetic impoverishment of small isolated populations possibly leads to declines and vulnerability to disease. Sport hunting is legal in 13 African countries, taking approximately 600–700 (mainly males) annually. CITES Appendix II, Red List VU (Global), EN (W Africa), CE (Asia).

African subspecies

LION

Male

Female

Cubs

Male

Asiatic
subspecies

Panther

HB ♀ 95–123cm, ♂ 91–191cm; T 51–101cm; SH 55–82cm; W ♀ 17–42kg, ♂ 20–90kg

Background colour varies from pale cream, through various shades of orange, to dark rufous-brown with white underparts, covered with rosettes, each a cluster of small black spots around a normally unspotted centre darker than the body colour. Forest Leopards tend to be dark, while those in arid areas are pale. Melanistic individuals ('black panthers') occur widely, usually associated with humid forests; they are most common in tropical Asia, e.g. outnumbering spotted Leopards in Malaysia. Size varies widely, correlated with changes in climate and prey availability. The smallest Leopards, from arid mountainous areas in the Middle East, are about half the weight of African woodland-savannah individuals. Leopards from an isolated population in the coastal Cape Mountains, S Africa, are also small, averaging 21kg (♀ s) to 31kg (♂ s). The largest Leopards (90kg) are recorded from E and southern African woodland and N Iran.

Distribution and Habitat

Widely distributed in southern, E and C Africa, reduced in W Africa and most of Asia, and relict in N Africa, the Middle East and Russia. Very wide habitat tolerance, ranging from Russian boreal forests with winter lows of –30°C, to desert with summer highs of 70°C. Reaches the highest densities in mesic woodland, grassland savannah and forest, and fairly common in mountains, scrub and semi-desert. Absent from open interiors of true desert, but occupies watercourses and rocky massifs in very arid areas. Tolerates human-modified landscapes provided cover and prey are available, e.g. coffee plantations and sugar-cane fields. Recorded exceptionally to 5638m (Mt Kilimanjaro, Tanzania).

Feeding Ecology

Extremely catholic and kills arthropods to adult male Elands (maximum weight 900kg), but prefers medium-sized ungulates weighing 15–80kg. Typical prey includes Steenbok, duikers, Impala, gazelles, Nyala, muntjacs, Chital, Roe Deer, Bushpig and Warthog, and young individuals of larger animals such as Gaur, wildebeest, oryx, hartebeest, Greater Kudu, Sambar and Wild Boar. Additionally, primates, hares, rodents, small carnivores and large birds are often important. Preys on livestock, occasionally entering corrals and settlements, and readily kills domestic dogs. Sometimes preys on humans. A consumate stalk-ambush hunter, approaching prey to as close as 4–5m before a final explosive rush. Hunting mainly nocturno-crepuscular, and most daylight hunts are unsuccessful. Hunting success estimates include 15.6% (Kalahari), 20.1% (Phinda, S Africa) and 38.1% (NE Namibia). Kalahari Leopards average 111 (♂ s) to 243 (♀ s) kills annually, with females making more smaller kills. Leopards hoist carcasses weighing up to 91kg into trees to avoid kleptoparasitism, and occasionally also cache in caves, burrows and kopjes. They typically pluck fur before feeding, usually starting at the underbelly or hind legs. They readily scavenge.

Social and Spatial Behaviour

Solitary and territorial. Adults defend a core area against same-sex conspecifics, but tolerate considerable overlap at the edges, with mutual avoidance and alternating use of shared areas. Territorial fights are uncommon, but may result in fatalities in both females and males. Males associate with familiar females and cubs for as long as 24 hours, but never form permanent family groups. Recorded territory size 5.6km² (♀ s; Tsavo NP, Kenya) to 2750.1km² (♂ s; Kalahari). Mean range size for mesic woodland, savannah and rainforest averages 9–27km² (♀ s) and 52–136km² (♂ s). Ranges are much larger in arid habitats, averaging 188.4km² (♀ s) and 451.2km² (♂ s) in northern Namibia, 488.7km² (♀ s) and 2321.5km² (♂ s) in Kalahari. A collared male in arid rocky habitat, central Iran, used 626km² in 10 months. Density varies from 0.5 Leopards/100km² (Etosha NP, Namibia), 1–1.4/100km² (Primorski Krai, Russia), 1.3/100km² (Kalahari), 4.6–12/100km² (Gabon rainforest), 11.1/100km² (Phinda-Mkhuze GRs, S Africa) to 16.4/100km² (southern Kruger NP). In African woodland savannah, average density under protection (10.5/100km²) is almost five times as high as outside protected areas (2.1/100km²).

Reproduction and Demography

Poorly known in its northern range (China, N Korea, Russia), where extreme winters might give rise to seasonality; otherwise aseasonal. Oestrus 7–14 days; gestation 90–106 days. Litter size normally 1–4, rarely 6 (captivity). Weaning begins at around 8–10 weeks and suckling typically ceases before 4 months. Inter-litter interval averages 16–25 months. Cubs independent at 12–18 months; earliest in which cubs survive 7 months. Female dispersers often inherit part of their mother's range, while males disperse more widely. Both sexes are sexually mature at 24–28 months; females first give birth at 30–36 months, and can reproduce to 16 years (19 in captivity). Males first breed at around 42–48 months. MORTALITY 50% (Kruger NP) to 90% (Kalahari) of cubs die in their first year, most killed by Lions. Estimates of adult mortality include 18.5% (Kruger NP) to 25.2% (Phinda). Aside from deaths attributable to humans, adults are killed primarily in territorial fights and by other predators, especially Lions and occasionally Tigers, Spotted Hyaenas, African Wild Dog and Dhole packs, baboon troops (rarely) and large crocodiles. Hunting accidents and deaths from disease are uncommon. LIFESPAN To 19 years for females, 14 for males in the wild; 23 in captivity.

Status and Threats

Surprisingly tolerant of human activity and persists where other large carnivores cannot. Even so, extirpated from approximately 37% of its African range and reduced elsewhere, with N African, Middle Eastern and Russian populations Critically Endangered. Loss of habitat and prey, closely followed by intense persecution in livestock areas, are the chief threats. Heavily hunted in south Asia for skins and parts supplying the Chinese medicinal trade, and killed for skins, canines and claws in W and C Africa. Bushmeat hunting, especially in tropical forest, competes directly for principal prey species and may drive extinctions even in intact forest. CITES Appendix I permitting 12 African countries to export sport hunting trophies (2010 total quota: 2648), in addition to small numbers of live animals, and skins sold commercially mainly as tourist souvenirs. Red List NT (Global), EN (Sri Lanka, C Asia), CE (Java, Middle East, Russia).

African
savannah
form

LEOPARD

Melanistic form

African
forest
form

Arabian
subspecies

Amur
subspecies

HB ♀ 116–219cm, ♂ 110.5–270cm; T 44–80cm; SH 68–75cm; W ♀ 36–100kg, ♂ 36–158kg

The world's third largest cat. Background colour varies from pale yellow through ginger to rufous-golden-brown, with white or cream underparts. Large block-like markings or rosettes usually enclose smaller black spots (lacking in similar Leopard). Melanism occurs, with the same pattern of markings apparent in oblique light; black individuals are most common in humid lowland rainforest. Size varies very widely; the smallest Jaguars occur in Central America and are about half the size of the largest individuals from the wet woodland savannah of Brazil and Venezuela.

Distribution and Habitat

N Mexico to N Argentina. Widely distributed in much of N and C S America, more fragmented in Mexico, C America, E Brazil, Argentina and S Bolivia. Resident breeding populations no longer occur in the USA, but individuals intermittently appear in Arizona and New Mexico from N Mexico. Broad habitat tolerance; while capable of occupying dry open savannah or desert habitats, the species is more commonly associated with the dense cover of tropical and subtropical lowland forest, typically below 2000m (exceptionally to 3800m). Strongly associated with water, and does well in well-watered habitats including flooded savannah (Pantanal and Llanos), swamps, riverine thicket and scrub, and mangroves. Excellent swimmer, capable of crossing large rivers.

Feeding Ecology

Diverse diet with at least 85 recorded prey species. Like all large cats, focuses on common large-bodied prey, but the natural absence of herds of large deer and wild cattle in S America results in the hunting of smaller animals more often than is the case for other big cats. Capybara and Collared and White-lipped Peccaries are frequently the most important prey species where they occur. Reptiles form a larger part of the diet than of any other large cat, especially caimans, as well as iguanas, freshwater turtles and tortoises, nesting marine turtles and large boas including anacondas. Can subsist on small abundant prey such as armadillos, pacas, brocket deer and/or coatis. Other relatively common prey includes White-tailed Deer, agoutis, marsupials and sloths. Occasionally kills very large prey including tapirs and Marsh Deer; Jaguars in the Brazilian Pantanal are recorded killing Freshwater Dolphins as they fish in shallow water. Readily kills domestic livestock; in ranching-dominated habitats such as the Pantanal and Llanos, introduced cattle form the major prey species. Less often, takes domestic pigs and dogs from villages. Almost never hunts humans; most recorded attacks result from extreme provocation, for example during Jaguar hunts, and verified unprovoked attacks are extremely unusual. Jaguars have proportionally the strongest bite force of all large cats, and kill either by crushing the back of the skull or with a typically feline suffocating throat bite. Hunting mainly nocturno-crepuscular and terrestrial, though Jaguars readily hunt prey, e.g. Capybara and caimans, in water. Often scavenges, including from cattle carcasses; the deaths of these cattle are often erroneously blamed on the cat.

Social and Spatial Behaviour

Solitary and territorial, but exclusive range use appears limited to small core areas; overlap between adults in some populations is extensive, possibly due to marked seasonal changes in the distribution of water and hence prey. Pantanal females establish largely exclusive ranges during the wet season, but overlap considerably in the dry season, while males overlap extensively in both. Similarly, Cockscomb (Belize) males overlap extensively. Adults engage in typically territorial behaviours such as roaring, scrapes and urine-marking, perhaps serving to foster avoidance rather than demarcate exclusivity. Aggressive interaction between adults appears to be rare, though there are records of fatal fights. Range size estimates include 28–40km² (♂s) in Belize; 38km² (average, ♀s) to 63km² (average, ♂s) in S Pantanal, Brazil; 47–83km² (♀s) to 93–108km² (♂s) in Venezuela, and 31–98km² (♀s) to 73–268km² (♂s) in N Pantanal, Brazil. A male in arid lowland desert and pine-oak woodland in Arizona used at least 1359km² in 2004–2007. Home range sizes in inundated habitat, e.g. Pantanal, often contract in the wet season, when flooded areas limit space available to prey. Density estimates include 1.1 Jaguars/100km² (Iguaçu NP, Brazil), 2.5/100km² (Atlantic forest, Brazil), 3.5/100km² (Corcovado NP, Costa Rica), 8.8/100km² (Cockscomb, Belize) to 6–11/100km² (Pantanal, Brazil).

Reproduction and Demography

Aseasonal. Oestrus 6–17 days; gestation 91–111 days (average around 101–105 days). Litter size 1–4, averaging 2 (captivity). Weaning begins at around 10 weeks, and suckling typically ceases by 4–5 months. Cubs independent at 16–24 months. Dispersal poorly known, but appears to be typically feline, in which females settle close to their natal range while males disperse more widely. Both sexes are sexually mature at 24–30 months; females first give birth at 3–3.5 years and can reproduce to 15 years. It is not known when wild males first breed. MORTALITY Rates poorly known, though humans remove significant numbers from some populations, e.g. an estimated 185–240 large cats (Jaguars and Pumas) were killed in 2002–2004 in a 34,200km² ranching landscape in Alta Floresta, Brazil. Adult Jaguars have no predators and are killed principally by humans and rarely by other Jaguars. Predators of cubs are poorly known, but infanticide by male Jaguars is recorded. LIFESPAN Poorly known from the wild, but unlikely to exceed 15–16 years; 22 in captivity.

Status and Threats

Extirpated from an estimated 45% of its historic range and extinct in El Salvador, Uruguay and the USA. Despite this, its remaining range is potentially still continuous, in part because the massive forested basins of S America have remained mostly inaccessible until recently. Habitat conversion for forestry, livestock and agriculture is the main threat, combined with intense persecution from ranchers and pastoralists in livestock areas, despite the fact that many cattle losses blamed on Jaguar predation occur from other factors. Hunting of prey is likely to impact Jaguar populations, and emerging evidence indicates that the species is hunted to supply the Chinese medicinal trade, though both are poorly quantified. Sport hunting is illegal in all range countries. CITES Appendix I, Red List NT.

JAGUAR

Central
American form

Melanistic
form

Brazilian
Pantanal
form

HB 55–80cm; T 19–30cm; SH 43–50cm; W 7.7–14kg
Smallest hyaena, superficially resembling Striped Hyaena but half its size and lightly built, with a narrow head and slender black muzzle. Jackal-sized in the field, but can appear much larger by erecting its dorsal mane when threatened. **Distribution and Habitat** Two disjunct populations in southern and E–NE Africa. Favours open habitat including semi-desert, grassland and woodland savannah, and does not require standing water, which it obtains from its diet. Avoids true desert, dense woodland and forest. **Feeding Ecology** Feeds almost exclusively on Snouted Harvester Termites (*Trinervitermes trinervoides*), lapping up as many as 300,000 (1.2kg) per night from dense feeding processions on the soil surface. Cold winters and high rainfall force these termites below ground, driving Aardwolves to switch to less social termite species that require more energy to find and consume. Under such food stress, adults can lose 25% of their weight and cubs are vulnerable to starvation. Foraging generally nocturnal, but may shift diurnally in winter. Reflecting the diet, the species has a very long, paddle-shaped tongue, copious sticky saliva and small, almost nonfunctional, peg-like cheek-teeth (it has retained the large canines, which are used for territorial defence and against predators). Does not eat meat, so does not take livestock or carrion. **Social and Spatial Behaviour** Monogamous (though extra-pair matings are common) and territorial. Forms breeding pairs that endure for 2–5 years. Pairs cooperatively maintain stable territories with very frequent scent-marking by both sexes, and share cub-raising duties, chiefly guarding the den against predators. Adults and independent cubs usually forage alone, but family members occasionally congregate at termite colonies. Territories 1–6km², depending on density of termite colonies. **Reproduction and Demography** Seasonal. Mates June–July. Cubs born October–December (E Cape). Oestrus 3 days; gestation 90 days. Litter size 2–4, exceptionally 5 in captivity. Cubs weaned at 12–16 weeks, by which time they begin to forage alone. Cubs are fully independent at 6–7 months, and most disperse before the following year's litter is born. MORTALITY Survival is tied to termite abundance; cub mortality is typically around 30%, but rose to 55% during a drought in one study. Adult mortality poorly known. Occasionally killed by all large carnivores, but the chief predator (of cubs) is Black-backed Jackal. LIFESPAN Unknown in the wild, 15 years in captivity. **Status and Threats** Generally widespread and secure. Harvester termites thrive in disturbed grassland habitats, including livestock areas, which are suitable for Aardwolves under enlightened management. Agricultural poisoning (usually for locusts) results in termite die-offs that trigger Aardwolf declines. Despite never eating meat, Aardwolves are erroneously persecuted for livestock losses. Hundreds are killed annually as 'by-catch' in jackal-control efforts in southern Africa. Red List LC.

STRIPED HYAENA *Hyaena hyaena*

HB 98–119cm; T 26–47cm; SH 60–74cm;
W ♀ 23–34kg, ♂ 26–41kg
The only hyaena whose range extends into Eurasia. Ash to straw coloured, with black stripes on the body and legs and a distinctive black throat patch. Tail typically lacks a black tip. Longest dorsal mane of any hyaena, which is erected defensively and becomes long and luxuriant in northern individuals in winter. **Distribution and Habitat** W, N and E Africa, the Middle East and Central Asia to India. Favours open semi-arid habitats with cover, especially dry woodland savannah, dry forest, semi-desert and mountainous terrain to 3300m. Absent from dense woodland, tropical forest and interiors of true desert. **Feeding Ecology** Principally a scavenger of dead wild and domestic ungulates. Scavenges the remains of kills made by large carnivores, and is often found close to livestock herds looking for dead animals. Hunting ability poorly understood. Reputed to hunt large prey and widely blamed for killing livestock, but there is little evidence for this. Opportunistically catches small mammals and birds, and capable of opening the carapaces of large land turtles and tortoises. Also eats a wide variety of vegetables, fruits and invertebrates; seasonally, these items may surpass carrion in importance in the diet. Raids fruit and vegetable crops in some areas, where it is treated as a pest, e.g. Israel. Typically forages alone, but may congregate loosely at food patches, including refuse dumps. Foraging mostly nocturnal, especially where it lives close to humans. Provided it is not persecuted, it is sometimes more crepuscular in winter or during overcast and rainy weather. Often brings prey remains back to the den, which is used by many generations, sometimes creating massive bone accumulations. **Social and Spatial Behaviour** The least-studied hyaena and poorly known. Inhabits an enduring home range, which is scent-marked regularly, but there is little evidence for territorial defence. In Central Asia reportedly forms monogamous breeding pairs that may endure for more than a season; cubs of previous litters sometimes remain with a pair, suggesting helpers (as for canids). In central Kenya, forms 'proto-social' groups with one female and up to three adult males that may or may not be related. Group members occupy the same shared range, and meet for social interaction and resting. Range size in central Kenya 36–101km² (♀s) and 52–115km² (♂s), with little overlap between females but high overlap among males. Only density estimates 2–3 hyaenas/100km² (E Africa). **Reproduction and Demography** Aseasonal. Oestrus reportedly only 1 day; gestation 90–91 days. Litter size typically 1–4, exceptionally 5 (captivity). Cubs begin to eat meat at around 1 month, but are suckled for 6–12 months. Females are sexually mature at 1 year and may give birth at 15–18 months, but usually first breed at around 24–27 months. MORTALITY Poorly known; 47% of adults in C Kenya survived a further 3 years from when first identified. LIFESPAN Unknown in the wild, 23–24 years in captivity. **Status and Threats** Although widespread, occurs naturally at low densities and much of its remaining range is fragmented into isolated populations. Often intensely persecuted for its perceived effect on livestock and widespread superstitions related to grave robbing (which occurs), witchcraft and folk medicine. In the Middle East, often killed on roads when searching for road-killed carrion. Red List NT.

Scent-marking

AARDWOLF

Central Asian
form, winter

Defensive posture

**STRIPED
HYAENA**

HB 110–136cm; T 18–27cm; SH 71.5–82cm; W ♀ 28–47.5kg, ♂ 35–49.5kg
Medium-sized hyaena covered in coarse, shaggy dark brown fur that fades to blond on the neck and shoulders. Dark stripes on a light background cover the legs. Males are only slightly larger than females. **Distribution and Habitat** Endemic to southern Africa. Independent of water and inhabits various arid to semi-arid habitats, from open desert to woodland savannah. Can occupy pastoral habitats close to people, e.g. Johannesburg's outer suburbs. **Feeding Ecology** Scavenger that feeds mainly on the carcasses of dead mammals. Opportunistically takes small vertebrate prey up to the size of antelope lambs, but kills comprise only 6–16% of the diet. Coastal Namibian individuals live almost exclusively on Cape Fur Seals that are mostly scavenged, though young pups are actively hunted. Also eats vegetables, fruits, invertebrates, eggs and human refuse. Predation on small livestock occasionally occurs. Foraging solitary, though it congregates at large carcasses, seal colonies and refuse dumps. Mostly forages at night, but more diurnal in cool conditions when protected. Covers prodigious distances when foraging, e.g. 34–89km/24hr in Namibia. Caches food and brings remains back to maternity dens to provision cubs. **Social and Spatial Behaviour** Forms small clans of 4–14 members centred around 1–5 related females that share a territory and raise cubs cooperatively. Unrelated adult males immigrate either permanently or as nomads that visit only for breeding. Clan members typically forage alone, but meet at carcasses and maternity dens. Territories overlap, but core areas are defended from intruders with ritualized aggression and extremely frequent 'pasting' with the anal gland. A single territory has up to 20,000 paste sites, which each adult marks up to 29,000 times anually. Individuals identify clan members from pastes. Average clan territory size estimates include 170km² (central Kalahari), 350km² (coastal Namibia, with seal colonies) to 1900km² (inland Namibia, with unpredictable food sources.) Occurs naturally at low densities, estimated at 1–2.9 hyaenas/100km². **Reproduction and Demography** Aseasonal. Oestrus 1 week; gestation 90 days. Litter size 1–5, typically 2–3. Cubs begin eating meat at 4 months and are suckled for 12–16 months. Females give birth alone, but often raise cubs in communal dens with other females. Females first breed at 35–36 months, and breed until at least 10 years. MORTALITY Appears moderate in protected populations: 89% of emerged Kalahari cubs reached independence and most adults died in old age. Starvation (in old age), Lion and Spotted Hyaena are the main natural causes. LIFESPAN At least 12 years in the wild, 13 in captivity. **Status and Threats** Widespread but naturally rare, and dependent on large areas with sufficient carrion. Livestock areas are important, but it is heavily persecuted (despite livestock depredation being infrequent) and threatened by habitat conversion to agriculture. Red List NT.

SPOTTED HYAENA *Crocuta crocuta*

HB 115–160cm; T 21–31.5cm; SH ♀ 73.5–88.5cm, ♂ 70–87cm; W ♀ 56–86kg, ♂ 49–79kg
Largest hyaena, heavily built with very strong forequarters, and covered with dark spots that tend to fade with age. Young cubs are dark chocolate-brown, developing adult colouration by 4–5 months. Females are slightly larger and heavier than males. **Distribution and Habitat** Endemic to Africa south of the Sahara, where it occurs in all woodland-savannah habitats. Largely absent from true forest, but lives in montane forest to 4000m (Abedares, Kenya) and penetrates rainforest-savannah mosaics along roads in C Africa. Cannot survive in hyper-arid desert interiors, but occurs deep in deserts along watercourses and massifs. **Feeding Ecology** Highly efficient scavenger and formidable predator whose own kills comprise 60–95% of the diet. Individuals can overpower adult wildebeest, and small groups kill Gemsbok, zebras and African Buffalo. Scavenged and killed ungulates dominate the diet, but it eats virtually anything organic including small mammals, birds, reptiles, fish, crabs, snails, insects, eggs, vegetables, fruits and human refuse. Readily kills livestock and domestic dogs, but very rarely humans. Appropriates kills from Leopards, Cheetahs and African Wild Dogs, often shadowing them when hunting. Large groups displace Lions from kills provided they have numerical superiority and adult male Lions are absent. Foraging is alone or in groups: clan members often hunt cooperatively or congregate on large carcasses. Usually nocturnal, especially near humans, but forages diurnally where it is protected. Rarely caches food; sometimes carries remains to dens, but does not provision cubs. **Social and Spatial Behaviour** Forms clans of 10–80 comprising inter-related females with their cubs, and unrelated immigrant adult males. Clan members occupy shared territory but are rarely all together; individuals move alone or in small subgroups that meet up frequently with ritualized greetings. Clans defend territories from intruders, but 'commuters' tracking migratory herds in E Africa are tolerated passing through. Clan ranges vary from 20km² where resident prey is abundant, e.g. Ngorongoro Crater, Tanzania, to over 1500km² in the Kalahari Desert. Density from 0.6–0.8 hyaenas/100km² (Kalahari and Namib Deserts), and 7–20/100km² (Kruger NP), 60–80/100km² (Masai-Mara-Serengeti ecosystem), to 170/100km² (Ngorongoro Crater). **Reproduction and Demography** Aseasonal. Oestrus 1–3 days; gestation 90–91 days. Litter size 1–3, usually 2. Cubs have a very prolonged suckling period of 13–24 months. Females usually give birth alone and suckle only their cubs, but bring them to communal dens to raise; up to 30 cubs of 20 litters were counted at a single den complex. Females are sexually mature at 24 months, but first breed at 3–6 years. MORTALITY Cub mortality in first year 40–50%, mainly from Lions, starvation when mothers are killed, occasional infanticide and siblicide (rarely). Adult mortality 13–15% in protected areas, from Lions, humans and occasional disease outbreaks (rabies and canine distemper). LIFESPAN Around 12–16 years, exceptionally to 20 in the wild; 41 in captivity. **Status and Threats** Relatively resilient and reasonably secure in southern and E Africa, but it has undergone significant declines everywhere outside protected areas due to very pervasive persecution by humans. Spotted Hyaenas are speared, shot, trapped and poisoned as livestock predators, and for superstitious beliefs mainly related to traditional medicine. They are highly vulnerable to snares due to their habit of scavenging: snaring kills 400 adults a year in the Serengeti ecosystem. Red List LC.

Cubs at den

BROWN HYAENA

SPOTTED
HYAENA

Greeting

Cub

SMALL INDIAN MONGOOSE *Herpestes auropunctatus* PLATE 20

HB 19.2–44.6cm; T 19.2–29cm; W 305–662g
Smallest Asian mongoose. Fur grizzled and colour varies widely, including pale yellow-grey, buff and rufous-brown, with pale buff around the mouth, chin and throat. **Distribution and Habitat** Southern Asia from E Iraq to Myanmar and S China; replaced by Small Asian Mongoose in Thailand and Indochina. Widely introduced, including to Bosnia-Herzegovina, Croatia, Fiji, Hawaii, Jamaica, Japan, Mauritius and many Caribbean islands. Occurs in virtually all habitats with cover, including those close to humans. **Feeding Ecology** Highly opportunistic and omnivorous, eating virtually any small vertebrate to the size of rats, and a wide variety of invertebrates, especially insects, arachnids and crabs, and fruits including berries. Kills poultry. Significant pest in its introduced range, where it has caused extinction of many endemic species, e.g. Hawaii. Foraging diurnal and solitary. Scavenges from carrion, handouts and human refuse. **Social and Spatial Behaviour** Solitary. Males apparently form coalitions in some introduced populations, e.g. Hawaii, possibly due to super-abundant food allowing greater sociality. Males have larger ranges that overlap multiple female ranges. Range size 0.014–1km². Density can exceed 300 mongooses/km² in introduced populations. **Reproduction and Demography** Breeding year-round, but birth peaks vary among populations. Gestation approximately 49 days. Litter size 1–5, averaging 2. Females have to 3 litters a year. MORTALITY and LIFESPAN Unknown. **Status and Threats** Common and adaptable. In its native range it is heavily hunted for meat and fur in some areas, but it is resilient to exploitation. Super-abundant in much of its introduced range, and has proved impossible to eradicate. CITES Appendix III (India), Red List NE.

SMALL ASIAN MONGOOSE *Herpestes javanicus*

Javan Mongoose
HB 30.2 –41.5cm; T 21–31.5cm; W 0.45–1kg
Very small mongoose, uniformly grizzled rufous-brown to dark brown. Formerly classified with Small Indian Mongoose, but molecular data indicate they are two valid species with a putative dividing line around the Salween River, Myanmar. Many records of this species actually concern Small Indian Mongoose outside the former's SE Asian range, where it is largely unstudied. **Distribution and Habitat** Thailand, Indochina to Peninsular Malaysia, Sumatra and Java; possibly extreme E Myanmar. Inhabits dry and wet forests, scrubland, brush and grassland. Occurs near humans, including in rice fields and cultivated areas. **Feeding Ecology** Poorly known; assumed to resemble Small Indian Mongoose, eating mainly small vertebrates and invertebrates. Bold and aggressive predator capable of killing large rodents and snakes. Raids domestic poultry. Foraging diurnal and solitary. **Social and Spatial Behaviour** Assumed to be solitary. **Reproduction and Demography** Thought to be aseasonal. Gestation approximately 7 weeks. Litter size 2–4. MORTALITY and LIFESPAN Unknown. **Status and Threats** Widespread, common and tolerant of anthropogenic habitats. Reaches high densities in suitable habitat; trapped as a pest by wildlife authorities in some areas. Red List LC.

SHORT-TAILED MONGOOSE *Herpestes brachyurus*

HB 35–49cm; T 19.3–24.5cm; W 2–3kg
Medium-sized mongoose, uniformly dark olive-brown with tawny speckling, especially on the head, neck and tail, and solid tawny around the mouth and chin. Short tail is bluntly conical in profile. Another described species, Hose's Mongoose *H. hosei*, from Borneo, is probably an aberrant Short-tailed Mongoose specimen and is not considered a valid species. **Distribution and Habitat** Borneo, Sumatra, Singapore, the Philippines and Peninsular Malaysia. Occurs primarily in intact lowland rainforest, usually close to streams and watercourses. Tolerates regenerating forest, and plantations close to forest with ground cover. **Feeding Ecology** Believed to eat small vertebrates, invertebrates, eggs and some fruits; has been captured in traps baited with chicken pieces and salted fish. Foraging solitary and strictly diurno-crepuscular. **Social and Spatial Behaviour** Based on the only radio-collaring study (five animals, Krau Wildlife Reserve, Malaysia), it is solitary and occupies mostly exclusive ranges. Male ranges overlap up to two female ranges, while female ranges overlap very little. Range size 1.15–1.5km² (♀s) and 2.24–2.5km² (♂s). **Reproduction and Demography** Unknown. **Status and Threats** Status poorly known. Appears relatively intolerant of habitat conversion, and much of its range is exposed to intense forest loss and hunting pressure. Eaten in some parts of Sarawak. Red List LC.

INDIAN GREY MONGOOSE *Herpestes edwardsii*

Common Grey Mongoose
HB 35.5–46cm; T 32–45cm; W 0.9–2kg
Small to medium-sized mongoose with grizzled tawny-grey to pale grey fur. Lower legs are usually darker than the body, ranging from rufous-brown to black. Tail never has a dark tip, but may be rufous in some individuals. **Distribution and Habitat** Indian subcontinent including Bhutan, Nepal and E Pakistan, and Sri Lanka; isolated populations in S Iran, Kuwait, Bahrain and Saudi Arabia. Occurs in dry forest, scrub and grassland. Common in cultivated areas and near rural villages. **Feeding Ecology** Mainly carnivorous, eating small mammals, reptiles, birds, eggs and a wide variety of invertebrates. Aggressive predator of snakes, including large venomous species; resistant to haemorrhagic snake venom, and the species commonly kept in India and Pakistan for staged mongoose-cobra fights. Apparently eats some fruits and vegetables. Foraging solitary and diurno-crepuscular. Readily scavenges from carrion and human refuse. **Social and Spatial Behaviour** Poorly known. Solitary. The only range estimate is for an adult male tracked for 3 months, 0.16km² (Nilgiri Biosphere Reserve, India). **Reproduction and Demography** Breeding year-round; births peak May–June and October–December. Gestation approximately 56–68 days. Litter size 2–4. Females have up to 3 litters per year. MORTALITY and LIFESPAN Unknown. **Status and Threats** Widespread, common and able to live in association with humans. Captured as pets for cobra-fighting shows, and for meat and hair, which is used in shaving and paint brushes. CITES Appendix III (India), Red List LC.

SMALL INDIAN
MONGOOSE

SMALL ASIAN
MONGOOSE

SHORT-TAILED
MONGOOSE

INDIAN GREY
MONGOOSE

COLLARED MONGOOSE *Herpestes semitorquatus* PLATE 21

HB 40–46cm; T 25.8–30.3cm; W 2–4kg
Large slender-legged mongoose with dark reddish-brown fur, dark brown legs and a cream tail. Throat tawny-yellow, becoming cream along the lower jaw. Has been considered the same species as Short-tailed Mongoose, but is now regarded as a valid species. **Distribution and Habitat** Endemic to Borneo; two equivocal records from Sumatra. Occurs in lowland rainforest, disturbed forest and plantations to 1200m.

Feeding Ecology Unknown. Camera-trap records indicate foraging is diurnal. **Social and Spatial Behaviour** Unknown. Camera-trap images suggest it is solitary. **Reproduction and Demography** Unknown. **Status and Threats** Status poorly known. Rare in camera-trap surveys and known only from lowland areas that are under intense pressure from forestry and hunting, but tolerance for disturbance and presence in better protected upland areas are uncertain. Red List DD.

BROWN MONGOOSE *Herpestes fuscus*

Indian Brown Mongoose
HB 33–48cm; T 20–33.6cm; W 1.1–2.7kg
Large stocky mongoose, uniformly grizzled dark brown with a slightly paler head and tawny chin and throat. Tail bushy with a tapering conical profile. **Distribution and Habitat** Endemic to W Ghats, SW India and W Sri Lanka. Inhabits mainly dense rainforest and adjacent habitats, including dense grassland as well as tea and coffee plantations. **Feeding Ecology** Unknown. Assumed to feed on a variety of invertebrates and small vertebrates. Recorded around refuse

dumps near research camps, and eats groundnuts (but not bananas or boiled chicken) left as bait during rodent surveys. Camera-trap surveys suggest foraging is nocturnal. **Social and Spatial Behaviour** Unknown. Camera-trap surveys record only solitary animals. **Reproduction and Demography** Unknown. **Status and Threats** Considered naturally rare with a very restricted range that is under considerable pressure from forest conversion to agriculture and pasture. Habitat loss is assumed to have resulted in a population decline of >30% in the last 15 years. Red List VU.

CRAB-EATING MONGOOSE *Herpestes urva*

HB 44–55.8cm; T 26–35cm; W 3–4kg
Large, fairly stocky mongoose, grizzled pale grey to dark greyish-brown, with dark legs and a pale tail ranging from tawny to white. Mouth and chin white, which extends as a distinctive stripe along the neck. **Distribution and Habitat** S China, Taiwan, Nepal, Bhutan, NE India, E Bangladesh and mainland SE Asia to Peninsular Malaysia. Inhabits forest, forest-swamp mosaics, scrubland and wetland to 2000m. Occurs in degraded forest, rice fields and cultivated areas, and near human settlements. **Feeding Ecology** Eats mainly water-associated invertebrates and vertebrates, particularly aquatic insects, crustaceans, rodents, earthworms, reptiles and amphibians. Birds, fish, arachnids, snails and fruits are eaten in small amounts. Occasionally

raids poultry and urban fishponds. Forages diurnally, usually near water; capable swimmer and dives for prey, at least in controlled conditions such as garden ponds. Hair of a Formosan Macaque found in a scat suggests that it scavenges. **Social and Spatial Behaviour** Poorly known. Mostly solitary but observed groups of 2–4 suggest limited sociality. **Reproduction and Demography** Poorly known. Gestation thought to be 50–63 days. Litter size 2–4. MORTALITY Unknown LIFESPAN 13.3 years in captivity. **Status and Threats** Widespread and tolerant of some habitat conversion. Hunted in Cambodia, China and Laos for the meat and pet trades, but apparently persists in areas of high hunting pressure if it is not combined with significant habitat conversion. CITES Appendix III (India), Red List LC.

RUDDY MONGOOSE *Herpestes smithii*

HB 39–47cm; T 35–41cm; W 1.75–2.7kg
Large mongoose with grizzled greyish-brown fur with a rusty tinge. Long tail ends in black-tasselled tip lacking in similar sympatric mongoose, e.g. Indian Grey Mongoose. **Distribution and Habitat** C and S India, and Sri Lanka. Inhabits dry forest, thorn scrub and dry grassland-forest mosaics; avoids evergreen forest. Occurs in disturbed forest, but avoids heavily modified habitats near humans. **Feeding Ecology** Known to eat rodents, birds and reptiles, and probably takes a variety of small prey. Foraging mainly diurno-crepuscular

and solitary. Largely terrestrial, but apparently a capable climber that hunts aboreally and carries prey into trees (unusual behaviour for mongooses, if true). Readily scavenges, including from road-killed carcasses. **Social and Spatial Behaviour** Poorly known. Largely solitary, but groups of 2–5 adult-sized individuals are fairly common; group composition unknown. **Reproduction and Demography** Unknown. **Status and Threats** Widespread but poorly known; common in some areas, e.g. C India. Habitat loss and localized hunting are main threats. CITES Appendix III (India), Red List LC.

STRIPED-NECKED MONGOOSE *Herpestes vitticollis*

HB 43–53cm; T 23–33.5cm; W ♀ 1.7–2.7kg, ♂ 2.6–3.4kg
Large distinctive mongoose with grizzled rufous-brown hindquarters, usually becoming grizzled grey-brown on the forequarters; some individuals are entirely rufous except for the head, which is always steely-grey, and an obvious black stripe along the neck. Tail has a conspicuous black-tasselled tip. **Distribution and Habitat** SW India and Sri Lanka. Inhabits evergreen and moist deciduous forests, often associated with swampy areas, watercourses and dense grassland clearings. Also occurs in teak plantations and rice fields. **Feeding Ecology** Eats small mammals, birds, reptiles,

eggs and presumably a wide variety of invertebrates. Preys on mammals to size of Black-napped Hare, and pursues young ungulate fawns. Foraging diurnal and solitary. Scavenges from carrion and refuse dumps. **Social and Spatial Behaviour** Poorly known; sightings are of singletons or pairs. **Reproduction and Demography** Unknown. Thought to have 2–3 young. LIFESPAN Almost 13 years in captivity. **Status and Threats** Restricted range and uncommon. Habitat loss and hunting are the main threats Habitat loss and hunting are the main threats, but poorly understood. CITES Appendix III (India), Red List LC.

COLLARED
MONGOOSE

CRAB-EATING
MONGOOSE

BROWN
MONGOOSE

RUDDY
MONGOOSE

STRIPED-NECKED
MONGOOSE

Small Grey Mongoose

HB 29–41.5cm; T 20.5–34cm; W 0.49–1.25kg

Small, pale to dark grizzled-grey with paler underparts and a bushy tail. Muzzle and lower limbs are dark grey. **Distribution and Habitat** S Africa, Lesotho and SE Namibia. Occurs in most habitats with cover, from sea level to 1900m; absent from very open and arid areas. Occurs on farmland and in urban parkland. **Feeding Ecology** Eats small vertebrates and invertebrates, especially small rodents and insects. Occasionally kills neonate Grysbok and juvenile Cape Porcupine. Foraging diurno-crepuscular, terrestrial and solitary. Scavenges from carrion (including roadkills) and from dumps. **Social and Spatial Behaviour** Solitary. Adult males occasionally associate loosely in pairs. Largely non-territorial; breeding females defend exclusive small areas. Range size 0.3km² (1 ♀) and 0.55–0.92km² (♂s). **Reproduction and Demography** Seasonal. Births peak August–December. Gestation approximately 60 days. Litter size 1–3. MORTALITY Prey of various larger predators, especially raptors. LIFESPAN 8.8 years in captivity. **Status and Threats** Common habitat generalist that adapts well to human presence. No significant threats. Red List LC.

SOMALI SLENDER MONGOOSE *Herpestes ochraceus*

Somalian Slender Mongoose

HB 25–29cm; T 22–27.3cm; W c. 0.3–0.75kg

Uniformly pale grizzled-grey to dark grey-brown. Tail lacks a black tip, distinguishing it from Common Slender Mongoose, with which it was formerly classified. **Distribution and Habitat** Somalia, E Ethiopia and NE Kenya. Inhabits semi-arid open woodland and hilly areas to 600m. **Feeding Ecology** Unknown. Assumed to resemble other slender mongooses. **Social and Spatial Behaviour** Unknown. Assumed to be largely solitary. **Reproduction and Demography** Unknown. **Status and Threats** Probably common with few threats, but very poorly known. Red List LC.

KAOKOVELD SLENDER MONGOOSE *Herpestes flavescens*

Black Slender Mongoose, Angolan Slender Mongoose

HB 31–35.5cm; T 31–37cm; W 0.55–0.9kg

Small mongoose with two colour phases: uniformly very dark reddish-brown to black, and pale tawny-red with a black tail-tip. Some authorities treat the black form as a separate species, Black Mongoose *H. nigrata*. **Distribution and Habitat** Endemic to SW Angola and NW Namibia. Occurs in arid habitats with cover; avoids true desert. Black form inhabits isolated granite kopjes and associated woodland. **Feeding Ecology** Small rodents, especially Dassie Rat, and insects are the main prey. Other prey includes birds, reptiles, arachnids, eggs and fleshy seeds. Foraging diurnal and terrestrial, though thought to raid White-tailed Shrike nests. Foraging mostly solitary; up to five adults congregate at carcasses but feed separately. Scavenges from carrion (mainly for carrion-eating flies), refuse and handouts at tourist lodges. **Social and Spatial Behaviour** Solitary, though males occasionally form loose pairs. Ranges overlap extensively, and adults use the same dens but not concurrently. Range estimates known only for males, 0.13–1.45km² (Erongo Mountains, Namibia). **Reproduction and Demography** Unknown. **Status and Threats** Considered relatively common. No serious threats. Red List LC.

COMMON SLENDER MONGOOSE *Herpestes sanguineus*

Slender Mongoose, Black-tipped Mongoose

HB 27.5–35cm; T 19.4–33cm; W 0.37–0.79kg

Very variable colour, but typically grizzled-tawny or reddish-brown with a black tail-tip. Melanistic individuals occur. **Distribution and Habitat** Ubiquitous in sub-Saharan Africa except the western Congo Basin, coastal Namibia and E/W Cape Provinces, S Africa. Occurs in most habitats (including anthropogenic ones) except true desert and rainforest. **Feeding Ecology** Insects, reptiles and small rodents are the main prey, as well as birds, amphibians, eggs, arachnids and wild fruits. Foraging diurnal, terrestrial and typically solitary; males sometimes travel (and presumably forage) in temporary pairs. Scavenges from carrion (mainly for sarcophagous flies), dumps and handouts. **Social and Spatial Behaviour** Solitary, but males may form groups of 2–4 that jointly defend territory from other males. Range estimates 0.25–1km². Density estimates 3–6 mongooses/km² (Serengeti NP, Tanzania). **Reproduction and Demography** Seasonal. Birth peaks coincide weakly with rainy periods: October–March (southern Africa), October–November and February–April (E Africa). Gestation 60–70 days. Litter size 1–4. MORTALITY Main predators are large raptors. LIFESPAN 8 years in the wild, 12.6 in captivity. **Status and Threats** Very widespread and common. No serious threats. Red List LC.

EGYPTIAN MONGOOSE *Herpestes ichneumon*

Ichneumon, Large Grey Mongoose

HB 50–61cm; T 43.5–58cm; W ♀ 2.2–4 kg, ♂ 2.6–4.1kg

Large mongoose, uniformly grizzled grey with a dark face, dark lower limbs and black-tipped tail. **Distribution and Habitat** Most of Africa except the Sahara, Congo Basin, arid southern Africa and NE Africa. Also Portugal and Spain (possibly introduced), and the Middle East. Inhabits woodland, grassland, wetland, semi-desert and montane areas. cultivated land and farmland. **Feeding Ecology** Eats vertebrates to the size of hares, invertebrates, fruits and fungi. Resistant to snake venom. Occasionally takes poultry. Foraging nocturno-crepuscular, terrestrial and solitary. Scavenges from carrion. **Social and Spatial Behaviour** Generally solitary; forms groups of male with 1–3 females and their offspring in Israel. Average range estimates (both sexes) 0.38km² (S Africa) to 3.1km² (Spain). Density estimates 0.1–2 mongooses/km². **Reproduction and Demography** Seasonal. Births peak October–December (southern Africa), September–February (E Africa) and May–July (Spain). Gestation about 63–70 days. Litter size 1–4. Females (Israel) communally raise kittens. MORTALITY Predation mainly from large raptors and Iberian Lynx (Spain). LIFESPAN 13 years in captivity. **Status and Threats** Widespread and common. Locally vulnerable to persecution and poisoning of rodent prey. Red List LC.

CAPE GREY
MONGOOSE

SOMALI SLENDER
MONGOOSE

KAOKOVELD
SLENDER
MONGOOSE

COMMON SLENDER
MONGOOSE

EGYPTIAN MONGOOSE

Water Mongoose

HB 44.2–62cm; T 25–41cm; W 2.4–4.1kg

Large dark mongoose with a distinctive blunt, triangular profile to the face. Shaggy fur uniformly dark reddish-brown to black, except for pale tawny around the mouth and chin. Toes long, slender and completely unwebbed, which is unique among mongooses; all other species possess a degree of webbing. **Distribution and Habitat** Sub-Saharan Africa; absent from arid NE Africa, and most of Namibia, Botswana and C South Africa. Inhabits dense wet habitat from sea level to 3950m, including streams, swamps, marshes, wet forest, mangroves, estuaries and coastal areas. Inhabits anthropogenic watercourses such as dams and canals provided there is cover and prey. **Feeding Ecology** Unique among mongooses in taking mainly aquatic prey, especially crabs, aquatic insects, molluscs and amphibians, as well as rodents. Takes small amounts of fish, birds, eggs and fruits. Occasionally kills larger terrestrial mammals, including other carnivores, e.g. Cape Grey and Yellow Mongooses. Sometimes kills poultry. Foraging nocturno-crepuscular and solitary. Excellent swimmer able to dive for prey for up to 15 seconds, but forages mainly in the shallows and on banks. Prey is located by sight or by feeling in mud and crevices with its long, extremely dexterous digits. Scavenges from carrion. **Social and Spatial Behaviour** Solitary. Occupies stable ranges that tend to be arranged along watercourses. Range size differs little between sexes, and ranges are scent-marked extremely regularly with obvious latrines. Range size 0.54–2.04km². The only density estimate is 1.8 mongooses/km² (KwaZulu-Natal, S Africa). **Reproduction and Demography** Aseasonal in much of its range; weakly seasonal in southern Africa, where breeding occurs in wetter periods, August–February. Gestation 69–80 days. Litter size 1–3. MORTALITY Poorly known; predation rarely recorded except by domestic dogs. LIFESPAN 19 years in captivity. **Status and Threats** Widespread and relatively common in much of its range. Vulnerable to destruction of watercourses by clearing, siltation and pollution, which causes local declines, and popular as bushmeat in W and C Africa. Red List LC.

LONG-NOSED MONGOOSE *Herpestes naso*

Long-snouted Mongoose

HB 44–61cm; T 32–43cm; W 1.9–4.5kg

Large dark mongoose with a long muzzle and prominent black nose. Fur grizzled dark brown with pale underfur, but the impression in the field is very dark brown or near-black. Sometimes classified in the genus *Xenogale* due mainly to its unique teeth characteristics; recent molecular evidence suggests that Marsh Mongoose is its closest relative. **Distribution and Habitat** Endemic to C Africa, from SE Nigeria to SE DR Congo, W of the Rift Valley. Inhabits rainforest, usually near streams, watercourses and swampy areas with dense tangled understory. Avoids open forest, but sometimes forages on burnt grassland close to forest edges. **Feeding Ecology** Omnivorous, eating mainly arthropods (particularly beetles, crickets, termites, ants and millipedes), rodents and shrews. Also eats snails, frogs, reptiles, birds, fish and fruits. Rarely takes prey larger than 4–5kg; Blue Duiker, primates, pangolins and Brush-tailed Porcupine occur in scats, but are probably scavenged. Foraging diurnal, terrestrial and solitary. **Social and Spatial Behaviour** Solitary. Both sexes establish stable ranges that differ little in size (based on one study with five radio-collared animals). Range size, including both independent subadults and adults, 0.1–1km². **Reproduction and Demography** Very poorly known. Young animals observed March–May (W Africa), and a litter of 3 is recorded. MORTALITY Poorly known; recorded in Leopard and Black-legged Mongoose scats (possibly scavenged in the latter). LIFESPAN 11 years in captivity. **Status and Threats** Regarded as rare, but poorly known. Common in some areas, e.g. Dzanga-Sangha forest, SW Central African Republic. Forest loss and bushmeat hunting produce local declines, e.g. Niger Delta, Nigeria. Red List LC.

WHITE-TAILED MONGOOSE *Ichneumia albicauda*

HB 47–71cm; T 34.5–47cm; W 1.8–5.2kg

Largest mongoose. Tall, long-legged and slender, usually pale grey to grizzled-grey with blackish lower legs and a bushy silvery-white tail. A dark morph occurs; a black tail and denser covering of black guard hairs over the body give it an overall darker appearance. **Distribution and Habitat** Widespread in the sub-Sahara except NE Africa, the Congo Basin and arid SW Africa; occurs on the southern Arabian Peninsula. Inhabits moist and dry woodlands, savannah, scrub and grassland; avoids open desert and dense forest. Tolerates anthropogenic habitats including farmland, plantations and orchards. **Feeding Ecology** Chiefly insectivorous, eating medium to large nocturnal arthropods, especially termites, ants, dung beetles, crickets, grasshoppers and their larvae. Additionally eats small mammals to the size of cane rats, amphibians, birds, and small amounts of fruits and vegetable matter. Occasionally raids poultry. Foraging nocturnal, terrestrial and mainly solitary, though pairs sometimes forage together, and up to nine congregate at food patches such as termite flushes. Scavenges from carrion and human refuse. **Social and Spatial Behaviour** Predominantly solitary. Occasionally forms loosely associated male-female pairs, and related females (usually mothers and daughters) sometimes share a range, forming small female clans with their offspring. Adults occupy enduring ranges that may overlap within and between sexes; aggressive behaviour between neighbours appears to be rare. Female ranges are slightly smaller than those of males; range size 0.4–4.3km², exceptionally to 8km². Density in very high-quality habitat (Serengeti, Tanzania) reaches 4.3 mongooses/km². **Reproduction and Demography** Seasonal. Births appear to coincide with wet periods: October–February (southern Africa), March–April and October–December (E Africa). Litter size 1–4. MORTALITY Occasionally killed by larger predators, but produces a nauseating anal secretion combined with an impressive threat display that probably deters predation. LIFESPAN 14 years in captivity. **Status and Threats** Widely distributed, common and present in many protected areas. Killed on roads, during predator-control operations and by dogs in rural areas, but not threatened. Red List LC.

MARSH MONGOOSE

LONG-NOSED MONGOOSE

Dark form, defensive posture

WHITE-TAILED MONGOOSE

Sokoke Bushy-tailed Mongoose

HB 40–50cm; T 18–30cm; W 1.3–2.1kg

Stocky dark mongoose with a prominent bushy tail. Colour typically dark blackish-brown with chestnut or tawny-brown underfur on the head, throat and body; tail and legs always blackish. Sokoke Bushy-tailed Mongoose from the coastal forests of N Tanzania and E Kenya is sometimes treated as a distinct species, *B. omnivora*; it has a paler body colour with more obvious light tawny underfur. **Distribution and Habitat** Eastern Africa, from S Kenya to N Zimbabwe and W to S DR Congo. Occurs mostly in dense woodland savannah, dry and wet forests, wooded grassland and hilly areas with rocky or shrubby cover. **Feeding Ecology** Eats largely insects, small reptiles, frogs and toads, land snails, and scorpions and other arachnids. Small rodents are recorded, but it appears to be relatively clumsy in handling vertebrates such as rats and large snakes. Captive animals refuse fruits. Foraging nocturnal, terrestrial and solitary. **Social and Spatial Behaviour** Poorly known. Solitary. No range or density estimates; the most frequently photographed small carnivore during camera-trapping surveys in Tanzania's Eastern Arc Mountains. **Reproduction and Demography** Poorly known. Litters recorded November and December in Kenya (based on very few records). Females have two pairs of teats, suggesting litter size of 2–4. **MORTALITY** Unknown. Confirmed predators include Crowned Eagle, Gaboon Viper and Spotted Hyaena. **LIFESPAN** Unknown. **Status and Threats** Rarely seen and has a patchy distribution, but widely distributed and appears common in certain areas. No significant threats at species level, but illegal logging combined with hunting represents a threat to forest populations, e.g. Arabuko-Sokoke and Zanzibar Island. Red List LC.

JACKSON'S MONGOOSE *Bdeogale jacksoni*

HB 50.8–57.1cm; T 28.3–32.4cm; W 2–3kg

Large, grizzled silver-grey mongoose with a bushy white tail and black or dark brown lower legs. Yellowish tinting on the cheeks, throat and sides of the neck. Considered by some authorities to be the same species as Black-legged Mongoose; that species is larger with a very robust skull, and the two species do not overlap in range. **Distribution and Habitat** Endemic to SE Uganda, and C and S Kenya, with an isolated population in the Udzungwa Mountains, Tanzania (900km to the south). Possibly occurs elsewhere in Tanzania's Eastern Arc Mountains. Recorded only in dense lowland forest, bamboo forest and montane forest to 3300m. **Feeding Ecology** Eats chiefly small forest rodents, e.g. vlei and marsh rats, soft-furred and brush-furred mice, and forest-floor insects, especially army ants, termites and beetles. Millipedes, snails, lizards and eggs are also recorded. Foraging nocturnal, terrestrial and solitary. Scavenges from carrion. **Social and Spatial Behaviour** Poorly known. Likely to be solitary; reports of pairs and groups numbering up to four are probably breeding pairs and mothers with grown litters. **Reproduction and Demography** Unknown. **Status and Threats** Dependent on forested habitat and limited to a series of isolated populations in a very restricted range. Ongoing forest clearing and degradation are the main threats. Probably also killed for bushmeat, though carnivores (apart from otters) appear not to be highly sought after in its range. Red List NT.

BLACK-LEGGED MONGOOSE *Bdeogale nigripes*

Black-footed Mongoose

HB 46–65cm; T 29–40cm; W 2–4.8kg

Large short-haired mongoose with silver-grey or yellowish-grey fur, and black legs. Tail moderately bushy and varies from bright white to cream. Erythristic individuals occur, in which the silver-grey colouration is replaced by brownish-red. Skull very robust, with short rounded ears and a large nose giving the head a blunt dog-like appearance. **Distribution and Habitat** C Africa, from the Rift Valley, E DR Congo to SE Nigeria and S Congo. Reports from N Angola are now rejected. Found only in rainforest to 1000m, mainly in dense groundcover associated with undisturbed forest. Rarely found in disturbed forest. **Feeding Ecology** Eats mainly terrestrial arthropods (chiefly ants, termites, beetles and grasshoppers) and small mammals, especially forest shrews and rodents; Brush-tailed Porcupine and Long-nosed Mongoose are also recorded as prey (possibly scavenged). Additionally eats snakes, lizards, frogs, toads and small amounts of fruits. Local people also report that it takes wild and cultivated fruits, including bananas and oil-palm fruits. Foraging nocturnal and solitary. Hunts mostly on the ground, but there is one observation of an individual hunting a young Potto with great agility in a tree 15m above the ground. **Social and Spatial Behaviour** Solitary. There are reports of pairs, probably breeding adults or mothers with large juveniles. **Reproduction and Demography** Poorly known. Records of litters cluster in November–January, suggesting that breeding occurs at the start of the dry season, but this is based on very few observations. Local people report litters numbering 1–2. **MORTALITY** Poorly known; most documented mortality is anthropogenic. **LIFESPAN** 15.5 years in captivity. **Status and Threats** Considered rare, though this is due in part to it inhabiting dense equatorial forest where it is poorly known. Widespread, with large parts of its range in relatively pristine condition, but thought to have undergone general decline due to forest loss and fragmentation. Also hunted as bushmeat, and one of the most frequently killed carnivores by people hunting with dogs, e.g. caught in 52% of hunts and represents about 25% of all carnivores caught by Bambuti pygmies in Ituri Forest, DR Congo. Red List LC.

BUSHY-TAILED
MONGOOSE

JACKSON'S
MONGOOSE

BLACK-LEGGED MONGOOSE

Suricate, Slender-tailed Meerkat

HB 24.5–29cm; T 17.5–24cm; W 0.62–0.97kg

Small mongoose with coarse yellow-brown to pale greyish-tan fur and dark eye patches. Back marked with short dark brown streaks, producing brindled (not striped) appearance. **Distribution and Habitat** Endemic to arid SW Africa, in Botswana, S Africa, Namibia and extreme SW Angola. Inhabits semi-arid desert, dry open savannah, dry scrubland and grassland. **Feeding Ecology** Eats chiefly arthropods on the soil surface and subsurface, especially beetles, millipedes, centipedes, termites, scorpions, spiders, and various insect eggs, pupae and larvae. Resistant to scorpion venom and the noxious secretions of millipedes. Also eats small rodents, herptiles, birds (mainly fledglings) and eggs. Not a poultry pest. Foraging strictly diurnal in large social groups. Prey is captured individually; not shared except with pups. **Social and Spatial Behaviour** Intensely social. Lives in family groups of 3–20 individuals (exceptionally to 49) comprising a dominant breeding pair, offspring from successive litters and some unrelated immigrant adults. Group members cooperate to rear young, avoid predation and defend communal range. Clashes over territory are sometimes fatal. Group ranges 2–10km². Density 0.32–1.69 Meerkats/km², fluctuating significantly depending mainly on rainfall (and therefore insect availability) and predation. **Reproduction and Demography** Seasonal. Breeding in wet summer, September–March. Dominant pair does most of breeding. Subordinate females produce about 25% of litters; most of the pups from these are killed by the dominant female or abandoned. Gestation about 70 days. Litter size 3–7. Females have up to three litters a year. All adults help raise pups; babysitters remain at the den when the group forages, and all adults provision pups. Unlike in Banded Mongoose, pups do not have an escort and beg from any adult. MORTALITY 80% (pups) and 32% (adults), mainly from predation and infanticide (on pups). LIFESPAN 8 years in the wild, 12 in captivity. **Status and Threats** Widespread, common and secure. No major threats at population level. Red List LC.

YELLOW MONGOOSE *Cynictis penicillata*

HB 27–46cm; T 25–29cm; W 0.44–0.9kg

Small mongoose with ginger-tawny to greyish-yellow fur. Southern individuals are larger, more reddish and have a white-tipped tail compared with smaller greyish animals in the north, which usually lack a white tail-tip. **Distribution and Habitat** Endemic to southern Africa from extreme S Angola through Namibia, Botswana and S Africa. Inhabits open semi-arid habitat including semi-desert, grassland, fynbos heath, scrubland and open bushland. Occurs in rangeland. **Feeding Ecology** Mainly insectivorous, preferring termites, beetles, grasshoppers and locusts, but eats a wide variety of invertebrates and small vertebrates to the size of large rats, as well as small amounts of fruits. Occasionally raids domestic poultry and eggs. Foraging mainly diurnal and usually solitary or in pairs. Scavenges from carrion and human refuse. **Social and Spatial Behaviour** Mostly travels and forages alone or in pairs, but lives in small family groups averaging 3–4 adults (exceptionally to 13) that reproduce cooperatively. Groups share a defined communal range, or individual group members occupy different areas of a large, loosely shared range with little overlap except at dens. Average individual range size 0.1–0.49km² (♀s) to 1.02km² (♂s). **Reproduction and Demography** Seasonal. Breeding in wet summer, August–February, with two litters produced in quick succession. All group females breed. Gestation 60–62 days. Litter size 1–5, averaging 2. Females den communally, and all adults provision and guard pups. MORTALITY Large raptors are the main predators. Rabies in S Africa is often fatal, though population effects are unknown. LIFESPAN 15 years in captivity. **Status and Threats** Widespread and common. Wholesale eradication campaigns to control rabies induce local declines; these programmes are now selectively applied to certain hotspot areas near humans. Red List LC.

SELOUS'S MONGOOSE *Paracynictis selousi*

HB 39–47cm; T 28–43.5cm; W 1.4–2.2kg

Slender medium-sized mongoose with a pale tawny-grey grizzled coat, darkening along the tail to the tip, which is greyish-white. Lower legs blackish-brown. **Distribution and Habitat** Endemic to SC Africa from S Angola to Malawi to N and E Botswana, NE S Africa and S Mozambique. Inhabits mainly dry and wet woodland savannahs, scrub, grassland and cultivated areas with cover. **Feeding Ecology** Based on limited records, eats invertebrates (especially locusts, grasshoppers, termites and beetles), small rodents, herptiles and birds. Foraging nocturnal and solitary. **Social and Spatial Behaviour** Poorly known. Most records are solitary; pairs are reported but composition is unknown. **Reproduction and Demography** Apparently seasonal, with births occurring mainly in the wet season, August–March. Litter size 2–4. MORTALITY Unknown; one record of predation by a Martial Eagle. LIFESPAN Unknown. **Status and Threats** Widespread, occuring in large areas of intact habitat. Considered uncommon compared with sympatric mongoose species based on mostly ad hoc survey efforts. Red List LC.

MELLER'S MONGOOSE *Rhynchogale melleri*

HB 44–48.5cm; T 28–41.2cm; W 1.7–3kg

Large mongoose with a distinctive blunt, slightly bulbous muzzle, and grizzled pale to dark brownish-grey fur. Tail colour can be black, brownish-grey or pale greyish-white within the same population. **Distribution and Habitat** C Tanzania, S DR Congo, Malawi, Mozambique, Zambia and NE S Africa. Inhabits mainly open woodland, savannah and grassland, but possibly has broader habitat tolerances; recently found in bamboo forest at 1850m (Udzungwa Mountains, Tanzania). **Feeding Ecology** Poorly known, but mainly insectivorous, apparently eating chiefly termites; almost always found in association with termite mounds. Also eats small vertebrates. Foraging thought to be nocturnal, terrestrial and solitary. **Social and Spatial Behaviour** Unknown. Believed to be solitary. **Reproduction and Demography** Believed to be seasonal based on a few records; births occur in wet summer, November–January. Litter sizes of to 3 are reported. MORTALITY and LIFESPAN Unknown **Status and Threats** Status poorly known; never properly surveyed. Widespread, occurring in many large protected areas. Red List LC.

Mobbing behaviour

MEERKAT

**YELLOW
MONGOOSE**

**SELOUS'S
MONGOOSE**

MELLER'S MONGOOSE

Savannah Mongoose

HB 25–33cm; T 16–23cm; W 0.3–0.4kg

Very small mongoose with grizzled dark brown fur, paler underparts and dark lower limbs. Head and neck dark brownish-grey and grizzled with greyish hairs. Similar to dwarf mongooses and sometimes classified with them in the genus *Helogale*; separated mainly on the basis of anatomical differences, especially in the skull and dentition. Lacks the groove in the upper lip present in both dwarf mongoose species. **Distribution and Habitat** NW Uganda, NE DR Congo, S Central African Republic, extreme southern Sudan and possibly Rep Congo. Known from dry savannah, savannah-forest mosaics and montane-forest grassland. **Feeding Ecology** Unknown. Has small, reduced dentition and front feet with very robust claws, suggesting that its feeding ecology is similar to that of dwarf mongooses, with a specialization mainly on surface invertebrates. One specimen had termites, millipedes and small seeds in its stomach. **Social and Spatial Behaviour** Unknown. Museum records and sightings indicate it is solitary, though it shares many evolutionary similarities with social mongooses such as cusimanses and dwarf mongooses. **Reproduction and Demography** Unknown. Litter of 4 reported from DR Congo. **Status and Threats** Status unknown. Pousargues's Mongoose is known only from 31 museum specimens and a handful of unconfirmed sightings; there have been no new records since the 1970s. Red List DD.

SOMALI DWARF MONGOOSE *Helogale hirtula*

Desert Dwarf Mongoose, Ethiopian Dwarf Mongoose

HB 20–27cm; T 15–18cm; W 0.22–0.35kg

Very small mongoose with longer, paler fur than Common Dwarf Mongoose. Colour warm tawny or grizzled pale grey-brown with contrasting dark brown digits. In the field, appears shaggier and less reddish than Common Dwarf Mongoose. **Distribution and Habitat** Somalia, SE Ethiopia and E Kenya; possibly Djibouti and extreme NE Tanzania (Mkomazi GR). Inhabits arid to semi-arid open woodlands, scrub and grassland to 600m, and thought to be largely independent of standing water. **Feeding Ecology** Unknown. Assumed to be similar to that of Common Dwarf Mongoose. Cheek teeth are more robust, suggesting small vertebrates might be relatively more important in the diet. **Social and Spatial Behaviour** Known to live in large groups similar to Common Dwarf Mongoose, but there are no details. **Reproduction and Demography** Unknown. **Status and Threats** Unknown. Said to be common in some areas, but its status has never been properly assessed anywhere, and its range is completely overlapped by that of Common Dwarf Mongoose, which might lead to misidentification. Red List LC.

COMMON DWARF MONGOOSE *Helogale parvula*

Dwarf Mongoose

HB 16–23cm; T 14.2–18.2cm; W 0.21–0.34kg

Smallest mongoose (both Pousargues's and Somali Dwarf Mongoose are comparable, but have very few verified measurements). Fur smooth, sleek and uniformly coloured. Colour variable, typically grizzled tawny-brown, red-brown or dark brown with warm-coloured underparts. Melanism occurs, in which most or all of a population is black, e.g. Erongo Mountains, C Namibia. **Distribution and Habitat** From the Horn of Africa, through E Africa (E of the Rift Valley), S DR Congo, Mozambique, Zambia and Angola to N Namibia, N Botswana and NE S Africa. Inhabits a wide variety of habitats with high densities of termitaria, including dry and wet woodland savannahs, brush, scrubland and open forest. Occurs in pastoral areas. **Feeding Ecology** Almost entirely insectivorous, with the diet dominated by surface and subsurface invertebrates, especially termites, beetles, grasshoppers, millipedes, centipedes and scorpions. Also eats small mammals, reptiles, frogs, birds, and small amounts of fruits including berries. Breaks open large eggs by throwing them backwards through its hindlegs at the ground or against a rock. Rarely kills poultry, but known to take the eggs of domestic fowl. Foraging strictly diurnal. Forages as a group, but spreads out individually to search for prey, most of which is caught and consumed alone. Large snakes are killed cooperatively (it is resistant to snake neurotoxins), and adults sometimes cooperate to kill large rats. **Social and Spatial Behaviour** Intensely social. Lives in complex extended groups of 2–32, averaging 9–12 adults and subadults that defend a stable group range. Each group is led by a dominant breeding pair, usually the oldest individuals of each sex, which bond for life. The group's remaining adults are made up of their grown offspring or unrelated immigrants. Group members are extremely social, engaging in constant contact and cooperative behaviours, including territorial defence, scent-marking, predator vigilance, raising young, grooming and sleeping together. Adults rescue pups from danger, and groups provision injured or invalid members. Group range size 0.27–0.96km². Density 3.9–30.9 mongooses/km². **Reproduction and Demography** Generally aseasonal, but births peak during wetter periods in strongly seasonal habitat, e.g. N Kenya. Group females synchronize oestrus, and all adults in a group mate; the alpha male dominates the alpha female, and she is often the only female to successfully reproduce. Despite mating, subordinate females are suppressed behaviourally and physiologically, and produce few litters. Of subordinate litters born, most do not survive due to infanticide by the alpha female; they are more likely to be tolerated during favourable conditions. Gestation 49–53 days. Litter size 2–6, averaging 2–3. Dominant female can produce up to four litters a year in favourable conditions. Subadults remain in the group, disperse voluntarily to seek breeding opportunities or are forced out following a takeover; females are more likely than males to remain in their natal group. MORTALITY Annually to 59% (emerged pups to 1 year), and 26% (adult ♀s) to 32% (adult ♂s). Major predators include large raptors, monitor lizards, puff adders, large mongooses and Honey Badger. LIFESPAN 10 (♂s) and 14 (♀s) years in the wild, 18 in captivity. **Status and Threats** Widespread, versatile and attains very high densities in some areas. Occurs in many protected areas and does not suffer major threats at population level. Red List LC.

POUSARGUES'S
MONGOOSE

SOMALI DWARF
MONGOOSE

Scent-marking
at den

COMMON DWARF
MONGOOSE

GAMBIAN MONGOOSE *Mungos gambianus*

PLATE 27

HB 30–45cm; T 23–29cm; W 1–2.2kg
Medium-sized stocky mongoose with coarse, grizzled reddish-grey-brown fur. Yellowish throat and chest demarcated by a distinctive black or dark brown streak running from the ear to the foreleg. Tail ends in a dark tip. Behaviour and ecology thought to parallel those of closely related Banded Mongoose, but have never been studied. **Distribution and Habitat** Endemic to W Africa from Gambia and S Senegal to the Niger River, Nigeria. Occurs in dry to semi-moist woodlands, open savannah, grassland and coastal scrub. **Feeding Ecology** Poorly known, but resembles Banded Mongoose in eating mainly arthropods on the soil surface, as well as small vertebrates including rodents, snakes and lizards. Forages in social groups and is strictly diurnal. **Social and Spatial Behaviour** Highly social. Lives in family groups typically comprising 5–15 individuals, but occasionally reaching 30–40 (reported from Côte d'Ivoire, Gambia and Senegal). **Reproduction and Demography** Poorly known. Young animals recorded January–February and September (Ghana) and June (Sierra Leone). MORTALITY Unknown; large raptors assumed to be the main predators. LIFESPAN Unknown. **Status and Threats** Status poorly known. Fairly widespread and common in some protected areas, e.g. Niokolo-Koba NP, Senegal and Upper Niger NP, Guinea. Sought after as bushmeat but effects are unknown. Red List LC.

BANDED MONGOOSE *Mungos mungo*

Striped Mongoose
HB ♂ 30–45cm; T 17.8–31cm; W 0.9–1.9kg
Medium-sized mongoose with short coarse fur. Colour grizzled pale grey-brown to dark brown with 10–15 dark, narrow transverse bands across the mid-back and rump. Juveniles lack the bands until around 8–10 weeks old and are easily mistaken for Dwarf Mongooses. **Distribution and Habitat** Widespread in sub-Saharan Africa; absent from forested W and C Africa, and arid southern Africa. Found in all types of woodland, savannah and grassland; does not occur in desert, semi-desert and montane habitat. Tolerates farmland and cultivated areas. **Feeding Ecology** Mainly insectivorous, eating arthropods on the soil surface or subsurface, especially millipedes, dung beetles, termites, ants, grasshoppers, crickets, their larvae and eggs. Also takes rodents, shrews, lizards, small snakes, amphibians, birds, fledglings, eggs and some fruits. Not considered a poultry pest. Foraging strictly diurnal and in large social groups; most prey is captured individually and hunting is not cooperative. Eagerly investigates ungulate dung, especially of elephants and rhinos, for dung beetles, and occasionally grooms Warthogs for ticks and other external parasites. Scavenges from human refuse in villages, tourist camps and refuse dumps. **Social and Spatial Behaviour** Intensely social. Lives in large, cohesive family groups typically of 12–20 individuals, which cooperate to raise pups, watch for predators and defend a communal range. Groups numbering up to 70–75 are occasionally recorded, usually under super-abundant food availability like that around refuse dumps in tourist lodges. Group adults are inter-related as both males and females may stay in their natal group; generally, group males are more closely related to other males, and females are more closely related to females. Groups are territorial and defend their ranges against other groups in clashes that are sometimes fatal. Group ranges 0.3–2km² (Queen Elizabeth NP, Uganda). Density 2.4–3 mongooses/km² (KwaZulu-Natal, S Africa, and Serengeti, Tanzania) to 18/km² (Queen Elizabeth NP). **Reproduction and Demography** Breeding occurs in wetter periods in regions with marked seasonality, otherwise aseasonal. Most adults in a group breed. Females synchronize oestrus and give birth within a few days of each other in a communal den; litters born out of synchrony are much less likely to survive. Gestation 60–63 days. Litter size 1–6, averaging 3. Males compete among each other for females, and dominant males sire most pups, but most males mate. All group adults contribute to pup care; 1–2 babysitters remain at the den when the group forages, and all adults groom, carry, play with and provision pups; females suckle any pup. When pups leave the den at 5 weeks, they are attached to an individual 'escort' that feeds and protects its pup until independence at around 3 months. MORTALITY 72% (birth to independence) and 14–33% (adults), mainly from predation and infanticide (on pups). LIFESPAN 17 years in captivity. **Status and Threats** Very widespread, common and secure. Disappears from intensively modified habitat, but not threatened. Red List LC.

LIBERIAN MONGOOSE *Liberiictis kuhni*

HB 42.3–46.8cm; T 19.7–20.5cm; W (single ♂) 2.3kg
Fairly large mongoose with a narrow head and long mobile snout. Uniformly dark brown with tawny-orange underfur, a pale throat and chest, and a distinctive dark neck stripe. Classification is disputed, but Banded and Gambian Mongooses are thought to be its closest relatives. **Distribution and Habitat** Endemic to Côte D'Ivoire and Liberia. Occurs in primary and secondary rainforests, usually associated with swampy areas and sandy stream beds where earthworms are abundant. **Feeding Ecology** Thought to be an earthworm specialist, with small amounts of other arthropods, small vertebrates and fruits consumed. Captives eat ground beef, commercial dry dog food, chicks, fish and live insects. Foraging diurnal and in small social groups. Uses its extremely well-developed front claws to excavate leaf litter and soil, and thrusts its long flexible nose in turned soil to locate prey. **Social and Spatial Behaviour** Social. Forms small groups typically of 4–6, though larger groups are sometimes observed. Adult males often travel alone and visit numerous groups for brief periods of 1–3 days, suggesting the social system might resemble that of coatis, with female groups and solitary males, but group composition is unclear. Range and density estimates unknown. **Reproduction and Demography** Poorly known. A few records suggest breeding occurs in the wet season, May–September. MORTALITY Unknown. Crowned Eagle is a known predator. LIFESPAN Unknown. **Status and Threats** Very restricted range and dependent on earthworm-rich wet forest. Habitat loss and bushmeat hunting are serious threats. Red List VU.

GAMBIAN MONGOOSE

BANDED
MONGOOSE

Grooming
Warthog

LIBERIAN MONGOOSE

Ansorge's Cusimanse
HB 32–36cm; T 20.8–22cm; W 0.6–1.5kg
Medium-sized cusimanse with shaggy dark brown fur and dense reddish-brown underfur. Lower face pale tawny and may have white flashes along the cheeks and eyebrows. Until 1984, only two specimens existed, one of which is still the only Angolan record from 1908. It is now known to be more widely distributed, but remains essentially unstudied. **Distribution and Habitat** Endemic to W and C DR Congo and NW Angola in two disjunct populations; not clear if it occurs in between. All records are from rainforest. **Feeding Ecology** Poorly known but assumed to resemble other cusimanses, with a diet of forest-floor invertebrates and small vertebrates. Captive individuals refuse fruits including berries, and mushrooms. Foraging likely to be diurnal, terrestrial and social. **Social and Spatial Behaviour** Highly social, living in groups with as many as 20 members, but no details are known. **Reproduction and Demography** Unknown. **Status and Threats** Rare compared to sympatric Alexander's Cusimanse based on lower frequencies in bushmeat surveys (10% of hunted carnivores compared to 42% for the latter). Unlike Alexander's Cusimanse, it is not recorded outside intact rainforest and may be forest dependent. Red List DD.

FLAT-HEADED CUSIMANSE *Crossarchus platycephalus*

Cameroon Cusimanse
HB 30–36cm; T 15.6–21cm; W 0.5–1.5kg
Small cusimanse similar to Common Cusimanse, but with a flatter, wider skull and paler colouration. Sometimes classified as Common Cusimanse, but they are generally regarded as distinct species separated by the Dahomey Gap. **Distribution and Habitat** S Nigeria through S Cameroon, SW CAR to Equatorial Guinea and NW Congo; probably occurs in N Gabon, and there are equivocal records from W Benin. Inhabits dense rainforest to 1600m, as well as forest-savannah mosaics and forest farmland. **Feeding Ecology** Poorly known, but resembles other cusimanses, with a diet of forest-floor invertebrates and small vertebrates. Forages in shallow water for freshwater crabs, which appear to be important prey. Foraging diurnal, social and mainly terrestrial; two adults were seen to pursue a large Black Cobra 3m into a tree (it escaped). **Social and Spatial Behaviour** Highly social but poorly known. Family groups are relatively small for cusimanses, typically comprising 5–8 members, though up to 25 are recorded. **Reproduction and Demography** Poorly known. Believed to be aseasonal, with litter size 2–5. **Status and Threats** Status poorly known. Relatively widespread and apparently persists in degraded habitat, e.g. Niger Delta, suggesting resilience to forest loss. Hunted for bushmeat. Red List LC.

COMMON CUSIMANSE *Crossarchus obscurus*

Long-nosed Cusimanse, West African Cusimanse
HB 30–37cm; T 14.6–21cm; W 0.45–1kg
Small stocky cusimanse with dark brown to blackish grizzled and shaggy fur with paler underfur. Fur on the head and face is shorter and paler than fur on the body, typically tawny or reddish-brown. **Distribution and Habitat** Endemic to coastal W Africa from W Guinea and Sierra Leone to the Ghana-Togo border. Inhabits mainly dense understory of rainforest and riparian forest to 1500m. Occurs in logged forest and plantations with understory. **Feeding Ecology** Eats mainly forest-floor invertebrates, especially millipedes, ants, termites, earthworms, beetles, grasshoppers, insect larvae, spiders and snails. Also takes rodents, small birds, herptiles and eggs; captives eat fruits including berries. Cusimanses forage diurnally and socially in large groups in which individuals mostly search for prey individually. Large prey, including cobras and Giant-pouched Rat, is hunted cooperatively, the latter by one or more cusimanses entering holes to flush it to the surface, where it is caught by the group. Foraging terrestrial; rarely pursues prey above the ground, but climbs vine tangles to 25m to den. **Social and Spatial Behaviour** Highly social. Lives in cohesive territorial groups with up to 20 members. Group structure unclear, but may be an aggregation of 2–3 family units (possibly inter-related), each comprising a mated adult pair and offspring from one or more litters. Group range size estimates 0.28–1.4km². Density in good-quality habitat 13.2–17 cusimanses/km². **Reproduction and Demography** Aseasonal in captivity, but most known wild births occur January–February and May–June. Gestation 53–60 days. Litter size 2–5. Breeding adults provide most of the care for pups, but non-breeding adults assist by guarding pups, carrying them between dens and provisioning them with prey. **MORTALITY** Poorly known; mortality of juveniles to 6 months estimated at 45.5% (based on few observations). Large diurnal raptors, especially Crowned Eagle, are major predators. **LIFESPAN** 13 years in captivity. **Status and Threats** Relatively restricted distribution that is under considerable human pressure, but not dependent on undisturbed forest and reaches high densities in suitable habitat. Vulnerable to bushmeat hunting. Red List LC.

ALEXANDER'S CUSIMANSE *Crossarchus alexandri*

HB 35–44cm; T 20.8–32cm; W 1–2kg
Largest cusimanse, with long, dark shaggy fur, conspicuous whorls on the neck and long dorsal hairs producing an inconspicuous crest. **Distribution and Habitat** Endemic to forested DR Congo and W Uganda; a disjunct population occurs on Mt Elgon on the Uganda-Kenya border. Inhabits lowland and montane rainforests, especially in swampy areas. Occurs near villages and cultivated areas. **Feeding Ecology** Poorly known but resembles other cusimanses, with a diet mainly of forest-floor invertebrates, small vertebrates and some fruits. Foraging diurnal, terrestrial and social. **Social and Spatial Behaviour** Highly social but poorly known. Family groups are relatively small for cusimanses, typically having up to 10 members. **Reproduction and Demography** Unknown. A pregnant female collected in Virunga NP, DR Congo, had six embryos. **Status and Threats** Believed to be widespread and relatively common based on high frequency in bushmeat surveys. Appears resilient to habitat modification, but severe degradation combined with hunting represents a local threat. Red List LC.

ANGOLAN
CUSIMANSE

FLAT-HEADED
CUSIMANSE

COMMON
CUSIMANSE

Foraging group

ALEXANDER'S CUSIMANE

FALANOUC *Eupleres goudotii*

PLATE 29

HB 45–65cm; T 22–25cm; W 1.6–4.6kg
Very distinctive euplerid with an extremely slender head that appears undersized, a stout body and a long bulbous tail. Some authorities recognize two species, Eastern Falanouc *E. goudotti* and a larger species, Giant Falanouc *E. major*, restricted to NW Madagascar, but this remains tentative. **Distribution and Habitat** Madagascar, in undisturbed humid and deciduous forests and associated marshland at sea level to 1600m. Sometimes observed in mesic, open grassy patches in these habitats. **Feeding Ecology** Almost exclusively insectivorous, with an exceptionally elongated snout and small conical teeth specialized for eating earthworms, slugs, snails and insect larvae. Captive animals eat frogs and small meat pieces, but otherwise refuse vertebrate prey. Long non-protractile claws are well suited for excavating prey. Accumulates up to 20% of its body weight as fat in the tail during the cold dry season, presumably to cope with reduced prey availability. Cathemeral. **Social and Spatial Behaviour** Virtually unknown. Probably solitary; all wild observations are of single animals or females with young. Both sexes mark prominent bushes and rocks with their ano-genital glands, but it is unclear if this demarcates territorial boundaries. **Reproduction and Demography** No information from the wild. Captive animals mate July–September; births November–January. All captive births are singletons, but twins reportedly occur. Kittens are surprisingly precocious, able to accompany the mother after 2–3 days, and weaned by 9 weeks. **Status and Threats** Secretive, solitary and poorly known, but thought to be naturally rare. Presumably vulnerable to the threats facing all of Madagascar's carnivores. CITES Appendix II, Red List NT.

FOSA *Cryptoprocta ferox*

Fossa
HB 70–80cm; T 61–70cm; W ♀ 5.5–6.8kg, ♂ 6.2–8.6kg
Madagascar's largest carnivore. Typically reddish to chocolate-brown; reports of melanism are unconfirmed. Kittens born with pearl-grey fur that darkens fully to adult colour by 5–6 months. **Distribution and Habitat** Madagascar. Inhabits forest and woodland to 2600m, from dry woodland savannah with <400mm rainfall to rainforest with over 6000mm rainfall. Reaches the highest densities in western dry deciduous forest. **Feeding Ecology** Formidable predator, able to kill all indigenous species including the largest lemurs (predation on adult Indris is equivocal). Primates frequently dominate the diet, but it opportunistically switches to other prey depending on availability. Rodents, tenrecs and other carnivores, including Narrow-striped Boky and Fanaloka, are important prey at some sites; birds, reptiles, amphibians and invertebrates are also eaten. Records of Bushpig and cattle are almost certainly scavenged. Raids domestic chickens. Foraging cathemeral, and both terrestrial and in trees, where it displays exceptional agility. Usually hunts alone, but cooperative hunting has been anecdotally reported for breeding pairs and mother-young groups; a direct observation exists of three adult males cooperating to drive a Verreaux's Sifaka to the ground, where it was caught and shared. **Social and Spatial Behaviour** Solitary and probably territorial; vigorously marks range borders with pungent ano-genital secretions, but territories overlap up to 30% and adults congregate at mating sites. Range size reaches 13km² (♀s) and 27km² (♂s), and is largest during the dry season, probably due to dispersed water and prey. Only density estimate is 5.5 Fosas/km from a high-density population in dry forest (Kirindy). **Reproduction and Demography** Likely seasonal. Mates September–December, usually in trees at traditional sites used by many females. Up to eight males attend an oestrous female, fighting for access, and females mate with multiple males. Gestation ~90 days. Litter size 2–4, rarely to 6 (captivity). Development of kittens is unexpectedly slow; weaning at around 4–5 months, and independence at around 12 months. MORTALITY Poorly known. Humans and domestic dogs are the only predators; people kill significant numbers in some areas. LIFESPAN 23 years in captivity. **Status and Threats** Restricted to peripheral forests of Madagascar, with an optimistic total population estimate of less than 2500. Forest loss and fragmentation are the key threats, combined with human hunting for meat (at high rates locally, especially for a top carnivore) and retributively for chicken depredation. Feral domestic dogs impact Fosa densities, and it is vulnerable to exotic diseases from dogs and cats, though impacts are unquantified. CITES Appendix II, Red List VU.

FANALOKA *Fossa fossana*

Spotted Fanaloka, Malagasy Civet, Malagasy Striped Civet
HB 40–45cm; T 21–26.4cm; W 1.3–1.9kg
Genet-like, with parallel rows of dark spots and blotches on a grizzled light brown background. Bushy tail is usually marked with brown bands, absent in some animals. **Distribution and Habitat** N and E Madagascar, primarily in humid and dry forests to 1300m. Most common in well-watered lowland rainforest, but also occurs in dry forested canyons of *tsingy* (limestone karst) with no permanent water at Ankarana. **Feeding Ecology** Hunts small prey such as rodents, tenrecs, small birds, reptiles, amphibians and invertebrates. In rainforest, hunts in shallow streams for freshwater crustaceans, eels and amphibians. Stores fat in its tail for the winter dry season, up to 25% of its body weight. Foraging mainly nocturnal and terrestrial, though it is an accomplished climber and sometimes hunts in small trees. **Social and Spatial Behaviour** Solitary. Spatial patterns are poorly known, but a brief study indicates that females and males inhabit discrete home ranges of 0.07–0.52km². Adults vigorously scent-mark ranges and males guard oestrous females, suggesting territoriality. In the same study, at least 22 adults were known from 2km², suggesting that it reaches high densities. **Reproduction and Demography** Seasonal. Mates August–September; births October–December. Gestation 80–89 days. All recorded litters comprise only 1 kitten, which is well developed at birth, able to walk at day 3 and weaned by 8–10 weeks. MORTALITY Poorly known; Fosa is a confirmed predator, and human hunters remove significant numbers from some populations. LIFESPAN Unknown in the wild, 11 years in captivity. **Status and Threats** Widespread within its limited range and seems to reach high densities. However, does not colonize human-modified habitats and is vulnerable to human hunting. Persecuted as a supposed predator of chickens (unsubstantiated but possible), and eaten by people in rural areas. CITES Appendix II, Red List NT.

FALANOUC

FOSA

FANALOKA

Broad-striped Malagasy Mongoose

HB 30–34cm; T 25–29.3cm; W 0.52–0.74kg

Smaller of two striped vontsiras, with broad dark stripes wider than the pale interlines. **Distribution and Habitat** E Madagascar in humid lowland forest usually to 700m (one record at 1500m); sometimes reported from degraded secondary forest. **Feeding Ecology** Unknown. Powerful skull and dentition suggest it mostly hunts small vertebrates;

invertebrates are almost certainly also eaten. Sometimes raids research camp stores, and reputededly kills chickens. Largely nocturnal and terrestrial. **Social and Spatial Behaviour** Observations and camera-trap images suggest adult pairs, but this is unsubstantiated. Rare in surveys, suggesting low densities. **Reproduction and Demography** Unknown. Females have a single pair of teats, suggesting 1–2 young. **Status and Threats** Unknown. Red List NT.

NARROW-STRIPED BOKY *Mungotictis decemlineata*

Narrow-striped Mongoose

HB 26.4–29.4cm; T 19.1–21.5cm; W 0.45–0.74kg

Distinctive grizzled grey fur with yellowish underparts and narrow orange-brown stripes. **Distribution and Habitat** SW Madagascar, restricted mainly to intact dry deciduous and baobab forests to 400m. Disappears from degraded forest. **Feeding Ecology** Feeds chiefly on insects and larvae in the soil, leaf-litter and rotting logs. Occasional preys on mouse and dwarf lemurs, rodents, tenrecs and small reptiles. Lemurs reportedly hunted cooperatively. Large prey including Malagasy Giant Rat (1–1.2kg) probably scavenged. Foraging

largely diurnal and terrestrial, though it climbs well. **Social and Spatial Behaviour** Gregarious with complex sociality based on groups of multiple adult females with young. Males are generally solitary but intermittently join groups, continually while breeding. Families occupy enduring ranges that are marked vigorously; interactions between groups mostly lack aggression. Group range estimates 0.13–0.25km². **Reproduction and Demography** Weakly seasonal. Mates July–December. Gestation 90–105 days. Group females may synchronize births. **Status and Threats** Very limited and fragmented range, but may reach high densities. Red List VU.

RING-TAILED VONTSIRA *Galidia elegans*

Ring-tailed Mongoose

HB 30–38cm; T 26–39.1cm; W ♀ 0.66–0.89kg, ♂ 0.9–1.09kg

Most conspicuous e_plerid. Rich chestnut-red with contrasting chocolate bands on the tail. Eastern animals are darker and more muted. **Distribution and Habitat** Madagascar, in most forest types to 1950m, as well as secondary forest and clearings close to cover. **Feeding Ecology** Feeds mainly on small rodents, tenrecs, reptiles, bird nestlings and eggs, and invertebrates, but capable of taking larger prey, e.g. Greater Dwarf Lemur (600g). Record of a 1.2kg Woolly Lemur was likely scavenged. Forages

terrestrially to 15m in trees, excavates rodents from burrows and fishes for aquatic prey in shallow water. Largely diurnal but occasionally hunts at night. **Social and Spatial Behaviour** Forms male-female pairs with dependent young. Males also occur alone; unclear if pairs form only for breeding, or if single males are dispersers. Family groups maintain ranges of 0.2–0.25km, which are vigorously scent marked. **Reproduction and Demography** Weakly seasonal. Mates July–November. Gestation 75 days. One kitten born October–December; families sometimes have different-aged litters, suggesting biannual breeding. **Status and Threats** Most common and widespread e_plerid. Red List LC.

BROWN-TAILED VONTSIRA *Salanoia concolor*

Brown-tailed Mongoose, Salano

HB 35–38cm; T 16–20cm; W c. 0.78kg

Uniformly dark chocolate-brown, and paler red-brown underparts without markings. A second species, Durrell's Vontsira *S. durrelli*, was first described in 2010 (page 9). **Distribution and Habitat** NE Madagascar from sea level to 1000m. Over 90% of records are from intact rainforest, but it sometimes inhabits secondary forest and cultivated land. **Feeding Ecology** Thought to be mainly insectivorous; the only direct observations involve individuals extracting larvae from rotting wood and leaf-litter. Foraging mainly diurnal

and terrestrial; climbs well and possibly forages for tree-living arthropods. Blamed by villagers for raiding poultry, which is questionable given its weak dentition. **Social and Spatial Behaviour** Poorly known. Groups with up to five individuals suggest adult pairs with dependent young. One sighting included an older kitten and very young juvenile, suggesting offspring from previous litters remain with family. **Reproduction and Demography** All records of kittens November–January, suggesting seasonal breeding. Observed litters comprise 1 kitten. **Status and Threats** Unknown. Locally common at Betampona Reserve. Red List VU.

GRANDIDIER'S VONTSIRA *Galidictis grandidieri*

Giant-striped Mongoose

HB 45–48cm; T 30–32.6cm; W 1–1.7kg

Larger of two striped vontsiras, with narrow dark longitudinal stripes on a pale grizzled grey-beige background. **Distribution and Habitat** SW Madagascar, restricted to only ~450km² on the edges of the hot arid Mahafaly Plateau in dry spiny forest. **Feeding Ecology** Eats unusually small prey for its size, chiefly hissing cockroaches, grasshoppers, arachnids and small vertebrates weighing <10g. Also eats small Malagasy iguanas, geckoes and snakes. Largest recorded prey weighs 140–200g, e.g. Hedgehog Tenrec and

Madagascar Turtle-dove, with one record of a 369g snake. Forages mainly on the ground in leaf-litter and rock crevices, but sometimes climbs small trees in pursuit of prey. Strictly nocturnal. **Social and Spatial Behaviour** Poorly known. Adults appear to be solitary and occupy defined home ranges, but occasional sightings of male-dominated groups of up to five suggest more complex social patterns. **Reproduction and Demography** Unknown. Thought to have 1–2 young. **Status and Threats** Has the most restricted range and habitat preference of any e_plerid; any threats to its habitat would be grave. Total population estimated at 2200. Red List EN.

BROAD-STRIPED
VONTSIRA

NARROW-
STRIPED
BOKY

RING-TAILED
VONTSIRA

BROWN-TAILED VONTSIRA

GRANDIDIER'S
VONTSIRA

HB 37.9–45cm; T 33–37.5cm; W 0.59–0.8kg
The two species of linsang are classified in their own family, the Prionodontidae. Banded Linsang is very small and slender, with a delicate elongated neck and head. Fur cream to yellow-buff with large dark bands across the spine, and interconnected blotches running in parallel bands lengthways along the body. Feet genet-like with fully protractile claws. **Distribution and Habitat** Java, Sumatra, Borneo, Peninsular Malaysia into S Thailand and S Myanmar. Occurs mainly in primary and secondary forests to 2700m, with a handful of records from degraded habitats such as plantations. **Feeding Ecology** Primarily carnivorous, taking small terrestrial and arboreal vertebrates including rodents, birds, reptiles and frogs; invertebrates such as large cockroaches are also eaten. Captive animals refuse fruits. Extremely agile and hunts both in the canopy (at least to 8m) and on the ground; descends trees rapidly head-first. Hunting mainly nocturnal. **Social and Spatial Behaviour** Very poorly known. Considered solitary; reports of pairs and trios are probably mother-young groups. Range size and density have never been studied. **Reproduction and Demography** Poorly known. Litters and pregnant or lactating females recorded April–October, but it is unknown if this represents seasonal breeding. Litter size 1–3. Juveniles reach adult size at 4 months. MORTALITY Unknown. LIFESPAN 10.7 years in captivity. **Status and Threats** Difficult to monitor and status is poorly known. Forest loss is considered the main threat, though the extent to which it can adapt to logged areas is unknown, and it occurs in better-protected higher altitudes. Appears occasionally in wildlife markets, probably reflecting its natural rarity. CITES Appendix II, Red List LC.

SPOTTED LINSANG *Prionodon pardicolor*

HB 31–45cm; T 30–40cm; W 0.55–1.2kg
Similar to Banded Linsang but less richly marked, with small, discrete round spots covering the body in roughly parallel rows. Spots along the back sometimes fuse over the spine, giving a similar appearance to the dorsal bands in Banded Linsang. The two species apparently do not overlap, but their ranges are poorly delineated and they may co-occur in S Thailand and S Myanmar. **Distribution and Habitat** Nepal, Bhutan, NE India, S China and SE Asia to S Cambodia and S Vietnam. Inhabits lowland, hill and mountain forests to 2700m (with equivocal records to 4000m), bamboo forest, forest-grassland mosaics and dense moist grassland scrub. Occurs in some disturbed habitats, including secondary forest and mosaics of pine plantations and cultivated areas. **Feeding Ecology** Similar diet to Banded Linsang's, comprising mainly small terrestrial and arboreal vertebrates such as rodents, birds, reptiles and frogs. Captive female in a semi-natural enclosure hunted terrestrial mice and voles by observing them from tree perches and descending quickly head-first to catch and kill them on the ground. She ate 4–6 small rodents a day. Mainly nocturnal. Three observations of scavenging from a Tiger's kill (Chitawan NP, Nepal). **Social and Spatial Behaviour** Virtually unknown. Usually reported as solitary. Captive female liberally marked her enclosure with urine and faeces in prominent places. **Reproduction and Demography** Poorly known. Very limited records of litters cluster in February–August, numbering 1–2 young. MORTALITY Virtually unknown; two adults drowned in an open well in Nepal, unlikely to be a regular occurrence. LIFESPAN Unknown. **Status and Threats** Status poorly known. Rarely seen or recorded in surveys and wildlife trade, but not necessarily due to rarity; semi-arboreal lifestyle makes it difficult to see and somewhat insulated from snaring. Main threats assumed to be habitat loss and hunting, both of which occur at high intensity throughout its range. Occurs in many protected areas in its range. CITES Appendix II, Red List LC.

SMALL-TOOTHED PALM CIVET *Arctogalidia trivirgata*

Three-striped Palm Civet, Javan Small-toothed Palm Civet

HB 43.2–53.2cm; T 46.3–66cm; W 2–2.5kg
Small civet with a long tubular tail exceeding the head-body length. Colouration very variable, including golden-yellow (especially in Java), various shades of grey, and tawny-brown to very dark brown. All forms typically have paler underparts, a dark tail and three dark dorsal stripes or parallel rows of interconnected spots (sometimes indistinct). Inside of the ear bright pink, contrasting conspicuously with the surrounding dark fur colour, especially while spot-lighting. There is evidence that the species should be reclassified as two species divided north and south of the Kra Isthmus (Thailand-Myanmar). **Distribution and Habitat** W Java, Sumatra, Borneo, mainland SE Asia to S China, NE India and E Bangladesh. Occurs in primary moist and dry evergreen forests at 150–1500m. Known from degraded forest in Laos provided canopy cover is intact; tolerance for more modified habitat is unclear. **Feeding Ecology** Omnivorous. Fruits, especially figs, are a key food, at least during seasonal fruiting peaks. Also eats small vertebrates such as tree squirrels, and invertebrates. Captive animals eat bananas, apples, oranges, grapes, tomatoes and meat. Foraging nocturnal and almost entirely arboreal in upper canopy; occasionally seen in understorey and (rarely) on the ground. Mostly forages alone, but often feeds in close proximity to other individuals in fruiting trees. **Social and Spatial Behaviour** Poorly known. Assumed to be solitary; sightings comprise around 82% single animals and 18% pairs, but pair composition is unknown. **Reproduction and Demography** Poorly known. In captivity, gestation is 45 days and litters number 1–3. Weaning occurs at around 2 months. Captive females can have two litters a year. MORTALITY Unknown. LIFESPAN 15.8 years in captivity. **Status and Threats** Status poorly known mainly due to highly arboreal lifestyle. Almost never appears in camera-trapping and other ground-based surveys; however, night spot-lighting shows it is locally common in some areas. Forest loss and hunting are the main threats; it is somewhat insulated from the latter except where night shooting is employed. Considered endangered in Java. CITES Appendix II, Red List LC.

BANDED
LINSANG

SPOTTED
LINSANG

SMALL-TOOTHED
PALM CIVET

HB 50.2–58cm; T 43.7–52.5cm; W c. 3.6kg
Medium-sized civet with a tail as long as the head-body. Uniformly yellow-beige to golden-brown, sometimes with a rich russet tinge, and paler underparts. Some individuals have three faint dorsal stripes. A dark brown form occurs, usually with a yellowish or white tail-tip; recently, some authorities have proposed this form comprises a separate species. **Distribution and Habitat** Endemic to Sri Lanka, where it is patchily distributed in discontinuous fragments of suitable habitat. Occurs in dry and wet forests from lowlands to cloud forest. Apparently tolerant of some habitat modification, but much less likely than sympatric Common Palm Civet to be near agriculture, plantations and settlements. **Feeding Ecology** Poorly known but believed to be omnivorous, feeding chiefly on fruits, as well as small vertebrates and invertebrates. Mainly nocturnal and believed to be chiefly arboreal, but it is trapped and camera trapped on the ground. **Social and Spatial Behaviour** Very poorly known. Most records are of solitary individuals. Range size and density have never been studied. **Reproduction and Demography** Poorly known. Litter size reported as 1–3, but there are few verifiable records. MORTALITY and LIFESPAN Unknown. **Status and Threats** Has a small and fragmented range estimated at less than 20,000 km², in an estimated 5–10 discontinuous populations. Can be locally common, especially in Sri Lanka's wet zone and highlands. Ongoing habitat loss, especially in highland areas, is the main threat. Red List VU.

BROWN PALM CIVET *Paradoxurus jerdoni*

Jerdon's Palm Civet
HB 51–61.5cm; T 44–50cm; W 2–4.3kg
Medium-sized civet with a long tail, typically very dark brown with slightly lighter sides and hindquarters and pale underparts. Some animals are dramatically lighter over the torso, ranging from light brown to pale tawny-grey and (rarely) golden, with only the head, forequarters and lower hind limbs retaining the dark brown colouration. Tail sometimes has a prominent yellowish or white tip, mainly in lighter individuals. **Distribution and Habitat** Endemic to W Ghats, S India. Inhabits evergreen forest at 500–1300m, most commonly above 1000m. Occurs in forest mosaics with highly modified habitat, including coffee and cardamom plantations. **Feeding Ecology** Mainly frugivorous, with the diet dominated by wild and cultivated fruits, seeds and flowers. Also eats invertebrates such as insects, centipedes, snails and crabs; small mammals, birds and reptiles are occasionally taken. Animal prey is taken mainly in the dry, non-fruiting season. Primarily arboreal, but readily travels and forages on the ground. Foraging nocturnal and alone; individuals sometimes feed close together in fruiting trees. **Social and Spatial Behaviour** Solitary. Ranges are small and overlap; conflict in shared areas is reduced by time-sharing, in which individuals largely avoid each other. Rests during the day in Giant Squirrel dreys, tree hollows or vine tangles, or on tree limbs. Male ranges are larger than female ranges. Limited range estimates (monitored for <1 year) 0.06–0.18km² (♀s) and 0.1–0.56km² (♂s). **Reproduction and Demography** Poorly known. Very limited observations of kittens report litter size as 1–2. MORTALITY Known mortality is mainly anthropogenic. LIFESPAN Unknown. **Status and Threats** Restricted distribution, but appears to be fairly common in much of its range. Also tolerates plantations provided the natural canopy is left intact, as is often the case with coffee, and persists in small forest fragments surrounded by modified habitat. Main threat is habitat loss and conversion to intensive plantation production such as tea, *Eucalyptus* and teak. Illegally hunted for meat. CITES Appendix III, Red List LC.

SULAWESI PALM CIVET *Macrogalidia musschenbroekii*

Giant Palm Civet, Celebes Palm Civet
HB 65–71.5cm; T 44.5–54cm; W 3.9–6.1kg
Large civet with a long tubular tail. Light chestnut to tawny-brown, with rich yellow-brown underparts clearly demarcated from the upper fur colour. Small faint brown spots on the upper hindquarters, and tail marked with alternating light and dark bands. **Distribution and Habitat** Endemic to Sulawesi, where it is confirmed from the North, Central and South-east peninsulas. Inhabits lowland forest, montane forest to 2600m and savannah-forest mosaics. Occurs in agricultural areas surrounded by forest. **Feeding Ecology** Omnivorous, with a diet dominated by rodents and fruits, especially of palm trees. Also takes Sulawesi Dwarf Cuscus and birds, including the large mound-building Philippine Scrubfowl and Maleo. Local people report that it takes piglets of Sulawesi Warty Pig, but this remains unconfirmed. Readily kills domestic chickens in villages, and captive animals take fruits such as bananas and papayas. Extremely agile and thought to forage both arboreally and terrestrially. Mainly nocturnal. **Social and Spatial Behaviour** Poorly known. All records are of solitary individuals. No range or density estimates. **Reproduction and Demography** Entirely unknown. Females have two pairs of teats, suggesting small litters comparable with those of other palm civet species. MORTALITY Adults do not have any predators except humans and their dogs. LIFESPAN Unknown. **Status and Threats** Status poorly known, but very restricted range combined with extremely high rates of forest clearing are reasons for concern. Over 50% of Sulawesi's forest was lost in 1985–1997, and the species is now absent from many lowland areas that previously provided suitable habitat. It is better protected in highland areas, which are more difficult to exploit. Apparently not sought after for food, but persecuted for raiding chickens. Red List VU.

GOLDEN
PALM CIVET

BROWN
PALM CIVET

SULAWESI
PALM CIVET

HB 50.8–87cm; T 50.8–63.6cm; W 3–5kg
Large palm civet with a highly variable body colour, typically greyish-brown but ranging from pale blond to dark reddish-brown. Most forms have a distinctive black-and-white facial mask, which also varies considerably; Sundaic animals often have a golden or pale grizzled face without a discrete mask pattern. **Distribution and Habitat** S and SE China, the Himalayan region from N Pakistan to NE India, the Andaman Islands, Indochina, Sumatra and Borneo. Introduced (presumably) in Japan. Occurs mainly in deciduous, evergreen and peat-swamp forests. Inhabits degraded forest, plantations and farmland with cover. **Feeding Ecology** Omnivorous, feeding mainly on fruits, small mammals and invertebrates. Diet shifts seasonally to track fruit availability, e.g. June–October in C China, while rodents and birds are mainly consumed November–May. Raids cultivated fruits including bananas, figs, kiwis and mangoes, and occasionally kills domestic poultry. Foraging mainly nocturnal, and both arboreal and terrestrial. Forages alone, but congregates amicably at dumps and feeding stations, where it scavenges food waste and handouts. **Social and Spatial Behaviour** Solitary. Adults are not strictly territorial, but avoid overlap in small core areas. Range size 1.9–3.7km² (♀s) and 1.8–5.9km² (♂s). No published density estimates, but very common in suitable habitat. **Reproduction and Demography** Reported as both seasonal and aseasonal, breeding once or twice a year, but there are few accurate data from the wild. Gestation 51–56 days. Litter size 1–4, averaging 2–3. MORTALITY Poorly known; humans are the main cause of mortality in most of its range. LIFESPAN 15 years in captivity. **Status and Threats** Widespread and common, but also exposed to very high levels of harvest. Common in Asian wildlife markets, where it is sold mainly for meat. Farmed in China, where wild populations have declined significantly from over-harvesting, and listed as NT. Carries the SARS coronavirus, but effects on wild populations are unknown. Red List LC.

COMMON PALM CIVET *Paradoxurus hermaphroditus*

Musang, Toddy Cat
HB 42–71cm; T 33–66cm; W 2–5kg
Medium- to large-sized palm civet, varying from tawny-grey to very dark grey, marked with longitudinal rows of small dark spots that merge into three narrow dorsal stripes. Most forms have a variable dark facial mask, dark legs and a dark tail. Population on the Mentawai Islands, Indonesia, is sometimes considered a separate species, the Mentawai or Siberut Palm Civet *P. lignicolor*. **Distribution and Habitat** Indian subcontinent including Sri Lanka, S China, Indochina, Sumatra, Java, Borneo and the Philippines. Inhabits forest, woodland and wooded scrub from lowlands to 2400m. Lives close to humans on plantations and farmland, and in villages. **Feeding Ecology** Mainly frugivorous, eating wild and cultivated fruits, nectar and sap. Often consumes coffee berries (passed beans are highly sought after for 'civet coffee', for which civets are also farmed, mainly in Indonesia). Also eats arthropods and small vertebrates, especially in non-fruiting periods. Often raids fruit plantations, and (rarely) kills poultry. Foraging solitary, nocturno-crepuscular and mainly arboreal. Scavenges food waste from village and rice-field dumps. **Social and Spatial Behaviour** Solitary. Marks prominent sites on branches and on the ground with piles of faeces, and ranges sometimes overlap extensively. Range sizes 0.06–1.4km² (♀s) and 0.17–4.2km² (♂s). **Reproduction and Demography** Thought to be aseasonal. Gestation 61–63 days (captivity). Litter size 2–5. MORTALITY Predators include large cats, and possibly large raptors and pythons. LIFESPAN 22.4 years in captivity. **Status and Threats** Widespread, common and tolerant of disturbed habitats. Hunted for meat (especially in China), captured as pets or rodent catchers, and killed by fruit farmers for raiding orchards: in concert, these threats may produce local declines. However, it is resilient, and persists in disturbed and heavily harvested areas. CITES Appendix III (India), Red List LC (VU on Mentawai).

BINTURONG *Arctictis binturong*

Bearcat
HB 61–97cm; T 50–84cm; W 9–20kg
Largest civet. Unmistakable, with black shaggy fur often tipped with silvery-white, giving a grizzled appearance. Face usually pale, varying from slightly grizzled to completely silvery-grey. Ears have a white rim and very long hair on the outer surface. Long bushy tail is prehensile. Subspecies on Palawan Island, the Philippines, is sometimes considered a separate species, Palawan Binturong *A. whitei*. **Distribution and Habitat** NE India, Bhutan, E Bangladesh, extreme S China, Indochina, Sumatra, Java, Borneo and Palawan (the Philippines). Presence in E Nepal uncertain. Inhabits mainly dense primary and secondary forests, and dry forest-grassland mosaics. Occurs in reduced densities in logged forest, and avoids heavily modified habitat. **Feeding Ecology** Poorly known in the wild. Omnivorous. Fruits are assumed to be the main food, with figs apparently favoured; often seen in fig trees. Also eats small vertebrates and arthropods. Captive animals eat a wide variety of fruits, vegetables, ground beef, eggs and milk. Not known to kill poultry. Foraging solitary and nocturno-crepuscular. Arboreal, but often comes to the ground to move between trees and feed on fallen fruits. **Social and Spatial Behaviour** Solitary. Ranges are stable and overlap considerably, suggesting lack of territoriality. Only range estimates are from Thailand, 4km² (1 ♀) and 4.7–20.5km² (♂s). **Reproduction and Demography** Aseasonal (captivity). Gestation 84–99 days. Litter size 1–6, averaging 2–2.5. Captive females may have two litters a year. MORTALITY Predation apparently rare, perhaps due to the combination of its mainly arboreal lifestyle, large size and ability to mount a formidable defence. LIFESPAN 22.7 years in captivity. **Status and Threats** Apparently adapts poorly to modified habitat, e.g. plantations, and estimated to have declined by more than 30% in the last 30 years as a result of forest loss in combination with heavy hunting for wildlife markets; both threats are rampant across its range, especially in SE Asia. Critically Endangered in China. CITES Appendix III (India), Red List VU.

MASKED
PALM CIVET

COMMON PALM
CIVET

BINTURONG

Owston's Palm Civet
HB 56–72cm; T 35–47cm; W 2.5–4.2kg
Slender long-tailed civet with a very narrow head, pointed muzzle and large ears. Fur creamy-white to buff with pale peach underparts. Four to six broad dark bands across the back, and small spots dotting the sides, neck and limbs. **Distribution and Habitat** Endemic to Vietnam, Laos and S China; uncertain in Cambodia. Occurs primarily in lowland and montane forests. Recorded in degraded forest near cultivation. **Feeding Ecology** Much-reduced dentition and a handful of records from the wild suggest that it eats primarily earthworms (which are relished in captivity), and other terrestrial invertebrates, including centipedes, grasshoppers, mantises and snails. Captive animals take meat, fruits, vegetables and small vertebrates (geckos, frogs and tadpoles), but largely refuse rodents. Blamed for taking poultry in villages, which is doubtful. Primarily terrestrial and nocturnal. **Social and Spatial Behaviour** Very poorly known. All camera-trap records are of solitary individuals or females with kittens. **Reproduction and Demography** Seasonal in captivity. Mates late January–mid-February; births April–May. Gestation 77–87 days. Litter size 1–3. Weaning at 12–18 weeks. Sexual maturity at 18 months. Captive males groom, sleep and forage with the kittens. MORTALITY and LIFESPAN Unknown. **Status and Threats** Very restricted distribution, which is under intense pressure from habitat clearing. Being ground-dwelling, it is vulnerable to very high levels of snaring and trapping throughout its range. CITES Appendix II, Red List VU.

BANDED CIVET *Hemigalus derbyanus*

Banded Palm Civet
HB 41–56.5cm; T 23.5–37.5cm; W 1–3kg
Very similar to Owston's Civet, but with smaller ears, darker buff to tawny-rufous fur with 4–8 bands, and no small spots; the two species do not overlap in their ranges. Distinctive orange-buff colouration on the underparts does not develop until adulthood. **Distribution and Habitat** Borneo, Sumatra (and associated small islands), Peninsular Malaysia, peninsular Thailand and possibly peninsular Myanmar. Inhabits forest mainly in lowland areas under 800m; occurs to 1200m on Borneo. Occurs in moderately disturbed habitats, including secondary forest and *Acacia* plantations near intact forest fragments. **Feeding Ecology** Poorly known, but available records indicate it is largely insectivorous, eating mainly earthworms and insects; occasionally takes crustaceans and small vertebrates including frogs, lizards and rodents. Fruits and plant matter are not recorded in the diet, though captive animals eat bananas. Foraging nocturnal. Dietary records suggest it forages mainly on the ground (though it climbs well), especially along rivers and riverine habitat. **Social and Spatial Behaviour** Virtually unknown. Sightings and camera-trap records are usually of solitary animals. **Reproduction and Demography** Poorly known. Litter size 1–2 (captivity). Juveniles reach adult size at 6 months. MORTALITY Virtually unknown; one record of predation by a Blood Python (Sumatra). LIFESPAN Unknown. **Status and Threats** Status poorly known, but close association with lowland forest and largely terrestrial behaviour make it very vulnerable to habitat loss and hunting; both threats are intense in its range. CITES Appendix II, Red List VU.

HOSE'S CIVET *Diplogale hosei*

HB 47.2–54cm; T 30–33.5cm; W (single ♀) 1.3kg
Similar shape and size to Banded Civet, but uniformly dark brown without markings, contrasting with bright white underparts, neck and lower face. Head narrow and elongated, with a slightly bulbous appearance to the muzzle. One of the least known viverrids. **Distribution and Habitat** Endemic to Borneo, where it is confirmed only from the N and NE of the island. Inhabits intact rainforest and montane forest at 450–1700m. **Feeding Ecology** Virtually unknown. Very long whiskers and partly webbed feet with fur between the pads, suggesting a semi-aquatic diet, but there are no records from the wild. The only animal ever kept in captivity, an adult female, ate fish, shrimp, chicken and processed meat; she refused fruits and boiled rice. All records (including behaviour in captivity) are nocturnal and terrestrial. **Social and Spatial Behaviour** Unknown. All records are of solitary animals. **Reproduction and Demography** Unknown. **Status and Threats** Status very poorly known. Probably declining due to forestry and habitat conversion to plantations and agriculture. Hunting is also a likely threat. CITES Appendix II, Red List VU.

OTTER CIVET *Cynogale bennettii*

Sunda Otter Civet
HB 57.5–68cm; T 12–20.5cm; W 3–5kg
Very distinctive semi-aquatic civet, grizzled dark brown with pale creamy-brown underparts. Small ears and a bulbous muzzle with very long white whiskers. Small obvious white patches above the eyes and on the cheeks. Feet have long dexterous digits and are partially webbed. Another species, Lowe's Otter Civet *C. lowei*, described from a skin found in N Vietnam, is now thought to be a *C. bennetti* specimen in trade. **Distribution and Habitat** S Thailand, extreme S Myanmar (unconfirmed), Malaysia, Sumatra and Borneo. Occurs primarily in wet lowland forest and swamp forest. Recent records from dry forest, bamboo and logged forest. **Feeding Ecology** Semi-aquatic and assumed to forage mainly along waterways for small aquatic prey, but there are no diet records from the wild. Has been observed in shallow water searching among stones with its stiff whiskers, and digging in moist leaf litter. Terrestrial and mainly nocturnal; recent camera-trap data from Sumatra show occasional diurnal activity. **Social and Spatial Behaviour** Unknown. All camera-trap records are of solitary individuals or females with kittens. **Reproduction and Demography** Virtually unknown. Litter size 1–3, based on very few records. **Status and Threats** Status poorly known, but it is nowhere common and severely threatened by forest loss and pollution of waterways, both of which are extreme in its range. Conversion of peat-swamp forest to oil-palm plantations is a particular concern. Hunting is also a likely threat. CITES Appendix II, Red List EN.

OWSTON'S CIVET

BANDED CIVET

HOSE'S CIVET

OTTER CIVET

HB 76–85cm; T 30–40cm; W 6.6–8kg

Large terrestrial civet with buff-grey fur marked with small spots that are most distinct on the sides and flanks, becoming paler and diffuse on the forequarters. Has the distinct black-and-white throat stripes, dark lower legs and banded tail of all *Viverra* species; the only similar sympatric species is Small Indian Civet. Formerly classified with Large-spotted Civet (which does not occur in India); the possibility of it being an introduced population of this species cannot be excluded. **Distribution and Habitat** Endemic to W Ghats, India. Known only from a handful of records in lowland forest and forest-plantation mosaics. **Feeding Ecology** Unknown. Assumed to be omnivorous with a diet similar to Large Indian Civet's. **Social and Spatial Behaviour** Unknown. Assumed to be solitary like other *Viverra* civets. **Reproduction and Demography** Unknown. **Status and Threats** Declared 'possibly extinct' in 1978, before being rediscovered in 1987; one fresh skin recovered since, in 1990. Otherwise, no recent records despite significant survey efforts. Its range is exposed to high levels of forest loss and hunting, and it may be extinct. Red List CE.

LARGE-SPOTTED CIVET *Viverra megaspila*

HB 72–85cm; T 30–37cm; W 8–9kg

Large civet with a pale buff-grey coat marked with large discrete spots. Erectile black dorsal crest runs from the nape to the base of the tail, continuing as a black dorsal stripe to the tail-tip (lacking in Large Indian Civet). Head heavy with a slightly bulbous muzzle. **Distribution and Habitat** S China and Indochina to Peninsular Malaysia. Inhabits mainly lowland primary forest under 300m (with one record at 520m in Laos). Occurs in logged areas, but disappears from small fragmented forest blocks. **Feeding Ecology** Unknown. Assumed to be omnivorous with a diet similar to Large Indian Civet's. Camera-trap records indicate it is nocturnal and terrestrial. **Social and Spatial Behaviour** Unknown. All camera-trap records are of solitary animals. Density unknown, but occurs in much lower frequencies than sympatric *Viverra* species during camera-trap surveys (except in S Laos). **Reproduction and Demography** Unknown. **Status and Threats** Thought to have undergone at least a 30% decline in the last 15 years. Its lowland forest habitat is exposed to very high levels of degradation and hunting: especially vulnerable to snaring due to its terrestrial habits. Red List VU.

MALAY CIVET *Viverra tangalunga*

Oriental Civet

HB 54–77.3cm; T 26–39.5cm; W 3–7kg

Smallest of the *Viverra* civets, with a slender build, narrow skull, and short and slender muzzle. Tail has 10–15 narrow dark rings compared with 4–8 broad bands in other *Viverra* species. **Distribution and Habitat** Peninsular Malaysia, Sumatra, Borneo, Sulawesi and the Philippines. Two records from Java, which are probably human assisted. Inhabits primary and secondary, lowland and montane forests to 1200m. Tolerates logged forest, plantations and agricultural areas near forests. **Feeding Ecology** Omnivorous. Eats small vertebrates, invertebrates (including toxic animals such as scorpions, millipedes and giant centipedes) and fruits. Civets in intact forest eat proportionally more fruits, whereas those in logged forest eat proportionally more rodents, insectivores, birds and herptiles, probably reflecting availability. Blamed for killing poultry. Foraging solitary, nocturno-crepuscular and terrestrial. Scavenges food waste and handouts from villages, camps and dumps. **Social and Spatial Behaviour** Solitary and apparently non-territorial. Ranges are stable and often overlap, in some cases extensively. Range estimates include 0.28–1.28km^2 (♀s) and 0.39–2.83km^2 (♂s). Ranges in logged forest slightly larger than in intact forest. Density estimates 0.9 civets/km^2 (logged forest) to 2.2/km^2 (unlogged forest). **Reproduction and Demography** Poorly known, but thought to be aseasonal. Gestation unknown. Litter size 1–2, rarely 3. MORTALITY Poorly known; roadkill is often the main cause of death in logging areas. LIFESPAN 5 years in the wild, 12 in captivity. **Status and Threats** Status poorly known, but it is widespread and tolerant of modified habitat. Less common in areas of human influence, but persists even in areas exposed to high rates of hunting. Hunted for meat and persecuted for killing poultry. Red List LC.

LARGE INDIAN CIVET *Viverra zibetha*

HB 75–85cm; T 38–49.5cm; W 8–9kg

Large civet with a typical *Viverra* appearance, but in which the spots are indistinct and interconnected, giving a mottled or marbled appearance. Dark bands on the tail are very broad, and each band may have a brownish inner ring. Sometimes regarded as a distinct species in Vietnam, the Thaynguyen Civet *V. tainguensis*, evidence for which is dubious. **Distribution and Habitat** S and C China, Nepal, Bhutan, NE India, Indochina to Peninsular Malaysia. Occurs mainly in evergreen and deciduous primary and secondary forests to 1600m. Does not tolerate heavily modified habitat, but inhabits plantations near forests. **Feeding Ecology** Omnivorous, with a similar diet to that of other *Viverra* species, but with more robust dentition suggesting greater carnivory. Diet includes small mammals, birds, herptiles, fish, eggs, insects, crabs and fruits. Kills domestic poultry. Foraging solitary, nocturno-crepuscular and terrestrial. Scavenges from villages and dumps. **Social and Spatial Behaviour** Poorly known. Solitary and apparently non-territorial. Only range estimate is 12km^2 (1 ♂; Thailand). **Reproduction and Demography** Poorly known but thought to be aseasonal, with two litters per year. Gestation unknown. Litter size 1–4. MORTALITY Unknown. LIFESPAN 20 years in captivity. **Status and Threats** Status poorly known, but the species is widespread and common in some areas. However, habitat loss coupled with high levels of hunting constitute serious threats, which have extirpated it from large areas, e.g. much of SE China. Large size and terrestrial habits make it especially vulnerable to hunting. Red List NT.

MALABAR CIVET

LARGE-SPOTTED
CIVET

MALAY
CIVET

LARGE INDIAN CIVET

Lesser Oriental Civet, Rasse
HB 45.5–68cm; T 30–43cm; W 2–4kg
Small terrestrial civet, buff-grey to tawny-brown, marked with small brown or black spots running in longitudinal lines along the body. Lower limbs uniformly black or dark brown. Tail marked with alternating light and dark bands ending in a pale tip, distinguishing it from similar terrestrial civets. The most commonly kept Asian civet (mainly in India) for perineal gland secretion used in perfume and traditional medicinal preparation. **Distribution and Habitat** S Asia, from E Pakistan through the Indian subcontinent, Sri Lanka, S and C China, Taiwan, Indochina, Sumatra, Java and Bali. Introduced in Madagascar, Zanzibar (Tanzania), Socotra (Yemen) and Comoros. Occurs in all kinds of forest, scrubland, grassland and riverine habitats. Thrives in disturbed and edge habitats, and is more common in degraded areas than in undisturbed closed-canopy forest. Lives in rural, agricultural and pastoral habitats. **Feeding Ecology** Omnivorous, eating a wide variety of mainly animal prey, especially small mammals, insects and earthworms. Also eats small birds, herptiles, eggs, crustaceans, arachnids, snails, fruits including berries, and young buds. Raids domestic poultry. Foraging strictly nocturno-crepuscular, with most activity taking place well after sundown. Forages alone, mostly on the ground; climbs well into the lower canopy to raid birds' nests and feed on fruits. Scavenges, including from carrion, villages and dumps. **Social and Spatial Behaviour** Solitary. Leaves faeces in prominent locations, but extent of territoriality is unclear; captive animals mark their surroundings prodigiously with the perineal scent gland, 6.3–8.2 times/2hr (non-breeding) to 11.5–19.5 times/2hr (breeding). Male range size 2.2–3.1km^2; female range size poorly known, but thought to be smaller. No density estimates, but often the most common small carnivore in secondary and degraded forests. **Reproduction and Demography** Poorly known in the wild. Captive animals (within their natural distribution) have two mating periods, February–May and August–December. Births occur November–February and May–July. Gestation 65–72 days. Litter size 2–5. Weaning at 2–3 months. MORTALITY Poorly known; most documented mortality is anthropogenic. LIFESPAN 9 years in captivity. **Status and Threats** Widespread, adaptable and common in altered habitats. In SE Asia and China very often killed in snares for meat and for scent. Farmed for scent in India, where thousands of animal are kept in captivity; farm stock is maintained largely by taking animals from the wild, and the species is resilient to harvest and remains fairly widespread in areas of collection. Frequently killed on roads and by domestic dogs, probably reflecting the species' ability to reach high densities in anthropogenic habitats. Red List LC.

AFRICAN CIVET *Civettictis civetta*

HB 67–84cm; T 34–47cm; W 7–20kg
Africa's only true civet species, and also its largest viverrid. Unmistakably marked with bold black spots and blotches on a buff to reddish-brown background, with dark lower limbs, a black facial mask and alternating black-and-white throat stripes. Black individuals occur, mostly in equatorial forest. Dark dorsal crest runs from the neck to the tail, which is raised in alarm. Kept in Ethiopia, Niger and Senegal for perineal gland secretion ('musk') used in perfume manufacture. Ethiopia produces around 90% of the world's supply from >3000 captive civets. **Distribution and Habitat** Widely distributed in most of sub-Saharan Africa except arid SW Africa. Inhabits most habitats with cover, including forest, wet and arid woodland savannahs, scrubland, wetland, mangroves and montane heathland to 5000m. Absent from very arid areas except along river courses. Occurs in modified habitats with cover, including plantations and farmland. **Feeding Ecology** Omnivorous. Catholic diet, but the most important food groups are terrestrial insects (mainly beetles, locusts, grasshoppers, termites and millipedes), fruits and small mammals, which fluctuate in relative importance depending on seasonal availability, e.g. insects peak during the wet season. Also eats birds, eggs, herptiles, arachnids, snails, crabs, molluscs and shallow-water fish, e.g. mudskippers. Occasional larger vertebrate prey includes lagomorphs, Spring Hare, small mongooses and one record of a domestic cat. Tolerant of noxious foods such as millipedes, unripe monkey-orange (*Strychnos*) fruits and highly decayed carrion. Takes domestic poultry, and cultivated fruits and grains; occasional claims of predation on lambs are equivocal. Foraging nocturno-crepuscular and solitary. Most hunting is terrestrial; climbs to feed on fruits, but is clumsy in trees. Scavenges, including from kills of larger carnivores, human refuse and handouts in tourist camps. **Social and Spatial Behaviour** Solitary. Deposits faeces in latrines or 'civetries', which may have territorial significance. Also scent-marks assiduously with its perineal scent gland on conspicuous natural and anthropogenic objects, including tree trunks, grass stems, rocks, fence posts and signposts. Scent marks are detectable to humans for up to 4 months in the dry season. Only range estimate 11.1km^2 (subadult ♂; Ethiopian farmland). **Reproduction and Demography** Thought to be aseasonal, with possible weak wet-season birth peaks in E and southern Africa. Captive females can have 2–3 litters a year, which likely applies in the wild. Gestation 60–81 days (captivity). Litter size 1–4, averaging 2–3. Weaning at 14–16 weeks. Females (captive) first give birth at 14–24 months. MORTALITY Confirmed predators include Lion, Leopard, Spotted Hyaena, African Wild Dog and domestic dogs. Rabies occurs, but population impacts are unknown. LIFESPAN 15 years in captivity. **Status and Threats** Widespread and relatively tolerant of habitat change. Sought after for meat in W and C Africa, and valued for traditional religious and medical beliefs in much of its range, which can denude populations locally, especially in concert with habitat loss. Civet farms rely on wild-caught animals, which may have local impacts, especially in Ethiopia. Often killed unintentionally during predator-control operations on southern African farmland, e.g. six civets killed for every jackal (the intended target) in S Africa. Red List LC.

SMALL INDIAN CIVET

Defensive posture

AFRICAN CIVET

Melanistic form

Ethiopian Genet
HB 40.8–43cm; T 38–40.3cm; W 1.3–2kg
Pale sandy-grey genet with contrasting dark spots that form two unbroken stripes on each upper side. Dark dorsal stripe often splits into two narrow parallel stripes with a pale centre. Highland animals are darker, with wider dark stripes and tail bands. **Distribution and Habitat** Eritrea, Ethiopia, E Sudan, Djibouti and extreme NW Somalia. Inhabits montane forest, heath and dense alpine grassland to 3750m; a few records from semi-arid lowland woodland savannah. **Feeding Ecology** Mainly carnivorous. Rodents (especially grass and brush-furred mice), small birds and insects comprise the main prey, based on a small scat collection in the dry season. Also eats fruits. Foraging generally nocturnal, but highland animals are also diurnally active, possibly because Afro-alpine rodents are mainly diurnal. **Social and Spatial Behaviour** Unknown. All records are of solitary adults. **Reproduction and Demography** Unknown. **Status and Threats** Status unknown. There are only 20 museum specimens and a handful of field sightings, but it has a wide altitudinal and habitat range. There is intense grazing and agricultural pressure in the highland areas of its range. Red List LC.

HAUSA GENET *Genetta thierryi*

Thierry's Genet
HB 44.3–45cm; T 40–43cm; W 1.3–1.5kg
Slender pale genet with small, widely spaced rufous-brown spots on a yellow-buff background. Dorsal stripe may be indistinct until the mid-back. Rufous colouring often infuses the dark bands on the tail, especially those closest to the body. **Distribution and Habitat** W Africa, from N Senegal to Nigeria and Cameroon. Occurs in rainforest, dry and moist savannah woodlands, and dry wooded steppe. **Feeding Ecology** Unknown. Assumed to be mainly carnivorous and nocturnal like other genets. **Social and Spatial Behaviour** Unknown. Assumed to be solitary. **Reproduction and Demography** Unknown. Subadults 8–10 months old recorded in November in S Mali, suggesting breeding coincides with the cool dry season, but this observation is based on very few records. Females have two pairs of teats. **Status and Threats** Considered rare based on a few sightings and records, but intensive surveys are lacking. Large areas of its range are exposed to severe forest conversion and hunting. Red List LC.

CRESTED GENET *Genetta cristata*

Crested Servaline Genet
HB 49.5–62.2cm; T c. 43cm; W c. 2.5kg
Slender dark genet with oatmeal-coloured fur and dense, evenly spaced black spots that become stripes on the neck and shoulders. Wide black dorsal stripe that can be erected as a crest. Head long and narrow, with large eyes giving a slightly bug-eyed appearance. **Distribution and Habitat** Endemic to SE Nigeria and SW Cameroon, with equivocal records from Gabon and Congo. Inhabits dense lowland and montane forests to 1000m. Occurs in secondary forest, forest-plantation mosaics and scrub near cultivated areas, but avoids human settlements. **Feeding Ecology** Small mammals and insects are the most important prey, based on a small number of records. Two captives allowed to roam freely also caught frogs and lizards. Nocturno-crepuscular. **Social and Spatial Behaviour** Unknown. Likely to be solitary. **Reproduction and Demography** Unknown. Records of pregnancies and kittens reported from late August–December. **Status and Threats** Very restricted range that is exposed to high levels of forest loss and bushmeat hunting; both probably constitute serious threats. Unknown if it occurs in any protected areas. Red List VU.

SERVALINE GENET *Genetta servalina*

HB 44.5–51cm; T 36.8–48.5cm; W c. 2.3–3kg
Dark genet, densely marked with small, tightly spaced block-like spots, and with dark lower limbs. Orderly rows of semi-connected spots along the spine that do not form a continuous dorsal stripe, and no dorsal crest. **Distribution and Habitat** Equatorial Africa, encompassing the Congo Basin eastwards to W Kenya, with four known discontinuous populations in Tanzania (Udzungwa, Uluguru and South Nguru Mountains, and Zanzibar Island). Occurs in rainforest, dense woodland savannah, thicket, bamboo forest and montane forest to 3500m. **Feeding Ecology** Chiefly carnivorous. Rodents, shrews and insects (mainly beetles, locusts, grasshoppers and termites) are the most important prey. Also eats small reptiles, birds and small amounts of fruits. Reportedly raids poultry on Zanzibar. Foraging thought to be nocturnal and mainly terrestrial. Has been observed scavenging from carrion. **Social and Spatial Behaviour** Unknown. Assumed to be solitary from sightings and camera-trap images. **Reproduction and Demography** Unknown. Kittens reported from Uganda in February–August (few records). **Status and Threats** Wide distribution, much of which comprises intact forest. Based on camera trapping, common in some sites, e.g. Udzungwa Mountains, Tanzania. Forest loss combined with bushmeat hunting can be expected to produce local declines. Red List LC.

GIANT GENET *Genetta victoriae*

HB 55–60cm; T 41.3–49cm; W 2.5–3.5kg
Dark-coloured genet with small, closely spaced spots that intermingle high on the sides to give an almost speckled appearance near the spine. Dark dorsal stripe discontinuous and can be erected as a crest. Dark tail bands very wide, with narrow pale bands in between. **Distribution and Habitat** Endemic to E DR Congo, possibly into extreme W Uganda and Rwanda. Occurs in rainforest and lowland deciduous forest. **Feeding Ecology** Unknown. Assumed to be similar to that of other genets. **Social and Spatial Behaviour** Unknown. Assumed to be solitary. **Reproduction and Demography** Unknown. Females have only one pair of teats, suggesting litters of 1–2. **Status and Threats** Very poorly known, but has a wide range, much of which comprises intact forest. Reported as locally common in some sites. Hunted for bushmeat and presumably vulnerable to forest loss. Red List LC.

ABYSSINIAN
GENET

HAUSA
GENET

SERVALINE
GENET

CRESTED
GENET

GIANT GENET

Common Genet

HB 46.5–52cm; T 42–51.6cm; W ♀ 1.4–2.3kg,
♂ 1.6–2.6kg

Pale grey to buff-grey genet, marked with small black spots that become blotches on the neck and shoulders. Black erectile dorsal crest. Tail has a white tip, distinguishing it from other sympatric genets. Namibian–S African population is sometimes considered a separate species, the Feline Genet or South African Small-spotted Genet *G. felina*. **Distribution and Habitat** Ubiquitous in Africa, except in the Sahara, Congo Basin and a narrow band across N Mozambique and Zambia; also occurs in Portugal, Spain, France (European populations are possibly introduced), SW Saudi Arabia, coastal Yemen and Oman. Occupies many habitats including forest, woodland, scrubland, wooded grassland and rocky areas. Tolerates modified habitats with cover, including plantations and agricultural areas. **Feeding Ecology** Eats mainly small rodents and insectivores, as well as birds, herptiles, arthropods, eggs and small amounts of fruits including berries. Sometimes raids poultry. Foraging solitary, mainly nocturnal, and both arboreal and terrestrial. Scavenges from carrion and food waste at tourist camps and villages. **Social and Spatial Behaviour** Solitary. Male ranges overlap multiple female ranges with low intrasexual overlap. Range estimates known mainly from Europe: 0.33–11.9km². Density estimated at 0.33–0.98 genets/km² (Spain). **Reproduction and Demography** Apparently weakly seasonal, with two birth peaks per year in E/N Africa and Europe (March–June and September–December), and one peak in southern Africa (September–February). Gestation 70–77 days. Litter size 1–4, averaging 2. MORTALITY Killed by many larger predators; Iberian Lynx predation (Spain) possibly reduces density. LIFESPAN 21.6 years in captivity. **Status and Threats** Widely distributed, common and occupies numerous habitats. Hunted in some areas for meat and fetishes, and disappears from areas of intense habitat conversion. Red List LC.

MIOMBO GENET *Genetta angolensis*

Angolan Genet

HB 44–48cm; T 38–43cm; W c. 1.5–2.5kg

Small slender genet, grizzled ochre tinged with brown or grey, and marked with brown to black spots. Black dorsal stripe and mid-dorsal crest, dark lower hind legs and a dark-tipped tail. Melanistic individuals occur. **Distribution and Habitat** Endemic to SC Africa, from S Tanzania/N Mozambique through Malawi, Zambia, S DR Congo and C Angola. Restricted to open moist miombo woodland and wooded grassland. **Feeding Ecology** Poorly known. One dead specimen (Kafue, Zambia) had eaten insects, fruits and grass. Small vertebrates likely to be important prey. All records nocturno-crepuscular. **Social and Spatial Behaviour** Unknown. Likely to be solitary. **Reproduction and Demography** Unknown. **Status and Threats** Status unknown, but widespread throughout Africa's miombo woodlands, which are relatively intact. Presumably killed for meat and fetishes (like all genets). Red List LC.

JOHNSTON'S GENET *Genetta johnstoni*

HB 47–51.4cm; T 46.2–49.5cm; W 2.2–2.6kg

Small slender genet with a narrow elongated face and slightly bug-eyed appearance. Coat yellowish-grey with tightly spaced dark brown or rufous-brown spots. Spots coalesce into 1–2 broken stripes along the upper sides, above which there is a black dorsal stripe. Lower limbs blackish. Tail has a dirty white tip with a faint dark band. **Distribution and Habitat** Endemic to coastal W Africa from S Guinea to extreme SW Ghana. Occurs in rainforest and associated dense habitat, including swamp forest, with one record from moist woodland savannah. **Feeding Ecology** Poorly known, but reduced dentition suggests arthropods are a chief component of the diet. Primarily nocturnal. **Social and Spatial Behaviour** Unknown. Likely to be solitary. **Reproduction and Demography** Unknown. Females have only two teats, suggesting small litters of 1–2. MORTALITY Crowned Eagle is a confirmed predator. LIFESPAN Unknown. **Status and Threats** Very restricted range and appears partially or wholly dependent on forest that is under intense pressure in W Africa from agriculture, logging and mining. Also severe hunting pressure across much of its range. Seen frequently in parks such as Tai NP (Côte D'Ivoire), the protection of which is increasingly necessary for its conservation. Red List VU.

AQUATIC GENET *Genetta piscivora*

Fishing Genet

HB 44.5–49.5cm; T 34–41.5cm; W c. 1.5kg

Unique genet with dense dull red to chestnut-red fur without markings, except it sometimes has a dark dorsal line. Bushy black tail and blackish lower limbs. Face has conspicuous white brows between the eyes, white patches under the eyes, and white cheeks, chin and throat. **Distribution and Habitat** Endemic to E DR Congo, between the Congo River and Rift Valley. May occur in W Uganda and Nyungwe Forest, Rwanda, but there are no unequivocal records. Only around 30 known records, all from rainforest at 460–1500m, mostly near very small streams with sandy bottoms. **Feeding Ecology** Believed to eat primarily fish. Stomachs of collected specimens contained only fish, and captive animals steadfastly refuse frogs, crustaceans and mice. Captives use a unique hunting method: they slap the surface of water and rest the tips of their long whiskers on it, thought to flush fish and detect vibrations from their movement. After several 'tests', they submerge the head to capture a fish with a swift open-mouthed lunge, taking barbell, catfish, squeaker and *Labeo* carp measuring up to 30cm. A captive kitten a few weeks old instinctively tapped the surface when first presented with a dish of water. **Social and Spatial Behaviour** Unknown. Thought to be solitary. **Reproduction and Demography** Unknown. Single record of a pregnant female, collected in December with one foetus. **Status and Threats** Considered naturally rare, based on reports from local hunters and trappers, and very low frequencies in which it appears in wildlife and market surveys. Opportunistically hunted in some areas for bushmeat, but extent to which this represents a threat is unknown. Fully protected in DR Congo, but protection is nominal in much of its range. Red List DD.

SMALL-SPOTTED
GENET

MIOMBO
GENET

JOHNSTON'S
GENET

AQUATIC GENET

Central African Large-spotted Genet

HB 41.1–52.1cm; T 39.5–54cm; W ♀ 1.3–2.5kg, ♂ 1.4–3.2kg

Sandy-grey to rufous-grey genet with large blotches, usually rufous-brown surrounded by a black border, or solid black with a rufous tinge. No dorsal crest. Tail has a black tip. **Distribution and Habitat** Sub-Saharan Africa, from S Burkina Faso and W Ghana, to Eritrea, south to C Namibia and KwaZulu-Natal, S Africa. Occurs in forest, woodland and moist savannah, and absent from very arid habitat. Inhabits cultivated areas, plantations and peri-urban areas. **Feeding Ecology** Mainly carnivorous, with rodents and insects comprising the most important prey, but eats a very wide range of small vertebrates and invertebrates. Rarely, kills mammals to the size of juvenile Red Duiker. Occasionally eats fruits and seeds; a population in the Shimba Hills, Kenya, atypically eats mainly fruits and seeds. Raids domestic poultry. Foraging solitary, nocturnal, and both arboreal and terrestrial. Scavenges, including from kills of large carnivores and tourist camps. **Social and Spatial Behaviour** Solitary. Adults mark their ranges assiduously, including at large latrines used by multiple individuals, but aggressive territorial defence is rare. Average range estimates 2.8km² (♀s) to 5.9km² (♂s) in Kenya. **Reproduction and Demography** Weakly seasonal. Breeding associated with the warm wet season, peaking August–March (southern Africa), and October–May (E Africa). Gestation 70–77 days. Litter size 2–5. MORTALITY Many mammalian and avian predators, but rates are unknown. LIFESPAN Unknown. **Status and Threats** Widely distributed, common and occupies many habitats, including anthropogenic areas. Hunted in some areas for meat and fetishes, and persecuted for killing poultry. Red List LC.

CAPE GENET *Genetta tigrina*

Large-spotted Genet

HB 43–56cm; T 39–46cm; W ♀ 1.4–1.9kg, ♂ 1.6–2.1kg

Very similar to Rusty-spotted Genet, with which it was formerly classified. Differs in having a mid-dorsal crest (apparent only when alarmed), and generally has larger, more widely spaced blotches. The two species overlap in S KwaZulu-Natal near the Eastern Cape provincial border. **Distribution and Habitat** Endemic to S Africa, from S KwaZulu-Natal to W Cape. Occurs in mesic habitats along coastal strip, including forest, woodland, dense grassland and fynbos. Tolerates anthropogenic habitats with cover. **Feeding Ecology** Very similar to Rusty-spotted Genet's; small vertebrates and invertebrates are the main prey. Raids domestic poultry. Foraging solitary, nocturnal, and both arboreal and terrestrial. Scavenges from dumps and at tourist camps; likely to scavenge carrion. **Social and Spatial Behaviour** Solitary. Less well known than Rusty-spotted Genet, but spatial patterns are essentially the same. Range estimates unknown. **Reproduction and Demography** Poorly known. Breeding records cluster December–February. Litter size 1–3. MORTALITY Many predators, but rates are unknown. LIFESPAN 9.5 years in captivity. **Status and Threats** Common and widespread in a limited range. Killed for taking poultry and hunted for bushmeat, but effects on populations are apparently minimal except in concert with severe habitat modification. Red List LC.

KING GENET *Genetta poensis*

HB c. 60cm; T c. 41.5cm; W c. 2–2.5kg

Dark genet with elongated, rectangular black or dark brown spots that often coalesce into long blotches, in closely spaced rows. Formerly classified with Pardine Genet. Known only from 10 museum specimens; no wild records since 1946. **Distribution and Habitat** Endemic to coastal W and C Africa from Liberia to Rep Congo; unequivocal presence limited to five sites in Liberia/W Côte D'Ivoire, S Ghana, Bioko Island (Equatorial Guinea) and coastal Congo, but distribution may be more continuous. All records are from intact rainforest. **Feeding Ecology** Unknown. Assumed to be similar to that of other genets. **Social and Spatial Behaviour** Unknown. Assumed to be solitary. **Reproduction and Demography** Unknown. **Status and Threats** Status unknown. Mostly known from hunters and markets, suggesting hunting is a threat, especially in W Africa. Large areas of rainforest are intact in much of its possible range, e.g. Gabon, where it would be considered secure if present. Red List DD.

PARDINE GENET *Genetta pardina*

West African Large-spotted Genet

HB 41–55.3cm; T 39–49cm; W to 3.1kg

Similar to both Rusty-spotted and Cape Genets, with which it was formerly classified. Small, rectangular dark brown or rufous brown spots, and no dorsal crest. **Distribution and Habitat** Endemic to W Africa from the Senegal–Mauritania border to W Ghana. Occurs in forest, woodland and moist scrubland. Inhabits plantations and cultivated areas with cover. **Feeding Ecology** Poorly known; rodents, invertebrates and fruits occur in the stomachs of dead specimens. Primarily nocturnal. **Social and Spatial Behaviour** Unknown. Likely to be solitary. **Reproduction and Demography** Unknown. Records of young suggest breeding January–February. **Status and Threats** Restricted range, but it is a habitat generalist and considered common. Frequently appears in bushmeat markets, which may constitute a local threat. Red List LC.

BOURLON'S GENET *Genetta bourloni*

HB c. 49.5cm; T c. 41cm; W c. 1.5–2kg

Similar to Pardine Genet but much darker, with dark spots that often fuse into long blotches, especially on the neck, shoulders and rump. Tail dark with broad bands connected by a dark dorsal line. Formerly classified with Pardine Genet, but recognized as a separate species in 2003 from 29 museum specimens. **Distribution and Habitat** Endemic to Liberia, E Sierra Leone, S Guinea and W Côte d'Ivoire, where it occurs only in rainforest. **Feeding Ecology** Unknown. Assumed to be similar to that of other genet species. **Social and Spatial Behaviour** Unknown. Assumed to be solitary. **Reproduction and Demography** Unknown. **Status and Threats** Very restricted range exposed to high levels of forest loss and bushmeat hunting. Red List NT.

RUSTY-SPOTTED
GENET

CAPE GENET

KING GENET

PARDINE
GENET

BOURLON'S
GENET

West African Linsang, Leighton's Linsang
HB 30–38cm; T 35–40cm; W 0.5–0.7kg
Very small, slender genet-like species, with soft yellow-buff fur fading to white or creamy white underparts. Marked with irregular, oval dark brown blotches that become small spots on the limbs and neck. Tail has 10–12 chevron-shaped dark rings. Has been considered a subspecies of Central African Oyan *P. richardsonii*, with which it does not overlap. **Distribution and Habitat** Endemic to W Africa, where confirmed only in SW Côte d'Ivoire and W Liberia. Presence in SW Guinea needs confirmation. All records are from rainforest. **Feeding Ecology** Unknown. Size and dentition suggest it eats small vertebrates such as rodents, birds and herptiles, as well as invertebrates. Arboreal with protractile claws, and thought to forage mainly in the canopy. **Social and Spatial Behaviour** Unknown. Assumed to be solitary. Local people (Liberia) report that it builds nests from leaves in the tree canopy similar to squirrels' dreys, which may be occupied by a number of individuals. **Reproduction and Demography** Unknown. **Status and Threats** One of the least known carnivores, confirmed from only 12 museum records. The most recent records are two skins collected in E Liberia in 1988–1989. Restricted to a very localized range that is exposed to high levels of forest loss and bushmeat hunting. Both are considered serious threats, but the species' current status is entirely unknown. Red List DD.

CENTRAL AFRICAN OYAN *Poiana richardsonii*

African Linsang, Richardson's Linsang
HB 32–40cm; T 34–40.2cm; W ♀ 0.45–0.5kg, ♂ 0.51–0.75kg
Very similar to Leighton's Oyan. Distinguished by slightly darker background colouration, yellow-brown with a grey or reddish cast, which is more densely marked with spots and blotches. Markings sometimes coalesce into elongated blotches or stripes along the back and dorsal area. Tail has 10–12 wide bands interspersed with narrow 'shadow' rings that are more distinct than in Leighton's Oyan. **Distribution and Habitat** Endemic to the Congo Basin, in S Cameroon, extreme S CAR, Equatorial Guinea (including Bioko Island), Gabon, Rep Congo and DR Congo to the Rift Valley. Occurs in lowland and montane rainforests. **Feeding Ecology** Unknown. Assumed to have a similar diet to that predicted for Leighton's Oyan. All records nocturno-crepuscular. Arboreal, but has been observed on the ground, where it probably also forages. **Social and Spatial Behaviour** Unknown. All sightings and records are of solitary animals. **Reproduction and Demography** Unknown. One record of a female lactating in October (Cameroon), and local people report it has two offspring, but there are no verified records. **Status and Threats** Status unknown. Assuming it occurs throughout the Congo Basin, there are large areas of intact forest where it is probably secure. Occasionally appears in the bushmeat trade, e.g. Bioko Is, though it is probably not especially sought after due to its very small size and arboreal nocturnal habits. Forest loss combined with hunting is likely to produce local declines. Red List LC.

AFRICAN PALM CIVET *Nandinia binotata*

Nandinia, Two-spotted Palm Civet
HB 37–62.5cm; T 34–76.2cm; W 1.2–3kg
Classified as the only species in the family Nandiniidae. Medium-sized genet-like species with a long tail that equals or exceeds the head-body length. Dense woolly fur is greyish-brown to rusty-brown with paler buff-yellow underparts, and lightly marked with small dark brown spots and a unique pale spot on each shoulder; limbs are unmarked. Tail has faint dark rings that can be difficult to distinguish in some individuals. Claws are partially protractile. **Distribution and Habitat** Sub-Saharan Africa from Senegal along coastal W Africa, through the Congo Basin to Uganda, Kenya, Tanzania (including Unguja, Zanzibar), Malawi, W Mozambique and extreme E Zimbabwe. Primarily a forest species, living in lowland and montane forests to 2500m, forest-savannah mosaics and moist woodland savannah. Occurs in logged and disturbed forests, and in forest patches near cultivated areas. **Feeding Ecology** Primarily frugivorous. Around 80% of the diet is made up of wide variety of wild and cultivated fruits, especially figs, African Corkwood, Sugar Plum, Myrianthus and the pulp of African Oil Palm nuts. Commonly eaten cultivated fruits include banana, passionfruit and pawpaw. Balance of the diet consists of small mammals, birds, fledglings, eggs and arthropods. Efficient nest raider, including of weavers' nests at the ends of very thin branches, which are negotiated easily. Kills mammals to the size of pottos and juvenile monkeys, though these appear rarely in the diet. Raids domestic poultry. Extremely agile in trees; able to hang from branches by its hind feet and descend trees rapidly head-first. Foraging nocturnal and solitary. Forages both arboreally and terrestrially; insects, fallen fruits and rodents are taken on the ground. Scavenges from carrion and village dumps, and known to drink fermenting sap from tapping vessels on palm trees. **Social and Spatial Behaviour** Solitary and territorial. Male ranges overlap multiple smaller female ranges. Fights between resident males are sometimes fatal. Unclear if territories are maintained only while breeding; there is evidence of seasonal nomadism during fruiting peaks, e.g. 12–15 individuals moving into 1km of forested valley to exploit localized fruiting of Corkwoods. Range estimates 0.29–0.7km² (♀s) and 0.34–1.53km² (♂s). Density estimated at 2.2–3.3 civets/km² (Bwindi, Uganda) and 5–8/km² (Gabon). **Reproduction and Demography** Thought to be weakly seasonal, with births peaking in wet periods, e.g. September–January (Gabon), and May and October (Uganda). Gestation ~64 days. Litter size 1–4, typically 2. Sexual maturity 1 year. MORTALITY Unknown. LIFESPAN 16.4 years in captivity. **Status and Threats** Widely distributed and reaches high densities in suitable habitat. Often considered the most abundant small carnivore in C African rainforest; this is supported by it being common in wildlife markets, e.g. Equatorial Guinea, but status in most of its range has never been assessed. Likely undergoing localized declines from a combination of forest loss and hunting. Red List LC.

LEIGHTON'S OYAN

CENTRAL
AFRICAN OYAN

AFRICAN
PALM CIVET

Descending head-first

Painted Dog, Cape Hunting Dog

HB 76–112cm; T 30–41cm; SH 61–78cm; W 17–36kg

Africa's largest canid, with coarse fur coloured a mottled patchwork of tawny, black and white. Tail always has a conspicuous white tip, extending variably along the length; some individuals have an entirely white tail. Colouration varies very widely within and between populations; southern African individuals tend to be tawnier, while E/NE African animals tend to be darker, but colour is not reliable for identifying origin. Colouration is unique to each individual, allowing identification for life. Males are slightly larger and heavier than females; there is little regional variation in size.

Distribution and Habitat

Sub-Saharan Africa, mainly in E and southern Africa, with scattered populations across Sahelian W/C Africa. Reaches highest densities in woodland savannah, but occurs widely in open grassland, semi-desert and scrubland. Absent from C African forest, but inhabits forest patches in E Africa. Penetrates deeply into true desert, but cannot permanently colonize very arid areas. Tolerant of habitat modification, but rarely inhabits pastoral landscapes due to intense persecution.

Feeding Ecology

Highly efficient pack hunter capable of taking prey as large as adult zebras and female African Buffaloes, but mostly kills medium-sized antelope species; each population focuses on 1–2 of the most common locally available species. Typical prey includes Impala, Nyala, Red Lechwe, Thomson's Gazelle and Blue Wildebeest (usually to subadult size). Largest preferred prey is Greater Kudu, in which adult females (135kg) are commonly killed (Namibia). Able to switch to smaller prey, especially where large species are absent or in low numbers, e.g. outside protected areas in N Kenya, Kirk's Dikdik (3–7kg) comprises 70% of prey. Bushbuck, duikers, Steenbok and Warthog are also important prey in some areas. Opportunistically kills smaller prey, including hares, small carnivores, e.g. Bat-eared Fox, and reptiles, but these form an insignificant proportion of the diet. Kills small livestock, but depredation is rare where wild prey is available, even when it is heavily outnumbered by stock. No records of predation on humans. Hunting almost always diurnal and highly social, often preceded by a frenzy of greeting between pack members. Reaches speeds of 66km/h and has terrific endurance, with chases extending for 2km. Hunts are highly coordinated and cooperative; pack members fan out and run in relays to maximize opportunities for capture, sometimes yielding more than one kill per hunt. Prey is killed cooperatively, usually by many pack members after capture by one dog; large and dangerous prey, e.g. Warthog and wildebeest, is often restrained by the head while other dogs disembowel and dismember it. Although this appears cruel (contributing to pastoralists' hatred for the species), the prey usually dies within 2–4 minutes. Hunts have high success rates, 42–70%, which increase with the number of adults present. Occasionally scavenges, including appropriating carcasses from other packs, Leopard, Spotted Hyaena and (very rarely) Lion.

Social and Spatial Behaviour

Intensely social, with pack members in almost constant association. Packs form around the dominant breeding pair, with up to 28 adults, but normally average 5–10 adults accompanied by yearlings and pups; including pups old enough to travel, the pack size occasionally exceeds 50. Same-sex adults in the pack are usually related and not related to opposite-sex adults. Packs usually form through interchange of same-sex subgroups, typically dispersing littermates that join opposite-sex subgroups. Both sexes may disperse, with females dispersing sooner but settling nearer to their natal range than males. Occupies enduring defined ranges that are often very large and overlap those of other packs. Active territorial defence is infrequent, but occurs in overlapping areas and around den sites, where inter-pack encounters are aggressive and sometimes fatal. Range size 150–2460km², averaging 423–1318km², dropping to 50–260km² when young pups are in the den. Due to wide-ranging behaviour, the species naturally occurs at low densities that fluctuate significantly depending on pup survival, e.g. 5 dogs/1000km² (semi-arid savannah, N Botswana), 2.8–22.5/1000km² (dry savannah, N Kruger NP), 16–24/1000km² (Selous GR, Tanzania) and 19–39/1000km² (mesic woodland, S Kruger).

Reproduction and Demography

Weakly seasonal. Pups born year-round, but most litters coincide with peak prey availability, e.g. March–June (Serengeti) and April–September (Kruger). Gestation 69–73 days. Litter size typically 10–11, exceptionally to 21. Pack's dominant female is usually the only breeder. All pack members cooperate to help raise pups by provisioning the mother at the den, regurgitating food to the pups and guarding the den. Helpers are essential for raising litters, and small packs (<4 adults) rarely reproduce successfully. Subordinate females occasionally breed, but their pups only survive when prey availability is high (they are sometimes 'stolen' and raised by the dominant female); otherwise, they are frequently killed by the dominant female or die from starvation brought about by harassment of subordinate mother/s. Weaning at around 8 weeks, and pups emerge from the den at around 12–16 weeks, after which they travel with the pack. Dispersal occurs most often at around 21–22 months for females (range 13–31), and 28 months for males (range 17–43). Inter-litter interval averages 12–14 months. Sexual maturity at around 2 years for both sexes, but breeding usually occurs with social dominance at around 4–5 years. MORTALITY 25% (Selous) to 65% (Kruger) of pups die in their first year. Adult mortality 23–28% for prime adults, rising to around 50% for older adults. Main natural causes are Lions, other African Wild Dogs and infectious diseases such as rabies and canine distemper. LIFESPAN 11 years in the wild, 16 in captivity.

Status and Threats

Species has undergone a drastic loss of range and is now extinct in 25 of 39 of its historic range countries. Total numbers are estimated at 3000–5500. African Wild Dogs were actively destroyed by wildlife managers until the 1970s; they are now protected throughout their range, though anthropogenic factors overshadow natural deaths, even in protected populations. Persecution, incidental killing in snares, roadkills and exotic disease transmitted by domestic dogs are the main causes. The species now persists mainly in areas with large parks; the main strongholds are the Okavango-Kaudom-Hwange ecosystem, Kafue NP and Luangwa Valley (Zambia), Selous GR-Mikumi NP and Ruaha-Rungwa (Tanzania), and Kruger NP (S Africa). Red List EN.

AFRICAN WILD DOG

Pack members
greeting

Simien Jackal, Abyssinian Wolf, Simien Fox

HB ♀ 84.1–96cm, ♂ 93–101.2cm; T 27–39.6cm;
SH 53–62cm; W ♀ 11.2–14.2kg, ♂ 14.2–19.3kg

Rich tawny-rufous with white underparts and bright white markings on the lower face, throat, chest and lower legs. Tail has a white base, darkening to a chocolate-brown tip. Hybrids with domestic dogs have a stockier build and lighter, duller coat. Despite its confusing array of common names, the species is most closely related to Coyote and Grey Wolf. **Distribution and Habitat** Restricted to seven isolated populations at 3000–4500m in Ethiopia. Inhabits open highland habitats, especially montane grassland, heath and shrubland. Avoids agricultural areas, which reach 3500–3800m in parts of its range. **Feeding Ecology** Feeds almost exclusively on small diurnal mammals, especially mole rats, rats and Starck's Hare. Infrequent prey includes Rock Hyrax, juvenile Grey Duiker, Reedbuck and Mountain Nyala, as well as birds including Blue-winged Goose goslings, francolins and eggs. Foraging largely diurnal and solitary, with most kills made by individual wolves stalking rodents or digging them from burrows. Small packs of 2–4 sometimes cooperatively pursue prey, especially hares and young antelopes. Rarely kills sheep lambs; does not kill cattle calves, and often forages among herds, which may assist hunting by flushing rodents and providing cover. Appropriates kills from raptors and scavenges, including from livestock carcasses. Caches surplus food in shallow holes. **Social and Spatial Behaviour** Forms packs of 2–13 adults that defend small stable territories from other packs. Pairs or small packs occur where prey availability is low. Males rarely disperse, so packs contain up to eight related adult males, as well as 1–3 adult females that may or may not be related; some females remain in their natal pack, while others disperse for breeding opportunities. Average territory size from 6km² in productive habitat to 13.4km² in poor habitat. Estimated densities include 0.1–0.25 wolves/km² in poor habitat or unprotected areas, to 1–1.2/km² in optimum protected habitat. **Reproduction and Demography** Seasonal. Mating August–November; births October–January. Gestation 60–62 days. Litter size 2–6. Reproduction is largely by the pack's alpha pair, but the dominant female also mates with visiting males from neighbouring packs. All pack members provision pups at the den, and subordinate females sometimes assist in suckling (it is unclear if extra-nursing females are pseudo-pregnant or absorb/abandon their own litters). Pups weaned from 10 weeks, and accompany the pack from 6 months. Sexual maturity at 18–24 months. **MORTALITY** Most mortality is anthropogenic and natural factors are poorly known; predation has not been observed, but may occur on pups by Spotted Hyena, Golden Jackal and large eagles. **LIFESPAN** 12 years in the wild. **Status and Threats** Endangered, with approximately 500 adults remaining in seven disjunct populations. Extreme pressure on habitat for agriculture and livestock is the chief threat, combined with exotic disease from domestic dogs; rabies epizootics reduce populations by up to 75%, e.g. in Bale. Roadkills, persecution and hybridization with dogs are lesser threats. Red List EN.

DINGO *Canis lupus dingo*

HB ♀ 70.3–101cm, ♂ 750–111cm; T 20–37cm;
W ♀ 8–17 kg, ♂ 7–22kg

Usually tawny-ginger; pale sandy, pure white (not albino) and black-and-tan variants occur. 'Sable' (Alsatian-like), brindled and piebald colouration indicates hybridization with dogs. Dingoes arose from Asian wolves in SE Asia some 6000–10,000 years ago, probably via domestication by humans; they colonized Australia with humans starting 3500–4000 years ago. **Distribution and Habitat** Australia (the only place where Dingoes live wild), SE Asia and New Guinea (only associated with humans). Inhabits desert, grassland, woodland savanna, wetland, alpine moorland and forest. Occurs in rural habitats, but avoids intensive agriculture. **Feeding Ecology** Very broad diet; 177 prey species recorded from Australia, with mammals comprising around 75%. At least one macropod (especially Red Kangaroo, Euro, and Swamp, Agile and Red-necked Wallabies) features prominently in the diet across its range. Other important prey includes wombats, brushtail possums, introduced European Rabbit and Magpie Goose. Other birds and reptiles form a small proportion of the diet. Kills sheep and cattle calves. Foraging mainly nocturno-crepuscular, but diurnal where it is free from persecution. Forages alone or socially; large prey such as kangaroos is usually hunted cooperatively in packs, which increases hunting success, e.g. from 5.5% (alone) to 19% (packs) when hunting Red Kangaroos. Scavenges, including from livestock carcasses and human refuse. **Social and Spatial Behaviour** Free from persecution, lives in stable packs of 2–12 adults and their pups in enduring home ranges. Territorial, but often shares important resources such as waterholes with neighbouring packs. Under persecution (most of its Australian range), social structure is fractured so that packs are smaller and less stable. Individuals associate in loose 'tribes', sharing a range that is not defended and tending to forage alone. In stable packs, breeding is usually restricted to the alpha pair, and other pack members help raise pups by provisioning and guarding. Size of pack territories from 4–55km² (moist cool forest) and 32–126km² (Simpson Desert), to over 300km² in SW Australian desert. **Reproduction and Demography** Breeding generally seasonal, most strongly in arid C Australia. Mating April–June; births June–August (births outside this period are attributed to the presence of hybrids). Gestation 61–69 days. Litter size 1–10, averaging 5. Females first breed at around 24 months; males sexually mature at 12 months, but breeding is limited by social dynamics. **MORTALITY** Most mortality is anthropogenic, including from introduced disease (especially distemper) and parasites (especially heartworm). Occasionally killed by Water Buffalo, Red Kangaroo and Wedge-tailed Eagle (pups). **LIFESPAN** 10 years in the wild, 13 in captivity. **Status and Threats** Widespread in Australia, but intense persecution in concert with hybridization from domestic dogs threatens the species. Pure Dingoes are most common in C and N Australia, rare/possibly extinct in S and NE Australia, and probably extinct in SE and SW areas. Protected in national parks, World Heritage areas and Aboriginal reserves, but legally regarded as pests elsewhere. Without intensive conservation effort, the pure Dingo is unlikely to persist. Red List VU.

Pack at den

ETHIOPIAN
WOLF

Black
form

Typical
form

DINGO

Pale
form

Timber Wolf, Arctic Wolf, Tundra Wolf

HB ♀ 87–117cm, ♂ 100–130cm; T 35–50cm; SH 66–81cm; W ♀ 18–55kg, ♂ 20–79.4kg
The world's largest canid, with significant variation in size and colouration. Largest individuals (Alaska and Canada) are 3–6 times as heavy as Middle Eastern and S Asian wolves. Typically pale to dark grey, but highly variable, e.g. ginger in E–C Asia ('Tibetan wolf'), brown in W–N Eurasia and white, especially in N Canada ('Arctic wolf'). Black colouration is rare outside forested N America, and traces to interbreeding with early domesticated dogs 10,000–15,000 years ago. Dingo is usually treated as a Grey Wolf subspecies; some authorities regard Red Wolf as a Grey Wolf-Coyote hybrid. Grey Wolf is the progenitor of the domestic dog, and they can interbreed. In early 2011, genetic analysis revealed the Golden Jackal in NE Africa is a Grey Wolf subspecies closely related to the Arabian or Indian Wolf from the Middle East to India.

Distribution and Habitat

N America and Eurasia, formerly with one of the most extensive distributions of any mammal. Widespread in Canada and Alaska, extirpated in the lower USA except in the N Rockies and N Midwest, and relict in SW USA–Mexico, where reintroduction has established a small population of Mexican Wolves. Widespread in Russia and C Asia, becoming more fragmented and reduced in SW Asia, the Middle East, and W and N Europe. Occupies many habitat types, including desert (to 50°C), open plains, steppe, mountainous areas, swamps, forest and Arctic tundra (to –56°C). Although tolerant of habitat modification, rarely inhabits agricultural and pastoral areas due to intense anthropogenic persecution.

Feeding Ecology

Highly opportunistic and proficient pack hunter. Diet varies extensively by region and season, but medium to large ungulates are usually the mainstay, e.g. Musk-ox, bison, Moose, Elk, Red Deer, Caribou, White-tailed Deer and Roe Deer, wild sheep, ibex and Wild Boar. Juvenile and debilitated individuals are most often killed, but even a single wolf is capable of killing healthy adults of these species, especially during winter. When ungulates are less vulnerable (spring–summer), the diet is more diverse with increased consumption of beavers, hares, rodents, waterfowl, fish, and fruits including berries. Kills smaller carnivores, especially Coyotes, and occasionally Pumas and young Black, Brown and (exceptionally) Polar Bears. Livestock is readily killed, especially during spring–summer, when wild prey disperses and stock occupies productive grazing areas; livestock (including the semi-domestic Reindeer) is the most important prey for many Eurasian populations. Unprovoked attacks on humans are very rare and usually attributed to rabies, e.g. two fatalities by healthy wolves in N America since 1900. Hunting cathemeral, with greater diurnalism where it is protected. Hunting highly social; pack members exchange the lead in chases and cooperatively bring down large prey. Can reach 64km/h and has extraordinary endurance, maintaining pursuit to 8km (with an exceptional record of 21km). Hunting success rates 10–49% for packs (N America). Scavenges, including from refuse dumps and appropriated carcasses from other carnivores.

Social and Spatial Behaviour

Highly social and territorial, living in packs numbering up to 42 but typically 2–15. Nucleus of the pack is a mated adult pair, accompanied by adult offspring. Pack size fluctuates depending on dispersal of offspring, which is affected by food availability. Dispersal is low in productive years, producing large packs with up to four generations of grown offspring. Most offspring ultimately disperse and seek non-related adults to form new packs. Recruitment of unrelated individuals (especially to replace a lost breeder) or small groups of dispersers occurs occasionally into established packs. Packs occupy enduring ranges that are usually defended aggressively from other packs. Range size 33–4335km², averaging 69–2600km². Ranges increase locally during winter and with increasing latitude; largest ranges recorded are from Alaska and the Canadian Arctic. Wolves following migratory herds have massive ranges of 63,000–100,000km² annually that are not defended. Density estimates from 5 wolves/1000km² (NW Alaska) to 92/1000km² (Isle Royale, Canada), but rarely exceed 40/1000km². Densities in arid areas, e.g. the Middle East, are likely to be very low.

Reproduction and Demography

Seasonal. Mating January–April depending on latitude, with pups born March–June. Gestation 60–75 days, typically 62–65. Litter size 1–13, averaging 4–7. Pack's dominant female is normally the only breeder, though multiple females (probably close relatives) occasionally breed under high prey availability. All pack members help raise pups by provisioning the mother and pups at the den, and by defending pups from predators. Weaning at around 8–10 weeks. Pups leave the den permanently and travel continually with the pack from 4–6 months by September–October. Dispersal is usually at 11–24 months, but is recorded at 5–60 months. Inter-litter interval typically 12 months; two litters a year produced on rare occasions. Sexual maturity at around 10 months for both sexes, though breeding opportunities rarely arise before 3 years. MORTALITY Annual pup mortality averages 34%, ranging from 9% (Denali NP, Alaska) to 61% (N Wisconsin). Adult mortality from 14%, to 44% where heavily exploited. Human hunting and trapping are often the main cause, but in unexploited populations starvation (mostly of pups) and aggression from other wolves are the main factors. Disease is an important cause of death, though population effects are poorly known. Wolves occasionally die in hunting accidents and are killed by bears, Puma and Amur Tiger. LIFESPAN 13 years in the wild, 17 in captivity.

Status and Threats

Widespread and stable in most of its northern range, especially in Alaska, Canada, Kazakhstan and Russia (total combined estimate 113,000–127,000 wolves). Nonetheless, it has lost an estimated third of its historic range, mainly in the USA/Mexico, W Europe and S Asia, where it is mostly threatened or endangered. Chief threat is persecution by humans, often as part of state-sanctioned control programmes. Diseases, especially canine parvovirus, mange and rabies, produce local declines. Reintroduction to Yellowstone NP in 1994–1995 has been highly successful, growing to approximately 1650 animals in 2010; reintroduction to the SW USA (Mexican wolf subspecies) has been less successful. Legally hunted and trapped in at least 15 countries, with the largest numbers killed in Canada (4000 annually), Mongolia and Russia (10,000–20,000 annually). CITES Appendix I (Bhutan, India, Nepal and Pakistan), elsewhere Appendix II, Red List LC.

GREY WOLF

Pack howling

Arabian form

Mexican form

Tibetan form

Eurasian form

DHOLE *Cuon alpinus* Plate 44

Asiatic Wild Dog
HB ♀/♂ 80–113cm; T 32–50cm; SH 42–55cm;
W ♀ 10–17kg, ♂ 15–21kg
Superficially resembles a large jackal. Pale tawny-brown to rich russet-brown with a dark-tipped tail. Northern temperate individuals are usually more reddish, with contrasting bright white underparts. **Distribution and Habitat** S China, Nepal, Bhutan, India, SE Asia, Sumatra and Java; patchily distributed in N China, uncertain in SE Russia and Mongolia. Inhabits forest, forest-grassland mosaics and montane scrubland. Avoids open habitat, and agricultural and pastoral areas. **Feeding Ecology** Pack hunter that mostly kills ungulates, especially Chital, Sambar, Blackbuck, Nilgai, Swamp Deer, Red Muntjac, Gaur, Asiatic Buffalo, Banteng, Markhor, Himalayan Tahr, gorals and Wild Pig. Juveniles are often selected, but packs are capable of killing adults of all but the largest species. Livestock is sometimes killed. Foraging usually diurno-crepuscular and cooperative. Scavenges, including kleptoparasitism from other carnivores, e.g. Leopard. **Social and Spatial Behaviour**

Lives in packs of 2–15 adults and their pups, exceptionally totalling 30 individuals. Packs have a dominant breeding pair and are biased towards males because females disperse more often. Packs occupy defined home ranges, though the extent of territorial defence is unknown. Pack range estimates 12–49.5km² (Phu Khieo WS, Thailand) and 40–83km² (dry forest, India). **Reproduction and Demography** Seasonal. Breeding October–April (India) and January–May (Java). Gestation 60–63 days. Litter size 4–12. Breeding usually restricted to the dominant pair, though multiple females occasionally breed and subordinate males sometimes mate alpha female. Pack members assist reproduction by guarding, provisioning at the den and regurgitating food to pups. MORTALITY Poorly known; Tiger and Leopard are known predators. LIFESPAN 16 years in captivity. **Status and Threats** Endangered and declining, chiefly from persecution, habitat loss and prey declines due to human hunting. Possibly extinct in its former E and C Asian range, and in SE Asia restricted to large protected areas. S/C India is the species' stronghold. CITES Appendix II, Red List EN.

COYOTE *Canis latrans*

Brush Wolf, Prairie Wolf
HB ♀/♂ 74–95cm; T 26–46cm; W ♀ 7.7–14.8kg (exceptionally to 25kg), ♂ 7.7–18.1kg (exceptionally to 34kg)
Uniformly coloured, from frosted grey to rufous-brown with a dark-tipped tail (rarely white-tipped). Varies widely geographically and seasonally; northern and winter individuals are generally larger and paler. Black, white and ginger individuals occur. **Distribution and Habitat** N Alaska and Canada (except NE Canada), throughout the USA and meso-America to C Panama. Inhabits virtually all habitats, including farmland, urban areas and agricultural land. **Feeding Ecology** Highly opportunistic generalist. Vertebrate prey is most important, ranging from rodents to juvenile ungulates, and carrion. Also eats invertebrates, fruits, seeds, vegetables and grains, including some crops. Takes poultry and small livestock. Foraging mainly diurno-crepuscular, but nocturnal near humans. Usually forages alone; hunts socially for larger prey. Scavenges, including from carnivores' kills, human refuse, pet food and birdfeeders. **Social and Spatial Behaviour** Sociality extremely flexible, changing regionally

and temporally depending on food availability. Basic social unit is a territorial monogamous pair that may breed for life. 'Associate' individuals, usually grown offspring of previous litters, remain with the pair under high food availability, forming packs numbering up to 10. Associates help raise pups and defend territories. Large packs occur where ungulates are the main prey, while pairs and trios are typical where prey is small. Some Coyotes never join packs and live as solitary nomads. Average territory size (excluding small breeding ranges) 2–3km² (SW USA) to 42–61km² (Minnesota). **Reproduction and Demography** Seasonal. Mates January–March; births April–June. Gestation 58–65 days. Maximum litter size 11, averaging 4–7. Pups weaned at 5–7 weeks; dispersal from 6 months. MORTALITY Humans are the main cause of death. Grey Wolf is the chief natural predator, e.g. Yellowstone NP. LIFESPAN 15.5 years in the wild (rarely beyond 8), 18 in captivity. **Status and Threats** Extremely widespread, common and very resilient to persecution. Very tolerant of habitat modification which, combined with Grey Wolf's extirpation, has enabled colonization outside its historic range, e.g. E USA and C America. Red List LC.

RED WOLF *Canis rufus*

HB ♀ 99–120cm, ♂ 104–125cm; T 30–46cm; SH 66–76cm; W ♀ 16–30kg, ♂ 21–41kg
Reddish fur, becoming pale ginger to cream on the lower limbs, and distinctive white throat and chest patches. Some authorities regard it as a Coyote-Grey Wolf hybrid, but recent genetic analysis suggests that it is a distinct species. **Distribution and Habitat** Extinct in the wild except for reintroduced population occupying 6000km² in E North Carolina, USA, which inhabits pine-forest wetland mosaics, marshland and agricultural land with cover. **Feeding Ecology** Eats mainly White-tailed Deer, raccoons, Marsh Rabbit and small rodents. Livestock is eaten, though recent records are all of carrion. Hunting nocturno-crepuscular, and usually in small packs or singly. **Social and Spatial Behaviour** Forms small family packs of 2–12 animals, comprising a dominant breeding pair and its offspring. Packs are territorial and occasionally kill unrelated intruders. Females disperse at higher rates than males, and some

individuals never disperse, remaining in the pack as non-breeding helpers. Pack range size averages 123.4km², ranging from 45 to 225.8km²; packs inhabiting agricultural habitats with high rodent densities have the smallest ranges. **Reproduction and Demography** Seasonal. Mating February–March; births April–May. Gestation 61–63 days. Litter size averages 3–5, exceptionally to 10. MORTALITY Rates low for canids: 32% (pups), 21% (yearlings) and 19% (adults) annually. Shooting and roadkills are the main factors; natural causes account for a quarter of deaths, mainly from intraspecific killing, sarcoptic mange and starvation (of pups). LIFESPAN 20 years in captivity. **Status and Threats** Extinct in the wild by 1980 from habitat loss, intense persecution and hybridization with Coyotes among a declining population. Reintroduced from captivity in 1987 and now numbers 100–130 including pups. Hybridization with Coyote is the main threat today; gunshot, roadkills and incidental trapping are also factors. Red List CE.

Northern
form

DHOLE

Southern
form

COYOTE

RED WOLF

Asiatic Jackal, Common Jackal
HB ♀ 74–100cm, ♂ 76–105cm; T 20–26cm;
SH 38–50cm; W ♀ 6.5–14.5kg, ♂ 7.6–15.5kg
The only jackal whose range extends into Eurasia. Uniformly coloured without distinct markings; bushy tail usually has a dark tip (never white). In parts of NE Africa, reclassified as a Grey Wolf subspecies in 2011. **Distribution and Habitat** N Tanzania through W and N Africa, the Arabian Peninsula, SE Europe and S Asia to Indochina. Inhabits semi-desert, grassland, dry woodland, forest, and agricultural and semi-urban areas. **Feeding Ecology** Omnivorous, eating mainly rodents, hares, lizards, snakes, birds and invertebrates; shifts to fruit and vegetable matter during the fruiting seasons. Small fawns are taken, and it can kill infirm ungulates to adult gazelle size. Foraging mainly nocturnal, but frequently crepuscular/diurnal where protected. Hunts singly or in small family groups, with larger prey more often taken by groups. Caches surplus food in shallow holes, and scavenges: congregations of up to 18 jackals are recorded at large carcasses and in dumps. **Social and Spatial Behaviour** Breeding pair is the main social unit, often accompanied by grown helpers from previous litters. Pairs are typically formed during the breeding season, but persist year-round in good conditions. Pairs defend a core territory around dens, and may cooperate to establish larger territories, depending on food availability. Territories 1.1–20km², with densities of up to 4 adults/km², e.g. Serengeti NP. **Reproduction and Demography** Seasonal, with births coinciding with peak food supply, e.g. December–March (E Africa), December–May (Israel) and April–June (India, C Asia). Gestation 63 days. Maximum litter size 8, typically 3–6. MORTALITY Poorly known; sometimes killed by large cats and domestic dogs. LIFESPAN 16 years in captivity. **Status and Threats** Generally common, though slowly declining outside protected areas in many countries. Tolerant of human activities and occurs in livestock and farming areas, but disappears under agricultural intensification (often with associated use of poisons) and urbanization. CITES Appendix II (India), Red List LC.

BLACK-BACKED JACKAL *Canis mesomelas*

Silver-backed Jackal
HB ♀ 66–85cm ♂ 69–90cm; T 27–38cm;
SH 38–48cm; W ♀ 5.9–10kg, ♂ 6.4–11.1kg
Recognizable by its dark-edged silver-grey saddle on a buff to rufous-brown body, with a dark, grizzled bushy tail. Usually the most conspicuous jackal where it occurs, due to preference for open areas and aggressive dominance over other jackals. **Distribution and Habitat** Two disjunct populations in southern and E Africa. Occurs in true desert, grassland, montane meadows, arid to mesic woodland savannahs, and agricultural habitats. **Feeding Ecology** Generalist omnivore. Common prey includes small rodents, Springhare, hares and young ungulates, though it is able to kill juvenile Cape Fur Seals, adult Springbok, Thomson's Gazelle and Impala. Birds, reptiles, eggs, invertebrates and carrion are also important food items. Readily takes poultry and small livestock. Foraging mainly nocturnal, especially where it is persecuted, but often crepuscular/diurnal where protected. Usually forages alone or in pairs, but up to 12 adults cooperate to kill large prey, and large congregations gather at carcasses and in food-rich patches like seal colonies. Readily scavenges, including from large carnivore kills and human refuse. **Social and Spatial Behaviour** Monogamous and territorial, forming breeding pairs that may endure for life and defend territories from other pairs. Up to three helpers accompany a pair and help to raise pups, including regurgitating food for them and the mother. Territories range from an average of 1.0km² (Hwange, Zimbabwe) to 24.9km² (coastal Namibia). Densities peak during the breeding season, e.g. 0.5–0.8 jackals/km² (non-breeding) to 0.7–1/km² (breeding; Hwange). **Reproduction and Demography** Seasonal. Mates May–August; births June–October. Gestation ~60 days. Maximum litter size 6, typically 3–4. Pups weaned at 8–9 weeks. Sexually mature at 11 months. MORTALITY Predators include large cats and hyaenas. LIFESPAN 14 years in captivity. **Status and Threats** Widespread and common in protected and pastoral areas. Persecuted intensely on farmland (especially southern Africa), but it is very resilient and anthropogenic population declines appear to be only temporary. Red List LC.

SIDE-STRIPED JACKAL *Canis adustus*

HB ♀ 65–76cm, ♂ 66–81cm; T 30–41cm;
SH 41–48cm; W ♀ 6.2–10kg, ♂ 5.9–12kg
Grizzled buff-grey with a characteristic pale stripe (often with a dark border) along the flanks and a distinctive white-tipped dark tail. **Distribution and Habitat** Southern, W and C Africa; replaced by Black-backed Jackal in the arid south and Golden Jackal in arid N Africa. Inhabits wooded grassland, woodland savannah, marshland, montane areas and forest edges. Avoids very open habitat, but can utilize agricultural areas with cover. **Feeding Ecology** Most omnivorous and least predatory jackal species, rarely killing prey larger than gazelle fawns. Diet is mainly a wide range of fruits, seeds and crops, as well as small rodents, hares, Springhare, small birds, insects and carrion. Poultry is sometimes raided. More strictly nocturnal than other jackals, perhaps to reduce competition, but becomes crepuscular when unmolested. Forages alone, but family groups gather at rich patches such as termite nests, and up to 12 from various families congregate to scavenge from carrion or dumps. **Social and Spatial Behaviour** Monogamous, forming mated pairs that may be lifelong. Resident pairs maintain exclusive use of a core territory with edges shared by neighbouring pairs. Yearling offspring often remain in the territory, sometimes forming groups with the resident pair numbering up to seven, though it is unclear if they act as helpers. Range size 0.15–0.56km² outside breeding, expanding to 0.55–1.6 km² during breeding. **Reproduction and Demography** Seasonal. Mating June–July; births August–November. Gestation 57–60 days. Litter size 4–6. Pups weaned at 8–10 weeks. Sexually mature at 11 months. MORTALITY Preyed on by larger carnivores, including domestic dogs near human habitation. Vulnerable to diseases such as rabies and canine distemper. LIFESPAN 10 years in captivity. **Status and Threats** Widespread and common. Tolerant of habitat conversion, and persists in suburban and agricultural habitats. Persecuted in human-dominated areas as a livestock predator, and many are killed by snares and vehicles, but human-caused deaths probably produce only local declines unless they are associated with poisoning or disease outbreaks. Red List LC.

GOLDEN JACKAL

Gathering at carcass

BLACK-BACKED
JACKAL

SIDE-STRIPED
JACKAL

Polar Fox, White Fox, Blue Fox

HB ♀ 50–65cm, ♂ 53–75cm; T 25–42.5cm;
W ♀ 3.1–3.7kg, ♂ 3.6–6.7kg

The northernmost canid and the only canid species to change colour seasonally. Two distinct colour phases: white winter phase moults to grey-brown with cream underparts in summer; blue phase is pale bluish-brown in winter and dark grey-brown in summer. White phase is more common, but blue dominates on islands and in coastal areas. **Distribution and Habitat** Circumpolar in the Arctic. Restricted to Arctic and tundra habitats, mostly north of the treeline in Canada, the USA (Alaska), Greenland, Russia, Finland, Norway, Sweden and Iceland. Inhabits most Arctic islands and winter sea ice to within 60km of the North Pole. **Feeding Ecology** Small rodents, chiefly lemmings and voles, are critical especially to inland populations, which fluctuate with rodent 'boom-bust' cycles. Other food includes fruits, eggs, birds to the size of Arctic Goose, ground squirrels, Arctic Hare and infrequent kills of Reindeer (Caribou) neonates. Coastal foxes also eat molluscs, crabs, fish, seabirds, and the carcasses of seals and whales; sometimes kills Ringed Seal pups. Blamed for killing domestic sheep lambs, though these are most probably scavenged. Foraging largely nocturno-crepuscular (light during the Arctic summer) and alone, but congregates at large carcasses. Caches surplus food; one larder contained over 500 eggs. Scavenges from bear and Grey Wolf kills, winter-killed Reindeer and human refuse. **Social and Spatial Behaviour** Usually solitary and monogamous, forming tight-knit breeding pairs that are territorial near the den. Pairs usually separate after raising pups, but remain in the territory year-round, pairing up again each spring to breed. Helpers from previous litters sometimes linger, forming extended family groups that persist if food is plentiful. Coastal home ranges are generally smallest (5–21km²), with high densities (4–8/100km²) due to more reliable food availability; inland ranges 15–60km² and densities 0.09–3 foxes/100km². Individuals may wander spectacular distances, possibly driven by rodent crashes, e.g. 2300km over 3 years by an Alaskan male. **Reproduction and Demography** Seasonal. Mates February–May; births April–July. Gestation 52–54 days. Litter size typically 3–11, exceptionally reaching 19 in rodent booms. Pups weaned at 7–8 weeks, reaching independence at 12–14 weeks. Most disperse in autumn, but they may over-winter on their parents' range; some (mainly females) become helpers. Both sexes mature at 10 months, but most individuals do not breed until their third year. MORTALITY Survival is tied to rodent abundance; up to 50% of adults and 80% of pups die in poor years, and populations collapse by up to 80%. Predators include Red Fox (especially of pups), large raptors, Wolverine, Grey Wolf and bears. Domestic dogs kill foxes near human settlements and may also transmit disease. LIFESPAN Maximum 10–11 years in the wild, typically under 5; 15 in captivity. **Status and Threats** Over 100,000 foxes are trapped annually for their dense fur, but populations appear to tolerate hunting if pressure is relaxed during poor food years. In Iceland, legally killed as pests by sheep farmers and eider-down collectors. Red Fox introductions from fur farms may elevate predation and competition, though population effects are unknown. Protected in Finland, mainland Norway and Sweden, where historic population crashes have not recovered; unprotected elsewhere. Red List LC.

RED FOX *Vulpes vulpes*

Cross Fox, Silver Fox, Common Fox

HB ♀ 45–68cm, ♂ 59–90cm; T 28–49cm;
W ♀ 3.4–7.5kg, ♂ 4–14kg

The world's most widespread and abundant wild carnivore. Typically various shades of rich red-brown, but highly variable including platinum-tipped black ('silver fox'), an intermediate phase called 'cross' and pale silvery-blond. All phases (except albinos) have black-backed ears, and white-tipped tails are typical but not ubiquitous. **Distribution and Habitat** N America, Eurasia and N Africa. Occurs in virtually all habitats, including farmland, suburbs and cities north of the Tropic of Cancer, except the northernmost Arctic and deserts of the SW USA. Introduced in Australia for recreational hunting in the mid-1800s, where it has serious impacts on native marsupials. **Feeding Ecology** Highly opportunistic, eating mainly reptiles, birds and small mammals to the size of hares; also eggs, amphibians, fish, invertebrates, fruits, acorns, fir cones, sedges, fungi and tubers. Kills juvenile small stock and poultry, and consumes crops such as wheat and corn. Hunting mostly nocturno-crepuscular, often more diurnal in winter and where it is undisturbed. Usually forages alone, but aggregates in food-rich patches, e.g. shorebird nest colonies and dumps. Caches surplus food for later use, and has an excellent memory for larder sites. Readily scavenges: ungulate carcasses (including of domestic livestock) can be especially important in winter, and scavenges from human refuse, birdfeeders and compost heaps. **Social and Spatial Behaviour** Usually solitary and monogamous. Mated pair is the main social unit, but sociality during breeding is very flexible. With sufficient food, may form groups comprising one male and up to five vixens (probably related). Group-breeding females may den alone or together; younger females are mostly non-breeding helpers from previous litters. Range size resource dependent, from 0.20km² (Oxford, UK) to 50km² (Oman). Density 0.1 fox/km² (Arctic tundra), 1–3/km² (temperate forest, Canada and W Europe), exceptionally reaching 30/km² with abundant food, e.g. urban areas where foxes are subsidized. **Reproduction and Demography** Seasonal. Mates winter, usually December–February, earlier in southerly latitudes (Australia: June–October). Births usually March–May. Oestrus 1–6 days; gestation 49–55 days. Litter size reaches 12 depending on food availability, typically 3–8. Pups weaned at 6–8 weeks, and most disperse from age 6 months before the next breeding season. Some young females remain as helpers. Sexually mature at 9–10 months. MORTALITY First-year mortality reaches 80% and averages 50% for adults, mainly due to humans. Large raptors, other carnivores and domestic dogs kill foxes, though humans are overwhelmingly their main predator. Major vector for rabies, with outbreaks causing intermittent population crashes. LIFESPAN Maximum 9 years in the wild, typically under 5; 15 in captivity. **Status and Threats** Remarkable adaptability and resilience enables it to tolerate intense persecution, with some 1–2 million wild individuals killed annually for the fur trade, and perhaps the same amount again by sport hunters and pest control. Unprotected in most of its range; exports of furs from India are restricted (CITES Appendix III); Red List LC.

White form, winter

White form, summer

ARCTIC FOX

Blue form, summer

Blue form, winter

Pale form

RED FOX

'Silver' form

Typical form

HB 45.5–54cm; T 25–34cm; W ♀ 1.6–2.2kg, ♂ 1.7–2.7kg

Smallest fox on mainland N America. Tawny-grey with ochre sides, neck and legs. Closely related to Swift Fox; hybrids occasionally occur in a narrow band of overlap in Texas and New Mexico. **Distribution and Habitat** W USA and N Mexico. Inhabits semi-arid to arid desert scrub and grassland. Can occupy urban and agricultural areas. **Feeding Ecology** Eats mainly mice, rats, kangaroo rats, ground squirrels, prairie dogs, lagomorphs and insects; also small birds, reptiles, carrion, wild cactus fruits and crops, e.g. tomatoes and almonds. Foraging mainly nocturnal and solitary. Scavenges from livestock carcasses and human refuse. **Social and Spatial Behaviour** Monogamous, normally mating for life. Helpers, usually grown daughters, often remain with the pair. Ranges stable, with exclusive core denning areas. Range size (both sexes) averages 2.5–11.6km². Densities fluctuate extensively depending on prey oscillations: 10–170 foxes/100km², averaging 44/100km². **Reproduction and Demography** Seasonal. Mating December–January; births February–March. Gestation 49–55 days (estimated). Litter size 1–7. Weaning at around 3 months and independence at 5–6 months. MORTALITY Average annual mortality 65% (juveniles) and 45% (adults), mainly from Coyote predation and starvation during prey shortages. LIFESPAN 7 years in the wild, 20 in captivity. **Status and Threats** Secure but has undergone significant local declines, e.g. California and Mexico, from habitat conversion and eradication of prey colonies, e.g. prairie dog and kangaroo rat towns. Red List LC (EN in California and Oregon, VU in Mexico).

SWIFT FOX *Vulpes velox*

HB 47.5–54cm; T 25–34cm; W ♀ 1.6–2.3kg, ♂ 2–2.95kg

Very similar to Kit Fox; smaller, more rounded ears, and a shorter tail. **Distribution and Habitat** Mid-western N America, from Alberta to New Mexico. Restricted to short- to medium-grass prairies and grassland. Tolerates dry-land agricultural areas. **Feeding Ecology** Diet dominated by prairie dogs, ground squirrels, mice and lagomorphs, plus wild fruits, seeds, insects, small birds, reptiles, eggs and carrion. Foraging solitary and mainly nocturno-crepuscular. **Social and Spatial Behaviour** Mated pairs typical, but sociality is flexible; trios and extra-pair breeding occur. Pairs/groups maintain exclusive core areas with overlapping edges; neighbours often related. Range size (both sexes) 7.6km² (Colorado) to 25.1km² (W Kansas). **Reproduction and Demography** Seasonal. Mating December (Oklahoma) to March (Canada); births March–May. Gestation 50–55 days. Litter size 3–6, exceptionally to 8. MORTALITY Rates average 67–95% (juveniles) and 36–57% (adults), mainly from Coyote predation. LIFESPAN 8 years in the wild, 14 in captivity. **Status and Threats** Extirpated from about 60% of its historic range, from massive prairie conversion and intense persecution of prey. Extirpated from Canada; now present as a small reintroduced population. Red List LC.

INDIAN FOX *Vulpes bengalensis*

Bengal Fox

HB ♀ 46–48cm, ♂ 39–57.5cm; T 24.5–32cm; W ♀ 2–2.9kg, ♂ 2.3–3.6kg

Slender fine-featured fox with a narrow face, yellow-grey to silvery-grey fur, and a black-tipped tail. **Distribution and Habitat** Endemic to the Indian subcontinent. Inhabits hot semi-arid grassland, plains, scrub and open dry forest. Occurs on agricultural land. **Feeding Ecology** Omnivorous, eating mainly small mammals and insects, supplemented by birds, reptiles, eggs, fruits, seeds and fresh shoots. Foraging solitary and nocturno-crepuscular,; often diurnal on cool or overcast days. Scavenges from carrion, but rarely because feral dogs dominate carcasses. **Social and Spatial Behaviour** Monogamous in mated pairs. Other adults sometimes associate with pairs, but do not help raise pups; their role and relatedness is uncertain. Range estimates average 1.6km² (♀s) and 3.1km² (♂s). Densities fluctuate due to rodent cycles and disease outbreaks, usually 1–15 foxes/100km², reaching 150/100km² in ideal conditions. **Reproduction and Demography** Seasonal. Mating November–January; births January–May. Gestation 50–53 days. Litter size 2–4. Both parents provision and guard pups. MORTALITY Disease is implicated in population crashes. LIFESPAN 8 years in captivity. **Status and Threats** Relatively widespread, but occurs at low densities in habitats that are under strong development pressure. Naturally vulnerable to population crashes, exacerbated by hunting pressure from people for food. Red List LC.

CORSAC FOX *Vulpes corsac*

Corsac, Steppe Fox

HB ♀ 45–50cm, ♂ 45–59.5cm; T 19–34cm; W ♀ 1.6–2.4kg, ♂ 1.7–3.2kg

Medium-sized fox, pale tawny-grey with silky frosted fur in winter, and a black-tipped tail. **Distribution and Habitat** C Asia from W Russia to NE China. Inhabits steppes, grassland, shrubland, semi-desert and desert. **Feeding Ecology** Omnivorous and opportunistic, exploiting seasonal fluctuations in food, especially of rodents, e.g. lemmings, voles, ground squirrels, gerbils and jerboas. Also consumes birds, reptiles, insects and carrion. Foraging solitary and usually nocturnal; diurnalism increases when feeding young pups and during food shortages. Scavenges, including from human refuse, Grey Wolf kills and winter-killed livestock. **Social and Spatial Behaviour** Forms monogamous breeding pairs, though it is unknown if they are permanent or territorial, and two females sometimes den together with their pups. Juveniles may become helpers. Range size (pairs) 3.5–11.4km² (C Mongolia), occasionally to 35–40km² in poor habitat. **Reproduction and Demography** Seasonal. Mating January–March; births mid-March–May. Litter size 2–10, averaging 5–6. Gestation 52–60 days. Both parents provision and guard pups; helpers sometimes assist. MORTALITY Average adult mortality 34% (protected reserve, C Mongolia), mainly from human hunting and Red Fox predation. LIFESPAN 9 years in captivity. **Status and Threats** Widespread and locally common, but hunting and rodent-poisoning campaigns (especially in China) are pervasive. Extirpated in many areas, especially in Russia and its former republics. Red List LC.

KIT FOX

SWIFT FOX

INDIAN
FOX

CORSAC
FOX

King Fox, Royal Fox, Afghan Fox

HB ♀ 34–45cm ♂ 38.3–47cm; T 26–36cm; W 0.8–1.6kg
Very small fox with distinctive dark 'tear' lines along the muzzle. Brown-grey fur is interspersed with long black guard hairs, and very bushy tail is usually black-tipped (rarely white). **Distribution and Habitat** E Egypt, the Arabian Peninsula and C Asia. Inhabits semi-arid to arid rocky desert and mountainous habitats. Independent of water. **Feeding Ecology** Largely insectivorous and frugivorous, eating mainly beetles, crickets, grasshoppers, ants, termites, scorpions, wild capers, olives, grapes and melons. Small rodents, birds and (rarely) reptiles are also hunted; newborn ibex records are probably carrion. Foraging usually solitary and nocturnal, with increased crepuscularity in winter. **Social and Spatial Behaviour** Forms territorial monogamous pairs

that cooperate to raise pups, but are fairly solitary outside the breeding period. Non-breeding yearling females (possibly from previous litters) are often tolerated by resident pairs. Range size (both sexes) 0.5–2km², averaging 1.6km². Density estimates include 0.5–2 foxes/km² (Israel). **Reproduction and Demography** Seasonal. Mating January–February (Israel); births late February–May. Gestation 50–60 days. Litter size 1–3. Males groom, guard and accompany pups (2–4 months) on foraging excursions. MORTALITY Rabies and old age appear to be the main mortality factors. Red Fox is a confirmed predator. LIFESPAN <5 years in the wild, 6 in captivity. **Status and Threats** Fairly widespread and common. Locally threatened by habitat development, especially in coastal areas, and incidentally killed by poison set for other species. CITES Appendix II, Red List LC.

PALE FOX *Vulpes pallida*

Pallid Fox, African Sand Fox

HB 38–55cm; T 23–28.5cm; W 1.5–3.6kg
Very small fox, uniformly pale sandy-cream except for dark-tipped tail (distinguishing it from Rüppell's Fox). Ears medium-sized in proportion to the head, lacking the over-sized appearance in Rüppell's and Fennec Foxes. **Distribution and Habitat** Sub-Saharan Africa, in a narrow band from W Senegal–Mauritania to Eritrea and Ethiopia. Inhabits very arid, sandy and stony deserts, and dry savannah. Occurs near human settlements. **Feeding Ecology** Poorly known. Invertebrates and small vertebrates are eaten,

but robust molars suggest fruits, seeds and plant matter are important. Largely nocturnal. **Social and Spatial Behaviour** Poorly known. Mostly seen in pairs and small groups, probably mated pairs with offspring, suggesting patterns similar to other small foxes'. **Reproduction and Demography** Probably seasonal. Gestation (captivity) 51–53 days. Litter size 3–6. MORTALITY Poorly known; occasionally killed by domestic dogs. LIFESPAN Unknown. **Status and Threats** One of the least known canids; status unknown. Occasionally killed near settlements, and used locally for traditional medicines, e.g. Sudan, for asthma. Red List DD.

RÜPPELL'S FOX *Vulpes rueppellii*

Sand Fox

HB 35–56cm; T 25–39cm; W ♀ 1.1–1.8kg, ♂ 1.1–2.3kg
Small fine-featured fox with large ears and a slender face. Body colour near white to greyish-brown, often with a silvery sheen due to dark guard hairs. Long bushy tail has a white tip. **Distribution and Habitat** N Africa (Sahara), the Arabian Peninsula and C Asia. Independent of water and inhabits semi-arid to very arid habitats, including sandy and stony deserts, rocky steppes, massifs, scrub and vegetated watercourses. Occurs near human habitation. **Feeding Ecology** Omnivorous. Eats small mammals, birds, lizards, insects and plant matter, including wild fruits (especially dates), desert succulents and grass (probably an emetic). Foraging solitary and nocturnal-crepuscular. Most hunting is terrestrial; climbs palm trees for dates. Scavenges from carrion and human refuse. **Social and Spatial Behaviour** Monogamous, forming

mated territorial pairs, but larger aggregations of up to 15 suggest more complex sociality. Territories are largely exclusive, though large ranges overlap significantly. Female range sizes 13.2km² (Saudi Arabia) to 53.8km² (Oman); male ranges 20.9–84.4km². Density 68–105 foxes/100km² (fenced reserve, Saudi Arabia). **Reproduction and Demography** Seasonal. Mating November–February (Saudi Arabia); births March–May. Gestation 52–56 days. Litter size 2–6. Pups independent at 4 months and disperse at 6–10 months. MORTALITY Rabies, distemper and predation are the main factors; Steppe Eagle and Pharoah Eagle Owl are confirmed predators. LIFESPAN 7 years in the wild, 12 in captivity. **Status and Threats** Widespread and quite common. Killed indiscriminately during poisoning campaigns, and absent in heavily grazed areas. Almost extinct in Israel due to pressure from increasing Red Fox populations. Red List LC.

TIBETAN FOX *Vulpes ferrilata*

Tibetan Sand Fox, Sand Fox

HB 49–70cm; T 22–29cm; W ♀ 3–4.1kg, ♂ 3.2–5.7kg
Distinctive stocky fox with a long narrow muzzle, small ears and small, wide-set eyes on a broad face. Body grizzled grey with white underparts, and a rufous cape, head and lower legs. **Distribution and Habitat** Restricted to the Tibetan Plateau in Ladakh (India), N Nepal, through S and C China. Presence in N Bhutan is unconfirmed. Inhabits remote, cold semi-arid to arid steppes, meadows, grassland and slopes at 2500–5200m. Tolerates ambient temperatures of –40°C to 30°C. **Feeding Ecology** Eats mainly small mammals, especially pikas, with which it is closely associated, and mice and zokors. Also eats hares, marmots, birds, lizards, insects and berries. Eats carrion, including scavenging Grey Wolf

kills, and follows foraging Brown Bears to mop up flushed rodents. Hunts alone and appears diurnal, possibly reflecting the activity patterns of pikas. **Social and Spatial Behaviour** Poorly known. Monogamous breeding pairs are typical, but adult trios with pups are recorded. Pairs are found in close proximity, especially in food patches like pika colonies, suggesting it is not strongly territorial. **Reproduction and Demography** Seasonal. Mating December–March; births February–May. Litter size 2–5. Gestation 50–55 days. MORTALITY and LIFESPAN Unknown. **Status and Threats** Widespread and inhabits remote areas, somewhat insulating it from human threats. Given its strong association with pika colonies, the gravest threat is state-sanctioned rodent poisoning affecting most of Tibetan plateau. Red List LC.

BLANFORD'S
FOX

PALE FOX

RÜPPELL'S
FOX

TIBETAN FOX

HB 33.5–39.5cm; T 12.5–23cm; W 0.8–1.9kg
Smallest canid, with proportionally the largest ears in the family. Pale cream to sandy-red with lighter underparts, a dark-tipped tail and a dark caudal spot. Soles of the feet are fully furred for traversing hot loose sand. **Distribution and Habitat** Restricted to N Africa; reports from the Arabian Peninsula are unproven. Extremely well adapted to deserts; water independent. Prefers stable sand dunes for burrows, but occupies all semi-arid/arid habitats. **Feeding Ecology** Eats very small prey, chiefly small rodents, lizards, geckos, birds to the size of sandgrouse, invertebrates and their larvae, eggs, fruits and tubers. Reportedly raids poultry coops, though evidence is anecdotal. Forages alone largely at night, becoming more crepuscular/diurnal in winter. Prodigious digger, catching most prey by speedy excavation after locating it with its extremely sensitive hearing. Caches surplus food in little excavations, and sometimes enters human settlements to scavenge. **Social and Spatial Behaviour** Poorly known. Thought to form territorial breeding pairs in enduring ranges; observed groups with up to 10 individuals suggest extended families with helpers. **Reproduction and Demography** Thought to be seasonal (but breeds year-round in captivity). Mates January–February; births March–April. Gestation 50–52 days. Litter size 1–4. Pups weaned at 61–70 days. MORTALITY Putative predators include large owls, Golden Jackal and domestic dogs. LIFESPAN 13–14 years in captivity. **Status and Threats** Status uncertain, but probably secure by virtue of inhabiting very remote regions. Main threat is trapping for the tourist pet and domestic fur trades, which drives local extinctions around settlements. CITES Appendix II, Red List LC.

CAPE FOX *Vulpes chama*

HB 45–61cm; T 25–40.6cm; W 2–3.3kg
The only small light-coloured fox in southern Africa. Body fur grizzled silver-grey, blending into pale tawny-reddish limbs, neck and head. Long bushy tail has a characteristic dark tip. **Distribution and Habitat** SW Angola, Namibia, Botswana and S Africa; possibly Swaziland and Lesotho. Favours semi-arid/arid habitats extending into moderately mesic scrub habitat in S Africa and Botswana. Occurs on farmland. **Feeding Ecology** Eats chiefly mice and gerbils; also invertebrates, small birds, reptiles and fruits. Largest kills are hares and Springhare. Very rarely kills newborn sheep and goats; most livestock consumed is scavenged. Forages alone, usually at night, with activity peaks after dusk and before dawn. Most prey is captured after prolonged listening at burrows or holes, ending with frenzied digging. Surplus food is cached. Eats carrion, though rarely visits large carnivore kills. **Social and Spatial Behaviour** Usually solitary and monogamous. Breeding pairs are typical, with both parents raising pups. Helpers occur rarely, and adult females (possibly related) occasionally den together. Solitary outside breeding, but pairs share the same range year-round. Range size of pairs 1–32.1km². Densities 0.05–0.3 foxes/km². **Reproduction and Demography** Broadly seasonal, weakly so in some regions. Seasonal populations mate from June; most births August–December. Gestation 51–52 days. Litter size 1–6, typically 2–4. Pups begin hunting at around 15–16 weeks, reaching independence at 5 months. MORTALITY Adult mortality 26–52% (S Africa), depending on densities of Black-backed Jackal, the most important predator. Other predators include large raptors, owls and large carnivores, including domestic dogs. Contracts rabies, but is apparently less susceptible than other canids. LIFESPAN <7 years in the wild. **Status and Threats** Many thousands of Cape Foxes are killed, mainly in S Africa and Namibia, for perceived livestock losses, deliberately and as 'by-catch' in trapping and poisoning campaigns targeting jackals and Caracal. Despite this, the species often prospers on farmland that lacks larger predators, and has expanded its range in some areas. Red List LC.

BAT-EARED FOX *Otocyon megalotis*

HB 46.2–60.7cm; T 23–34cm; W 3.4–5.4kg
Grizzled smoky-grey with black legs and a bushy black-edged tail. Unique dentition with a range of 46 to 50 total teeth, the most for any placental land mammal. **Distribution and Habitat** Two disjunct populations in southern and E Africa. Favours arid or semi-arid grassland and open woodland savannah. Inhabits farmland and degraded habitat provided insecticide use is limited. **Feeding Ecology** Almost entirely insectivorous, eating particularly two termite genera with which its distribution overlaps almost completely. Also eats other invertebrates, small reptiles, rodents and some fruits, especially in the dry season. Does not kill livestock or poultry, but eats insect larvae in livestock carcasses, sometimes incurring blame for depredations. Forages mainly by sound, typically in family groups. Feeding aggregations of up to 15 individuals from different families occur when termites are abundant. Foraging chiefly nocturnal, but shifts diurnally in winter, reflecting changes in termite activity. Does not cache food and rarely scavenges. **Social and Spatial Behaviour** Monogamous pair is typical, but sometimes forms extended groups with one male and 2–3 related females that breed communally and allo-suckle pups. Insectivorous diet precludes female provisioning at the den, so the male is essential for guarding pups and allowing mothers to forage. Ranges overlap and territorial defence is limited to the den area. Range size 1–8 km². Densities fluctuate depending on food and season, e.g. 2.3 foxes/km² (non-breeding) to 9.2/km² (breeding: Mashatu GR, Botswana), and 0.7–14/km² (Kalahari, S Africa). **Reproduction and Demography** Seasonal. Pairs mate for life. Mates July–September; births October–December. Gestation 60–75 days. Litter size 1–6, averaging 3. Pups weaned at 10–15 weeks, but remain with the parents until the following June–July; many disperse, though some stay as helpers. MORTALITY Adult mortality 25–30% (S Africa), mainly from predation by Black-backed Jackal and disease episodes (rabies and canine distemper). Other predators include large raptors, owls and larger carnivores, but groups vigorously mob predators, often deterring predation. LIFESPAN 9 years in the wild, 13 in captivity. **Status and Threats** Reasonably widespread and tolerant of habitat conversion, but mistakenly persecuted as a livestock predator and cannot persist where insecticide kills off prey. Disease outbreaks associated with domestic dogs sometimes trigger severe die-offs. Red List LC.

FENNEC FOX

CAPE FOX

BAT-EARED FOX

Listening for termites

Island Gray Fox, Channel Islands Fox

HB 45.6–63.4cm; T 11.5–32.2cm; W 1.07–2.7kg

Smallest North American canid, very similar to Gray Fox but much smaller. An insular dwarf form of Gray Fox, separated for 10,400–16,000 years, and genetically a distinct species. **Distribution and Habitat** Restricted to six Channel Islands off the California coast, USA. Occupies all island habitats, including grassland, chapparal scrub, woodland, coastal scrub and dunes. Avoids degraded areas such as overgrazed pasture. **Feeding Ecology** Omnivorous, focusing on the most abundant food source, which differs by habitat and island. The most important food types are small mammals (especially deer mice), insects and fruits (especially Prickly Pear, and also berries). Also eats seeds, acorns, birds, nestlings, eggs and crustaceans. Cathemeral, with nocturnal activity peaks. Forages on the ground, climbs shrubs for fruits and birds' nests, and scales cliffs for seabird eggs and chicks. **Social and Spatial Behaviour** Monogamous and territorial, living as mated pairs with offspring, which usually remain until their second year; adult offspring are often tolerated even after they are living independently. Resident males chase and fight with other males, mainly during the breeding season. Ranges among the smallest for any canid: 0.15–0.87km². **Reproduction and Demography** Seasonal. Mates January–March; births February–May (peaking April). Gestation 50–53 days. Maximum litter size 5, typically 1–3. Pups weaned at 6–8 weeks and forage with adults from 2 months. MORTALITY Recent Golden Eagle colonization on northern islands has led to hyperpredation, prompting population crashes. Exotic disease (especially canine distemper) reduced the Santa Catalina population by 95% in 1998–2000. LIFESPAN 10 years in the wild, 15 in captivity. **Status and Threats** Critically Endangered and restricted to small isolated populations that are greatly vulnerable to random events. Golden Eagle predation and exotic disease (and less so roadkill) have produced calamitous declines on four of the six islands where it is found, reducing the total population from approximately 6000 to 1500 in 2002. Red List CE.

GRAY FOX *Urocyon cinereoargenteus*

Northern Gray Fox, Tree Fox

HB ♀ 52.5–58cm, ♂ 56–66cm; T 28–44.3;
W ♀ 2–3.9kg, ♂ 3.4–7kg

Grizzled grey body colour, bordered by a sharp edge of rufous on the neck, sides and legs, changing to white underparts, chest and cheeks. Long bushy tail has a black dorsal stripe and tip. **Distribution and Habitat** Extreme SE Canada, most of the USA, meso-America, extreme N Colombia and N Venezuela. Inhabits temperate and tropical forests, woodland, brush, semi-arid scrubland, agricultural habitats and peri-urban areas. Avoids very open areas such as grassland and prairie. **Feeding Ecology** The most omnivorous North American fox. Hunts mainly rodents and rabbits in winter, but greatly expands its diet in other seasons as different foods become available. Can be almost exclusively insectivorous in summer, while fruits, seeds and nuts comprise up to 70% of its diet in autumn. Also eats small herptiles, birds, eggs and carrion. Foraging mainly nocturnal and solitary. The most arboreal of canids, and forages both terrestrially and in trees up to 18m. **Social and Spatial Behaviour** Monogamous and territorial, living as mated pairs with offspring. It is unclear if pair bonds are permanent or if offspring remain as helpers. Territory size differs little between sexes, ranging from 0.13km² (Wisconsin) to 27.6km² (Alabama), averaging 1–6.7km². Densities 0.4 foxes/km² (California) to 1.5/km² (Florida). **Reproduction and Demography** Seasonal. Mates January–April; births March–early June. Gestation 60–63 days. Maximum litter size 10, typically 3–5. Pups weaned at 6–8 weeks and accompany foraging adults from 3 months. MORTALITY Most important predator is Coyote, which suppresses populations in some areas. Distemper and rabies produce local crashes. LIFESPAN 4–5 years in the wild, 14 in captivity. **Status and Threats** Widespread and common. Legal trapping is the main source of mortality; trapping is not regarded as a threat, though it probably contributes to population impacts in combination with disease epidemics. Red List LC.

RACCOON DOG *Nyctereutes procyonoides*

Tanuki

HB 49–70.5cm; T 15–23cm; W 2.9–12.5kg

Grizzled dark grey to buff-grey with a black facial mask, and a black chest, legs and feet. An ancient canid lineage with no close relatives, and classified in its own genus. **Distribution and Habitat** Japan, SE Russia, W Mongolia, E China, Korea and extreme N Vietnam. Introductions and escapes from fur farms have established populations throughout N, E and W Europe. Inhabits a variety of forest types, shrubland, farmland and urban areas. **Feeding Ecology** Omnivorous. Rodents are the mainstay, supplemented by small herptiles, birds, eggs, fish, crustaceans and carrion. Fruits including berries, and seeds are important in late summer–autumn before hibernation. New leaves and flowers are consumed mainly in spring, and insects peak in the diet in summer. Eats crops including oats, corn, maize, watermelon and fruits, and urban populations exploit garden fruits such as gingko and persimmon, e.g. Japan; rarely raids poultry. Foraging mostly terrestrial and nocturnal. Adult pairs forage together, but often some distance apart. The only canid that hibernates, in November–March in areas with severe winters. **Social and Spatial Behaviour** Strictly monogamous, forming permanently mated pairs that share a territory, usually moving and denning together. Territory cores are exclusive, especially while breeding, but range edges overlap. Territory sizes 0.07km² (urban, Japan), 6.1km² (subalpine habitat, Japan), to 20km² (SE Russia, introduced). **Reproduction and Demography** Seasonal. Mates February–March; births April–June. Gestation 59–70 days. Litter size averages 4–9, exceptionally reaching 19. Pups weaned at 5 weeks, and forage with adults shortly thereafter. MORTALITY Sarcoptic mange, distemper and rabies cause population declines, but impacts appear temporary. LIFESPAN 5 years in the wild, 13 in captivity. **Status and Threats** Abundant in much of its original range, and widely considered a pest where it has been introduced. Very tolerant of suburban and agricultural habitats. Up to an estimated 370,000 are killed annually on roads in Japan: it is unclear if this produces population declines. Red List LC.

ISLAND FOX

GRAY FOX

Tree climbing

RACCOON
DOG

Argentine Gray Fox, Southern Gray Fox, South American Gray Fox

HB 50.1–66cm; T 11.5–34.7cm; W 2.5–5kg

Small fox, pale grizzled grey with rufous-buff lower legs and a rufescent head tinged with grey. Lower thighs have a distinctive dark patch, and the lower jaw is conspicuously dark. **Distribution and Habitat** Restricted to Argentinean and Chilean Patagonia. Inhabits grassland, scrubland and steppes in lowlands and Andean foothills (rarely to 3500m). Tolerates ranching, agriculture and plantations. **Feeding Ecology** Eats mainly small rodents, European Hare and carrion. Where small mammal availability declines, the diet includes more fruits, arthropods, reptiles and birds. Kills domestic poultry and (very rarely) lambs. Foraging largely nocturno-crepuscular, and solitary or as loosely associated pairs with offspring during breeding. Buries excess food. **Social and Spatial Behaviour** Forms monogamous breeding pairs, occasionally with female helpers, which help raise pups. Pairs associate loosely outside breeding season. Range size 1.4–2.8km². **Reproduction and Demography** Seasonal. Mates August–September; births October. Gestation 53–58 days. Litter size 4–6. Males help raise pups and provision mother at den. Two mothers (possibly related) may cooperate to raise litters. **MORTALITY** Killed by domestic dogs and canine diseases, but impacts are unknown. **LIFESPAN** Unknown. **Status and Threats** Widespread and locally common, especially in its southern range. Once heavily hunted for fur; now apparently better regulated. Killed for depredation and in the false belief that it transmits disease to livestock. CITES Appendix II, Red List LC.

DARWIN'S FOX *Pseudalopex fulvipes*

HB 48–59.1cm; T 17.5–25.5cm; W ♀ 1.8–3.7kg, ♂ 1.9–4kg

Small stocky fox, dark grey-brown with contrasting white underparts, rufous extremities and a dark bushy tail. **Distribution and Habitat** Endemic to Chile, on Chiloé Island, and with a single mainland population 600km N in Nahuelbuta NP. Relies on dense southern temperate ('Valdivian') forest, but tolerates forest mosaics with dunes, beaches and pasture patches. **Feeding Ecology** Eats mainly small vertebrates, insects, crustaceans, fruits and seeds. Largest recorded prey, Pudu (10kg) and Magellanic Penguin, is probably scavenged. Foraging solitary, but congregates at food patches. Cathemeral; foraging mainly nocturnal in Nahuelbuta, perhaps to avoid larger Chilla (absent from Chiloé). Scavenges, including from fishing waste, pet food and carrion. **Social and Spatial Behaviour** Solitary. Forms monogamous pairs, which associate mainly for breeding on Chiloé, but persist year-round in Nahuelbuta. Range size (both sexes) averages 1.5–3km². Densities 0.95–1.14 foxes/km². **Reproduction and Demography** Poorly known. Seasonal. Observed litters occur October–January. Gestation unknown. Litter size estimated at 2–3. Both parents help raise pups. **MORTALITY** Rates in Nahuelbuta are 7–16% annually (juveniles and adults combined). Puma is a confirmed predator; no natural predators on Chiloé. Killed by domestic dogs at both sites. **LIFESPAN** Unknown. **Status and Threats** Restricted to two populations with less than 250 adults; 90% occur on Chiloé. Strictly protected, but threatened by continued logging in unprotected areas, and by domestic dogs. Red List CE.

SECHURAN FOX *Pseudalopex sechurae*

Sechura Desert Fox, Sechura Fox

HB 50–78cm; T 27–34cm; W 2.6–4.2kg

Slender head with a narrow long muzzle, large rufous-backed ears and a rufous ring around the eyes. **Distribution and Habitat** Coastal NW Peru and extreme SW Ecuador. Inhabits desert, associated beaches and sea cliffs, dry forest and Andean foothills to 1000m. Occurs in agricultural areas. **Feeding Ecology** Omnivorous and opportunistic. Fruits and seeds comprise much of the diet; also small rodents, birds, reptiles, insects, scorpions and carrion. Coastal individuals eat crabs, and seabirds (probably scavenged) and their eggs. Occasionally takes poultry and domestic guinea pigs; blamed for killing goats (unlikely). Foraging primarily nocturnal and solitary; congregates in small groups at large carcasses. **Social and Spatial Behaviour** Poorly known. Sightings largely of single adults or females with pups. **Reproduction and Demography** Poorly known. Births probably peak October–January. **MORTALITY** Roadkills common in N Peru; impacts unknown. **LIFESPAN** Unknown. **Status and Threats** Very limited range; status poorly known. Appears to tolerate rural and agricultural areas. Vulnerable to persecution, and religious ceremonial uses as amulets and handicrafts. Red List NT.

CULPEO *Pseudalopex culpaeus*

Andean Fox

HB 44.5–92.5cm; T 31–49.5cm; W ♀ 3.9–10kg, ♂ 3.4–13.8kg

Largest South American fox, powerfully built with a robust head. Light to dark grey with tawny extremities and pale underparts. **Distribution and Habitat** Extreme S Colombia to southern tip of S America. Inhabits all Andean habitats, from dry desert to temperate rainforest, from coast to 4800m (higher than other S American canids). Occurs on ranchland. **Feeding Ecology** More carnivorous and predatory than other S American foxes. Eats small rodents, introduced lagomorphs, and wild and domestic ungulates (most as carrion); also fruits, birds, insects and herptiles. Kills small livestock, generally young lambs, but occasionally adults, e.g. a 3.6kg juvenile fox killed a 24kg goat by suffocation. Generally forages alone and is cathemeral; strictly nocturnal where hunted. Scavenges. **Social and Spatial Behaviour** Solitary and territorial. Forms pairs for breeding; male assists in caring for pups. Range estimates (both sexes) typically 4.5–8.9km², but as large as 800km². **Reproduction and Demography** Seasonal. Mating August–October; births October–late December. Gestation 55–60 days. Litter size 3–8, averaging 5. **MORTALITY** Rates for juveniles 20% (unhunted) to 92% (hunted), and 31% (unhunted) to 49% (hunted) for adults. Chief predators are domestic dogs and Puma. **LIFESPAN** 11 years in the wild. **Status and Threats** Widespread, common and resilient. Benefits from pastoral/agricultural conversion with introduction of exotic lagomorphs. Intensely persecuted as a livestock pest and extirpated from extensive sheep-ranching areas. CITES Appendix II, Red List LC.

CHILLA

DARWIN'S
FOX

SECHURAN
FOX

CULPEO

HOARY FOX *Pseudalopex vetulus*

Plate 52

Hoary Zorro, Small-toothed Dog
HB 49–71cm; T 25–38cm; W 2.5–4kg
Small slender fox, grizzled grey with buff lower legs and underparts, and with a less crisply contrasting chest and throat patches than in similar foxes. Smallest canid in its range. **Distribution and Habitat** Endemic to Brazil. Inhabits mainly open cerrado savannah, and occasionally associated woodland, forest and floodplains. Occurs in pastoral, agricultural and plantation habitats. **Feeding Ecology** Largely insectivorous, with a diet dominated by harvester termites, dung beetles and grasshoppers. Also eats small rodents, birds, reptiles, wild fruits and grasses. Evidence for poultry depredation is equivocal. Foraging largely nocturnal, singly or as pairs with offspring. Most prey is taken on the soil surface; flips cattle dung for termites and dung beetles. **Social and Spatial Behaviour** Forms monogamous breeding pairs that inhabit a defined range and cooperate to raise pups. Unclear if pairs are permanent, and there is no evidence of extended family groups with helpers. Range size 3.8–4.6km². **Reproduction and Demography** Seasonal. Mates late May–June; births July–August. Gestation 50 days (estimated). Litter size 3–5. Weaning at around 3 months. Males help groom, guard and chaperone pups on foraging excursions. MORTALITY Maned Wolves are putative predators, and killed by domestic dogs. LIFESPAN 8 years in captivity. **Status and Threats** Widespread, common and tolerant of some habitat conversion. Persecuted (probably mistakenly) for poultry depredation, and roadkills are fairly frequent. Rabies and sarcoptic mange are confirmed, though population impacts are unknown. Red List LC.

PAMPAS FOX *Pseudalopex gymnocercus*

Azara's Fox
HB ♀ 50.5–72cm, ♂ 60–74cm; T 25–41cm; W ♀ 3–5.7kg, ♂ 4–8kg
Medium-sized fox, grizzled grey with reddish ears, neck and lower limbs. Throat, chest and lower hind limbs creamy-white. Smaller than sympatric Culpeo. **Distribution and Habitat** Extreme SE Brazil, E Bolivia, NE Argentina, Paraguay and Uruguay. Optimal habitat is pampas grassland, but also inhabits scrub, open woodland, pasture and agricultural land. **Feeding Ecology** Adaptable omnivore with a diet that shifts depending on availability. Most important food items are small rodents, European Hare, grassland birds, insects and fruits. Eats carrion, especially livestock carcasses. Rarely kills newborn lambs. Foraging solitary, but congregates at large carcasses. Cathemeral, becoming largely nocturnal where persecuted. Scavenges. **Social and Spatial Behaviour** Solitary. Forms monogamous pairs, but appears to associate only during the breeding season. Range estimates 0.4–1.8km². Reported densities typically 1–3 foxes/km², peaking at 5.85/km² in optimal pampas habitat free of persecution. **Reproduction and Demography** Seasonal. Mates July–August; births September–December. Gestation 55–50 days. Maximum litter size 8, typically 3–4. Pups weaned at 2 months. Both parents guard pups, and males provision females and pups at the den. MORTALITY Predators include Puma and domestic dogs. LIFESPAN 14 years in captivity. **Status and Threats** Widespread, common and tolerant of agricultural/pastoral conversion, but legal control and bounties for perceived livestock depredation kill many tens of thousands, resulting in local population declines. Red List LC.

CRAB-EATING FOX *Cerdocyon thous*

Crab-eating Zorro
HB 57–77.5cm; T 22–41cm; W 4.5–8.5kg
Medium-sized fox, dark grizzled grey with coarse fur, giving a dark bristly appearance. Muzzle and lower limbs usually conspicuously dark. **Distribution and Habitat** N Colombia, Venezuela, E and S Brazil, E Bolivia, Paraguay, N Argentina and Uruguay. Occupies all kinds of forest, woodland, grassland and marshland, as well as pastoral and agricultural habitats. **Feeding Ecology** Omnivorous, with a catholic diet, especially fruits, insects and small mammals. Also eats birds, reptiles, eggs, amphibians, land crabs, insects and carrion. Raids poultry and kills small lambs (rarely). Foraging mainly nocturno-crepuscular and as pairs with pups, or alone. Congregates in larger groups on turtle-nesting beaches and at carcasses. **Social and Spatial Behaviour** Monogamous, forming mated pairs occupying exclusive territories. Yearling offspring often remain with the resident pair, forming family groups of up to seven. Dispersers may settle near their parents and interact amicably as adults. Range size for adults 0.5–10.4km², averaging 2.2–5.3km². Densities of 0.55 foxes/km² (Brazilian scrub savannah) to 4/km² (Venezuelan Llanos). **Reproduction and Demography** Possibly seasonal with dry-season breeding peaks, though births occur year-round in some areas. Gestation 52–59 days. Litter size 3–6. Pups weaned at 12 weeks. MORTALITY Killed by larger carnivores like domestic dogs; common as roadkill. LIFESPAN 9.2 years in the wild, 11.5 in captivity. **Status and Threats** Widespread, common and adaptable. CITES Appendix II, Red List LC.

SHORT-EARED DOG *Atelocynus microtis*

Small-eared Dog
HB 72–100cm; T 25–35cm; SH c. 35cm; W 9–10kg
Medium-sized with short, sleek uniformly coloured fur ranging from very dark brown to rufous-grey. Face long and slender, and ears conspicuously small. Classified in its own genus; most closely related to Bush Dog. **Distribution and Habitat** Restricted to W lowland Amazonia, in Colombia, Ecuador, Brazil, Peru and Bolivia. Inhabits undisturbed primary forest usually associated with rivers. Avoids human presence and disturbed habitats. **Feeding Ecology** Poorly known; fish dominate the diet in Cocha Cashu, Peru. Other food includes agoutis, small marsupials, rodents, birds, crabs, frogs and fruits. Has been observed hunting in waterholes and swimming after prey; its elongated shape, sleek fur and partially webbed toes may be adaptations for semi-aquatic hunting. **Social and Spatial Behaviour** Poorly known. Most sightings are of individuals, suggesting it is solitary, but adult pairs have been observed foraging together. **Reproduction and Demography** Poorly known. Observed litters cluster in the dry season (May–December). MORTALITY Unknown. LIFESPAN 11 years in captivity. **Status and Threats** Naturally rare and relies on intact forest. Amazon deforestation is a serious threat, and domestic dog diseases were implicated in an apparent population decline in Peru. Red List NT.

HOARY
FOX

CRAB-EATING
FOX

PAMPAS
FOX

SHORT-EARED
DOG

BUSH DOG *Speothos venaticus*

HB 57.5–75cm; T 11–15cm; SH c. 30cm; W 5–8kg
Small stocky dog unlike any other canid, with a long body, short legs and short bushy tail. Broad head is bear-like with small eyes and short rounded ears. Body colour varies from blond to dark brown; usually tawny-blond on the neck and head, and dark brown on the legs and tail. **Distribution and Habitat** N Argentina and N Paraguay, through much of Brazil into N S America, N Ecuador to C Panama. Records from Costa Rica are equivocal. Closely tied to intact forest and forest savannah, including well-vegetated cerrado and pampas grassland. Occurs on ranchland and in agricultural areas, but is dependent on forest fragments in disturbed habitats. **Feeding Ecology** Eats mainly small mammals up to its own weight, especially Paca, agoutis, Nine-banded Armadillo, Brazilian Rabbit, opossums and rats. Large terrestrial reptiles, e.g. tegu lizards, and birds, e.g. tinamous, are also taken. Packs supposedly kill large prey, including Capybara, brocket deer and Rhea, by biting the legs until the quarry tires: a report exists of a pack of six harassing and badly wounding an adult Brazilian Tapir. Sometimes eats *Cercropia* fruits. Local people report that it occasionally takes chickens. Foraging mostly diurnal, and it sleeps in burrows overnight; there are a handful of observations of nocturnal hunting. Hunts socially, mainly by prolonged pursuit through thick vegetation, assisted by its squat long body shape. Readily takes to water during pursuits, and enters and excavates burrows in pursuit of prey. **Social and Spatial Behaviour** Most social of small canids, living in small packs of 2–12 (generally 2–6) that are in constant close contact, including sleeping in groups. Pack composition is thought to comprise a monogamous breeding pair with its grown offspring. Pack members help raise pups by provisioning mothers at the den during nursing. Territorial and ranging behaviour poorly known; a pack inhabiting a mosaic of ranchland, soy plantations and cerrado fragments in Brazil covered at least 145km². Ranges in intact habitat likely to be smaller. **Reproduction and Demography** Aseasonal, though birth peaks possibly occur in the wet season. Gestation averages 67 days (65–83 days in captivity). Litter size 3–6, exceptionally to 10. Pups weaned from 4 weeks. Sexual maturity at 10 (♀s) to 12 (♂s) months. MORTALITY Poorly known; predation by large cats is reported by local people. LIFESPAN 10.4 years in captivity. **Status and Threats** Widespread but nowhere abundant, and appears to be naturally rare. Conversion of forested habitat for livestock and agriculture is the main threat. Occasionally killed on roads and as a perceived predator of poultry. Captive individuals are vulnerable to canid diseases such as distemper and parvovirus. CITES Appendix I, Red List NT.

MANED WOLF *Chrysocyon brachyurus*

HB 95–115cm; T 28–50cm; SH 70–74cm; W 20.5–30kg
Tall, very long-legged canid with tawny-rufous fur, dark fur on the neck and shoulders, black socks and a bushy white-tipped tail. Face fox-like with very large ears, a white throat and a black muzzle. Despite its common name, it is not closely related to wolves and belongs in its own genus; its closest relative (but distant nonetheless) is thought to be Bush Dog. **Distribution and Habitat** C and S Brazil, N Bolivia, E Paraguay and patchy occurrences in N Argentina and NW Uruguay. Inhabits primarily cerrado and pampas savannah, woodland-savannah mosaics, shrub forest and seasonally flooded wetland such as the Pantanal. Occurs on ranchland and in agricultural areas with cover. **Feeding Ecology** Omnivorous, with a very broad diet of both plant and animal matter; at least 102 fruits and 157 animal species are recorded in Brazil alone, though small vertebrate prey is most important by weight. Common prey includes mice, spiny rats, cavies and small armadillos, as well as small birds, arthropods and (less so) reptiles, including poisonous snakes. Occasionally kills tamandua and brocket deer; one record of predation on an adult Pampas Deer, which was killed by a throat bite. Wolf Apple or Lobeira (*Solanum lycocarpum*) is consumed year-round and is the most important fruit in the diet. Diet expands in the wet season to include more fruits, such as bell peppers, coffee and papaya, as well as new grass. Readily kills chickens, and possibly takes juvenile small stock, though reports are equivocal. Foraging cathemeral and solitary. Scavenges, including from road-killed carcasses and human refuse. **Social and Spatial Behaviour** Basic social unit is a monogamous pair that shares and defends a common territory from other pairs. Most behaviour, including hunting, is solitary; pairs occasionally rest and travel together, but longer associations occur only when breeding. Territories stable, marked constantly with urine and faeces, and used equally by both pair members. No difference in range size between males and females, except that female ranges decrease during breeding. Pair territories average 49–57km² (Brazilian cerrado), with a range of 4.7–114.3km². Density estimated at 5.2 wolves/100km² (Emas NP, Brazil). **Reproduction and Demography** Seasonal. Mates April–June; most births occur in the dry season, June–September. Gestation 56–67 days. Litter size 1–7, averaging 3. Weaning at around 15 weeks, but pups may accompany the mother from 7 weeks. Role of male in raising pups is unclear; captive males regurgitate food and groom pups, and wild males often accompany mothers and young pups. Sexual maturity at 12 months, but first breeding is probably during the second year. MORTALITY Annual adult and subadult mortality 35% and 36% respectively (Emas NP). Most mortality is anthropogenic; Puma and domestic dogs are confirmed predators. LIFESPAN 15 years in captivity. **Status and Threats** Lives at naturally low densities, and its habitat is under intense pressure from agricultural development. In Brazil, roadkills are one of the leading causes of death, and domestic dogs on ranches often kill Maned Wolves; dogs may also be a source of exotic disease, though effects on populations are unknown. Frequently killed in retribution for depredation on poultry. CITES Appendix II, Red List NT.

Family group

BUSH DOG

Hunting
rodents

MANED
WOLF

HB 120–180cm; T 8–16cm; SH 71–86cm; W ♀ 70–100kg, ♂ 85–125kg

Unmistakable and probably the most recognizable mammal on Earth. The distinctive white-and-black colouration is not apostematic as in skunks and other black-and-white carnivores. It has been suggested that it helps pandas locate one another in dense habitat during the breeding season, though they clearly utilize scent-marking and calls to find mates, as do most carnivores. Cubs are tiny at birth, weighing only around 100g, relatively much smaller compared with the mother's size than in all other bears. Cubs have sparse white fur: black markings appear gradually by 3–4 weeks. The forefoot has a 'false thumb', actually a greatly modified sesamoid bone (normally tiny in bears) with its own pad, which helps in grasping and manipulating bamboo, the species' main food source.

Distribution and Habitat

Restricted to six mountain ranges in the central Chinese provinces of Sichuan (which has about 75% of the population), Gansu and Shaanxi. Formerly widespread across C and SW China into N Myanmar and N Vietnam. Dependent on temperate montane forest with abundant bamboo at altitudes of 1200–4100m. Suitable habitat now occurs only on steep rugged slopes that are inaccessible for agriculture.

Feeding Ecology

The most specialized and herbivorous of bears, and indeed of carnivores, with over 99% of the diet made up of bamboo. Eats over 60 bamboo species, moving seasonally between altitudes to exploit the availability of different types as they germinate and grow. Most parts of the plant are eaten, but it prefers leaves and shoots, which are high in protein and easier to digest than stems and branches. Although it is a capable climber it eats on the ground, usually in dense stands of bamboo, in a sitting or reclining position. It occasionally eats other items, particularly in years of bamboo die-off when it is forced to seek other food, including leaves, shoots, roots and the bark of other plants, crops (including wheat, kidney beans and pumpkin) and fruits. Recorded occasionally scavenging from carrion and human refuse. Foraging occurs day and night, with about 50–55% of the time spent eating or gathering bamboo; adults eat 12–15kg a day. Unlike in other bears, foraging patterns do not change during the year to maximize periods of high food abundance. As food is available year-round, pandas do not hibernate. Females fast for a limited period of 2–3 weeks when they give birth.

Social and Spatial Behaviour

Solitary and non-territorial, with considerable overlap in home ranges. Individuals may remain in small core areas ≤3km² for extended periods; these overlap little between same-sex adults, though it is unknown if they are actively defended as territories. Range size 1–60km², averaging 5–15km², and changes seasonally as different bamboo species flower and pandas migrate vertically to track food. Usually stays in high altitudes during summer and spring, and descends to lowlands in winter to forage and avoid deep snow. Unusually among bears, it appears that young females disperse and young males settle close to their natal range. Density estimates poorly known and controversial; 48 pandas/100km² estimated for Mabian Nature Reserve.

Reproduction and Demography

Pandas have a reputation for poor reproduction, but this arises from the difficulties of breeding them in captivity. Wild pandas have a similar reproductive output to other bears. Seasonal. Mates March–May. Oestrus 12–25 days; gestation 97–163 days (averaging 145–146 days in the wild), with delayed implantation. Cubs born August–September. Litter size 1–2 (very rarely 3); with twins, the mother often ignores the second cub, which dies. The only bear that regularly gives birth to more cubs than it raises, the reasons for which are unknown. Weaning at 8–9 months. Independence at around 1.5 years. Inter-litter interval 2–3 years, averaging 2.2. Sexually maturity at 4.5 years for both sexes: earliest breeding 5–7 years for females and probably similar for males. Females can reproduce until their early 20s, males until age 17. MORTALITY Post-emergence (i.e. excluding abandoned twins) most cubs survive, with mortality estimated at 10–30%. Adult mortality normally ≤10% for both sexes, except in occasional years of mass bamboo die-offs, when starvation is a serious threat, e.g. at least 138 adults starved in a die-off event in the mid-1970s. No definite records of predation: panda remains have been found in Leopard scats, though it is not clear if the bears were killed or scavenged. Brown Bear, Asiatic Black Bear and Dhole co-occur and could plausibly kill subadults and perhaps adults. Cubs are helpless for a prolonged period and could be killed also by Snow Leopard, Indochinese Clouded Leopard, Asiatic Golden Cat and Yellow-throated Marten. LIFESPAN Maximum 26 years in the wild, 34 in captivity.

Status and Threats

Numbers in the wild estimated at 1500–2000, though the species' shyness and rugged habitat makes accurate population estimates challenging. Strictly protected in China and now rarely killed by humans; occasionally killed unintentionally in snares set for ungulates. Continued destruction of habitat for forestry and agriculture is a grave threat. Now restricted to six mountain ranges, each of which is an isolated island surrounded by deforested areas and cultivated land. Within each, the habitat is further fragmented by cultivation and forestry, so that most populations number under 50 adults. An estimated 71% of the habitat and 54% of the population are protected in 59 reserves, but both protected and non-protected areas are vulnerable to human activities. As well as ongoing habitat conversion, pressures include bamboo-shoot collection by local people, and infrastructure development that further fragments habitat, especially the building of highways and dams. CITES Appendix I, Red List EN.

Reclining
feeding
posture

GIANT PANDA

Handstand
urination

Female with newborn

Himalayan Black Bear, Moon Bear, Tibetan Black Bear

HB ♀ 110–150cm, ♂ 120–189cm; T <12cm; SH 70–100cm; W ♀ 40–140kg, ♂ 60–200kg

Solid black, often paler on the muzzle and face, and usually with a characteristic crescent-shaped cream patch on the chest. A rare chocolate-brown phase exists, and a highly variable blond phase is known from Cambodia, Laos and Thailand. **Distribution and Habitat** Southern Asia in a narrow band mostly associated with mountains, from SE Iran across central Asia into SE Asia (except Malaysia) and S China, including Taiwan, NE China, the Russian Far East, S Korea, N Korea and Japan. Lives primarily in temperate and tropical forested habitats in hilly and mountainous terrains to 4300m. Avoids open country, but enters plantations, agricultural fields and open alpine meadows in forested habitat. Westernmost population in southern Pakistan/Iran lives in arid thorn forest. **Feeding Ecology** Chiefly herbivorous, with 80–90% of the diet comprising plant matter. Diet varies seasonally as it moves between habitats and elevations, tracking food availability. In spring, eats mainly young succulent vegetation, including grass, leaves, forbs and bamboo shoots, switching to fruits including berries in summer, and hard mast such as oak acorns, beechnuts, walnuts, chestnuts and hazelnuts in autumn. Invertebrates are consumed when available, typically peaking in summer. Vertebrates to the size of small (10–20kg) ungulates such as muntjacs and serows are also eaten. Active hunting occurs, though most meat is probably scavenged. Both sexes hibernate in winter in the northern part of its range (October–May; Russia), while only pregnant females hibernate in the tropics; southern bears not hibernating eat mainly hard mast and fruits. Feeds in fruit orchards, plantations (where it ring-barks trees to access the tender cambium) and crops, especially corn and oats. Humans are rarely attacked, and probably never as prey. Foraging mainly diurnal, though nocturnal activity increases in autumn when hard mast is abundant e.g. Taiwan. Readily scavenges, especially from mammals caught in hunters' traps. **Social and Spatial Behaviour** Solitary and appears non-territorial. Home range size averages 26km² (♀s) to 66km² (♂s) in Japan, and 117km² (1 ♀) and 27–202km² (♂s) in Taiwan. Ranges overlap extensively, but females and young males avoid productive areas where adult male ranges concentrate during autumn (Taiwan). **Reproduction and Demography** Likely seasonal, especially in temperate areas; poorly known from the tropics. Mates May–July. Gestation 6–7 months (captivity), with delayed implantation. 1–3 cubs (most often 2) born December–March (Russia). Cubs accompany the mother for 2–3 years. Sexual maturity 3–4 years for both sexes (captivity). MORTALITY Humans are the main cause of death in many populations. Known natural predators of adults include Brown Bear, Tiger and (rarely) other Asiatic Black Bears. LIFESPAN Unknown in the wild, 36 years in captivity. **Status and Threats** Chief threats are habitat loss, combined with intense human hunting feeding a massive commercial market for bear products, mainly bile used in the Asian medicinal trade and paws for luxury restaurants. Bear farms are common in China and Vietnam for bile production. Some are ostensibly self-supporting, but many do not breed bears, and wild bear products are more valuable, fueling ongoing hunting and capture of wild bears. Sport hunting is legal only in Russia (75–100/year) and Japan (around 500): combined illegal and nuisance killing numbers an additional 500 and 1000–4000 respectively. CITES Appendix I, Red List VU.

ANDEAN BEAR *Tremarctos ornatus*

Spectacled Bear

HB 130–190cm; T <10cm; SH 70–90cm; W ♀ 60–80kg, ♂ 100–175kg (exceptionally to 200kg)

The only bear in South America. Typically black, occasionally reddish-brown, with pale facial 'spectacles' that usually extend onto the throat and chest, and are unique to individuals. **Distribution and Habitat** Endemic to the tropical Andes in Venezuela, Colombia, Ecuador, Peru and Bolivia; possibly NW Argentina (where it occurred historically) and Darien, Panama. Optimum habitat is high-elevation humid forest and paramo (humid alpine meadows), but it inhabits a variety of forests, woodland and grassland at 250–4750m. An isolated population lives in coastal desert scrub forest around Cerro Chaparri, NW Peru. **Feeding Ecology** Perhaps the most herbivorous of bears after Giant Panda, eating mainly bromeliads and fruits (especially of the fig and avocado families). Opportunistically eats cacti, moss, orchids, bamboo, tree wood, palms, honey, invertebrates, birds and small mammals. Individuals occasionally kill livestock, but most carcasses are probably scavenged. Eats crops, particularly corn, and is considered a major crop pest in some places. Foraging primarily diurnal (06.00–21.00h), though probably more nocturnal where it is persecuted. Given that its preferred food is above ground, it is highly arboreal. Builds a large nest-like 'feeding platform' by pulling fruiting branches into a bunch to support its weight while feeding; this possibly doubles as a rest bed. Readily scavenges, especially from livestock carcasses. With food available year-round, does not hibernate. **Social and Spatial Behaviour** Solitary, but reported to feed in groups (of up to nine) in cornfields and cactus groves. Spatial patterns poorly known, but like other bears it is unlikely to be strongly territorial. Home ranges of radio-collared bears in Ecuador overlap considerably and average 34km² (♀s) to 150km² (♂s). **Reproduction and Demography** Likely seasonal; pairs observed March–October during the fruiting season. Gestation 160–255 days (captivity), probably with delayed implantation. 1–4 cubs (usually 2) born December–February. They emerge at around 3 months, coinciding with the following fruiting season, and accompany the mother for up to 14 months (possibly longer). First reproduction 4 years for both sexes (captivity). MORTALITY Poorly known. Humans are the main cause of death in many populations, e.g. 11 bears killed in 13 months around Antisana Ecological Reserve, Ecuador. LIFESPAN Unknown in the wild, 36 years in captivity. **Status and Threats** Reduced to an estimated 42% of its historical range in over 100 fragments scattered along the Andes. The largest, most intact sites are in Peru and Bolivia. Habitat destruction from forestry and agriculture is the chief threat, though human hunting is equally grave for many populations. Killed mainly for crop raiding and perceived cattle killing, as well as for fur, meat and some traditional medicinal use. CITES Appendix I, Red List VU.

**ASIATIC
BLACK BEAR**

Golden
form

ANDEAN BEAR

Honey Bear, Malayan Sun Bear

HB 100–140cm; T 3–7cm; SH 70cm; W ♀ 25–50kg, ♂ 34–80kg

Smallest bear, with short, velvety jet-black (occasionally chocolate-brown) fur, very small ears and a pale muzzle and chin. Named for its distinctive orange, yellow or cream chest patch, which is highly variable in shape and sometimes absent altogether. **Distribution and Habitat** From Bangladesh throughout Indochina to Malaysia, Sumatra and Borneo. Recorded in Yunnan, S China, but possibly extinct there now. Optimum habitat dense lowland dry and wet forests, as well as montane evergreen forest and swamp forest, from sea level to 2800m. Extends into marginal habitats such as mangroves and plantations, provided dense forest is nearby. **Feeding Ecology** Omnivorous, with a narrow diet dominated by insects, honey and fruits. Eats over 100 species of insect, especially social species such as ants, termites and bees. Breaks into nests with its powerful claws, and uses its extremely long tongue to consume adult insects, larvae, eggs, honey, honeycomb, beeswax and nesting resin. Fruits are also extremely important, with over 40 species eaten, especially figs. Occasionally eats other invertebrates, such as earthworms and scorpions, reptiles (including small turtles), rodents and birds' eggs. Eats little vegetation, but apparently relishes coconut-palm growth shoots ('hearts'); their extraction kills the trees and creates conflict in plantations. Also may raid fruit orchards and crops such as sugar cane, potatoes and manioc. Livestock depredation is almost unknown, except for exceptional attempts at raiding poultry coops. Scavenges from human dumpsites, especially when natural food is scarce. Foraging mainly diurnal when undisturbed, but almost exclusively nocturnal when close to humans. Highly arboreal, adeptly climbing trees to forage for fruits and insect nests, building large nest platforms from branches for resting. Does not hibernate. **Social and Spatial Behaviour** Solitary; reported to occasionally gather around fruiting trees. Spatial patterns poorly known; unlikely to be strongly territorial, but four collared males in Borneo occupied small exclusive core areas of less than 1km^2 with larger overlapping ranges of 6.2–20.6km^2. Two females in poor habitat had small ranges of around 4km^2. **Reproduction and Demography** The only bear species that is not seasonal at all (though reproduction is very poorly known in the wild). Gestation short, 95–97 days (captivity), suggesting no delayed implantation. One cub (exceptionally 2) born in a tree cavity or hollow log. First reproduction 3 years for females, probably later for males. MORTALITY Poorly known. Humans cause most deaths in many populations. Vulnerable to starvation in poor fruiting years, when conflicts with humans (and related retributive killings) also rise. Confirmed predators include Tiger and Reticulated Python. LIFESPAN Unknown in the wild, 24–25 years in captivity. **Status and Threats** Numbers have declined an estimated 30% in the last 30 years, chiefly due to commercial hunting, and deforestation from logging and conversion to plantations. Hunted mainly for gall bladders for traditional Chinese medicinal use, and for bear-paw soup. Hunting reduced one Thai Sun Bear population by an estimated 50% in 20 years. CITES Appendix I, Red List VU.

SLOTH BEAR *Melursus ursinus*

HB 140–190cm; T 8–17cm; SH 60–92cm; W ♀ 50–100kg (exceptionally to 120kg), ♂ 70–150k (exceptionally to 190kg)

Solid black, rarely chocolate- or reddish-brown, with a crescent-shaped white-cream patch on the chest, and a pale muzzle and face. Fur long and shaggy. **Distribution and Habitat** Restricted to Sri Lanka, India, Nepal and Bhutan; formerly Bangladesh, where it is now possibly extinct. Lives primarily in dry or moist forest, scrubland, savannah and grassland, mostly in lowlands under 1500m, but occasionally to 2000m, e.g. W Ghats, India. **Feeding Ecology** The only bear species adapted for myrmecophagy, with flexible protrusible lips and nostrils that it can seal when sucking termites and ants. Switches mainly to fruits, especially figs, during the fruiting season; mostly eats fallen fruits rather than foraging in trees, though it is a capable climber, for example to eat honey. Eats tubers, roots and some flowers, but otherwise consumes little vegetation. Rarely eats meat, aside from exceptional meals of small vertebrates and carrion. Feeds on crops, including sugar cane, maize, rice, potatoes and sweet potatoes. Livestock depredation is virtually unknown. Has a reputation for unpredictability and aggression, and may attack people, for example 735 attacks (48 fatal) in Madhya Pradesh, India, in 1989–1994. Attacks occur mostly in fields and adjacent forest where people and bears forage for the same food items; humans are rarely hunted as prey. Foraging mainly nocturnal. Pregnant females den for up to 2 months, though it is unclear if they actually hibernate; other cohorts are active year-round **Social and Spatial Behaviour** Solitary and non-territorial. Home ranges often overlap extensively, though females appear to maintain small exclusive core areas. Home range size averages 2km^2 (♀s) to 4km^2 (♂s) in Sri Lanka, to 25–100km^2 in Panna NP, India. Density estimates include 6–8 bears/100km^2 (dry-forest habitat, Panna) to 27/100 km^2 (productive Terai grassland and forest, Royal Chitwan NP, Nepal). **Reproduction and Demography** Seasonal. Mates May–July. Gestation 4–7 months (captivity), with delayed implantation. Litter size 1–3 (usually 2), with cubs born November–January. Cubs emerge from the den at around 6–10 months, and ride on their mothers' backs for up to 6 months, probably as a defence against predation. Weaning at 12–14 months, and cubs accompany the mother for 1.5 or 2.5 years, depending on food availability. First breeding at 4 years for females, probably later for males. MORTALITY Poorly known. Humans are the main predator; natural predation of adults is rare, but is known by Tiger and Leopard. LIFESPAN 40 years in captivity. **Status and Threats** Reasonably secure inside protected areas, but outside suffers heavily from habitat fragmentation and anthropogenic killing. Killed retributively and preventively in conflict situations, as well as for commercial trade in parts, particularly gall bladders. In some parts of India, females are killed and cubs captured to be trained mainly by Kalandar gypsies as 'dancing bears', now illegal but still widely practised. CITES Appendix I, Red List VU.

SUN BEAR

SLOTH BEAR

Female
carrying cub

HB ♀ 120–168cm, ♂ 129–188cm; T <12cm; SH ♀ 63–85cm, ♂ 72–103.5cm; W ♀ 40–150kg (exceptionally to 190kg), ♂ 60–300kg (exceptionally to 400kg)

Smallest of America's bears and endemic to that continent. Colour highly variable, most often black, but also many shades of brown ('cinnamon' common in the W USA); a grey-blue phase called 'glacier' in Alaska and NW Canada; and a cream phase called 'Kermode' that occurs only on the British Columbian coast. Except in the Kermode phase, the muzzle is usually a contrasting shade of buff-brown, useful for distinguishing it from Brown Bear. Cubs of any colour may be born into the same litter, and brown cubs often darken after their second year, occasionally becoming black by adulthood. Up to 80% of cubs have a white chest patch that usually disappears but is retained in some adults.

Distribution and Habitat
Widespread and abundant in most of Canada, and the E and W USA (including Alaska), becoming patchy and fragmented in the S USA and N Mexico. Adaptable, and occupies a wide range of forested habitats from sea level to 3500m. Generally avoids areas without cover, though a unique population has colonized open coastal tundra in NE Canada since the recent local extinction of Brown Bears. Readily occupies human landscapes, e.g. farmland and peri-urban areas, provided there is cover.

Feeding Ecology
Omnivorous, feeding mainly on vegetation, with a relatively small proportion of the diet consisting of animal matter. As in other bears, the diet shifts seasonally reflecting available food sources. Eats mainly new grasses, buds, shoots and forbs in spring, fruits and soft mast in summer, and hard mast such as hazelnuts, oak acorns and Whitebark Pine nuts, and berries, huckleberries and buffalo berries, in autumn. Kills are made opportunistically in any season, but especially of winter-debilitated and newborn ungulates in spring. A capable hunter recorded killing animals to the size of Moose cows, but most kills are the size of White-tailed Deer fawns or smaller, e.g. rodents, birds, reptiles, fish and invertebrates. Carrion eater, especially of winter-killed ungulates after emerging from hibernation in spring. Readily exploits foods from human sources, especially during autumn hyperphagia. Occasionally kills livestock and eats crops such as apples, oats and corn, and can become a pest raiding domestic beehives, suburban birdfeeders, refuse bins, dumpsters, camp grounds and refuse dumps. Humans are sometimes killed: although rare, kills are thought to be mainly predatory rather than in defence of food or cubs. There are very few records of female Black Bears attacking humans while defending cubs. Foraging mostly solitary and diurnal, with early morning and late afternoon peaks; becomes more nocturnal in human-dominated areas. Forms tolerant congregations at food-rich sites like refuse dumps. Hibernates throughout its range for 3–7 months; in its southern range, where food is available year-round, only pregnant females and mothers with yearling cubs hibernate.

Social and Spatial Behaviour
Mostly solitary, except mothers with emerged cubs that stay in family groups for 16–18 months. Non-territorial, with enduring home ranges that overlap considerably with those of other adults. Females sometimes establish exclusive areas in their home range that they defend from other females, probably related to raising cubs. Excursions up to 200km outside its range are fairly common in periods of food scarcity, especially during autumn hyperphagia; individuals typically return to their own range to hibernate. Depending on food availability, range size is 2–1100km², exceptionally reaching 7000km² in the NE Canadian tundra, where food is especially scarce. Ranges expand in seasons, years and regions of low food availability. Male ranges typically 2–10 times the size of female ranges: average range sizes include 6.9km² (♀s) and 51.2km² (♂s) in Tennessee; 17.1km² (♀s) and 22.4km² (♂s) in California; 28km² (♀s) and 318km² (♂s) in Massachusetts; 48.8km² (♀s) and 112.1km² (♂s) in Idaho, and 295km² (♀s) and 495km² (♂s) in Manitoba. Females mostly settle near their natal range; males disperse more widely. Dispersal distance typically 10–220km. Density estimates include 20 bears/100km² (south-central Alaska), 40–50/100km² (N Montana) to 120–150/100km² (SW Washington).

Reproduction and Demography
Seasonal. Mates mid-May–July, occasionally as late as September in southern areas. Oestrus is 'seasonally constant', lasting until conception or the end of the breeding period. Gestation 200–253 days, with delayed implantation. Cubs born January–February (rarely December). Litter size 1–5, average 2–2.5. Weaning at 5.5–7 months. Independence at 16–17 months. Inter-litter interval usually 2 years, longer when food availability is low. Sexually mature at 2.5–3 (♀s) and 3.5 (♂s) years; earliest breeding at 3–8 (♀s) and 3.5–8 (♂s) years. Reproduction by females is possible until at least age 20. MORTALITY On average, 40% of cubs die in their first year, but cub mortality ranges from 0% (e.g. wilderness, Nevada) to 83% (urban area, Nevada). Mortality is highest following independence, mainly from starvation and human causes: 30–48% of subadult females and only 14–17% of subadult males survived to adulthood in one Alaskan population. Adult mortality 13–21% (♀s) and 12–41% (♂s). Even in unhunted populations, humans are often the main cause of death, especially from roadkills and problem animal control. Adults have few predators except Brown Bears and Grey Wolf packs, typically of hibernating individuals killed in their dens. Cubs are additionally vulnerable to predation by male Black Bears, and occasionally by Bobcats and Coyotes; there is at least one record of Golden Eagle predation. LIFESPAN Maximum 24 (♀s) and 20 (♂s) years in the wild, 32 in captivity.

Status and Threats
Widespread and globally secure, with a total population estimated at 900,000. Has disappeared from large areas of its southern and mid-western US historic range, where remaining populations are patchy, and some are threatened (Florida and Louisiana) or endangered (Mexico). Vulnerable to severe habitat loss and fragmentation, but thought to be recolonizing or increasing in much of its range. As many as 40,000–50,000 are legally killed annually by sport hunters and trappers in the USA and Canada. Poaching, illegal killing and roadkills are serious local threats for some populations. CITES Appendix II, Red List LC.

**AMERICAN
BLACK BEAR**

Typical
form

Kermode
form

Cinnamon
form

Glacier
form

BROWN BEAR *Ursus arctos* Plate 58

Grizzly Bear

HB ♀ 140–228cm, ♂ 160–280cm; T 6.5–21cm; SH 90–152cm; W ♀ 55–277kg, ♂ 135–725kg

Second largest terrestrial carnivore after the closely related Polar Bear. Massively built with a distinctive shoulder hump, distinguishing it from similarly coloured phases of American Black Bear. Colour variable, ranging from light blond through various shades of brown to near-black. Some individuals have blond tips to the hairs, giving a grizzled appearance, hence the name Grizzly Bear (used only in N America). Cubs often have a white or blond collar that usually fades after their first year, but sometimes persists in adults, especially in Eurasian populations. Size varies significantly, seasonally within populations and regionally between populations: smallest bears are W Eurasian, e.g. Syrian bears, while bears with access to abundant salmon, e.g. in coastal habitats of Alaska, British Columbia and eastern Russia, are largest.

Distribution and Habitat

The most widespread bear, ranging from NW N America, the Russian Far East through northern Asia to Fenno-Scandinavia, south to the Himalayas in the east, and SW Iran in the west. Isolated populations occur in W Europe and SW Central Asia. Inhabits a wider variety of habitat types than any other ursid, including all types of temperate forest, coastal habitats, meadows, grassland, steppes, tundra and semi-desert, from sea level to 5000m.

Feeding Ecology

Highly omnivorous and opportunistic, with the broadest diet of any bear species, comprising all types of fungi, plant matter, invertebrates, fish, reptiles, birds, eggs and mammals. In most populations, the diet changes seasonally as bears move between habitats and elevations, reflecting food availability. In spring and early summer, they eat mainly grasses, shoots, sedges and forbs, and actively hunt newborn ungulates such as Elk, Moose and Caribou/Reindeer calves. Fruits including berries become increasingly important during summer and early autumn; roots and bulbs are also eaten, and may become critical in autumn for some inland populations if fruit crops are poor. Super-abundant salmon is a key food source for many coastal populations, especially in autumn during the spawning run. Invertebrates and hard mast are also important in autumn, given their high fat content, e.g. Rocky Mountain bears in high talus slopes eat up to 40,000 Army Cutworm Moths daily, and harvest Whitebark Pine nuts from Red Squirrel middens almost exclusively. Meat dominates the diet at the end of winter as the bears emerge from hibernation to scavenge carcasses of winter-killed ungulates, and hunt weakened Moose, Elk, bison, Caribou/Reindeer, Wild Boar and various deer species; adult Musk-ox occasionally recorded. Occasionally kills livestock, raids crops and ransacks domestic beehives, typically peaking during autumn hyperphagia and especially when natural food sources are poor. Humans are sometimes killed, usually in defence of carrion or cubs, but occasionally as food. Loses 5–43% of its body mass during winter hibernation. Foraging mostly solitary, but large seasonal congregations gather at food-rich sites like spring meadows, salmon runs, alpine-moth aggregations and refuse dumps. May cache food, especially large carcasses, by covering it with dirt and vegetation, and often lies close to the pile. Readily scavenges, including from human refuse.

Social and Spatial Behaviour

Solitary, but emerged cubs (≥3 months old) accompany their mothers for up to 3 years, and adults congregate amicably at seasonal feeding sites. Not classically territorial. Adults maintain enduring home ranges that often overlap extensively; subordinate bears give way to dominant individuals in well-defined dominance hierarchies that segregate individuals and prioritize access to food and mates. Range size smallest in populations with concentrated dependable food sources such as salmon, e.g. 24–89km² (♀s) and 115–318km² (♂s) in coastal Alaska and British Columbia. Ranges generally increase inland where food sources are more dispersed, e.g. 281km² (♀s) to 874km² (♂s) in Yellowstone NP; up to 2434km² (♀s) and 8171km² (♂s) in the central Arctic (Canada). Subadult bears occupy areas of up to 20,000km², e.g. central Canadian Arctic, before settling into their adult range. Females tend to settle near their natal range, while males disperse more widely. Density estimates include 1–3 bears/1000km² (Norway), 3–4/1000km² (Arctic Wildlife Refuge, Alaska), 14–18/1000km² (Yellowstone NP), 47–80/1000km² (Glacier NP, USA), 135–190/1000km² (Abruzzo NP, Italy) to 191–551/1000km² (coastal Alaska).

Reproduction and Demography

Seasonal. Mates April (Eurasia) to mid-May (N America) until July. Oestrus 10–30 days; gestation 210–255 days, with delayed implantation. Cubs born January–March. Litter size typically 1–3, rarely to 6. Adoption and exchange of emerged cubs sometimes occurs. Weaning at around 18 months, independence at 2–3 years, occasionally as long as 4.5 years. Inter-litter interval averages 3.5 years, range 2–6, depending on food availability. Sexual maturity at 3.5 (♀s) and 5–5.5 (♂s) years; earliest breeding at 4–8.1 (♀s) and 5–8 (♂s) years. Reproduction by females possible until age 28. **MORTALITY** 13–44% of cubs die in their first 1.5 years. Adult mortality ≤10% (♀s) and 6–38% (♂s), higher rates occurring in hunted populations. Even in unhunted populations, humans are often the main cause of death. Adults have few predators except other Brown Bears and (rarely) Tigers (Russia). Both sexes (but usually males) occasionally kill cubs, and males sometimes kill other adults. Cubs are occasionally killed by Grey Wolf and Golden Eagle (one confirmed record). **LIFESPAN** Maximum 25–30 years in the wild, reportedly 47 in captivity.

Status and Threats

Still widespread and relatively common in much of its northern range, with large numbers in western Canada (~25,000), Alaska (~32,000) and Russia (100,000–125,000). Southern range has undergone significant contraction and fragmentation into numerous very small and isolated populations. Extirpated from Mexico, most of the lower 48 United States, and most of its former range in Europe and the Middle East. Formerly occurred in N Africa, perhaps until the mid-1800s in Morocco and Algeria, the only modern bear species to inhabit Africa. Globally secure, but small isolated populations are vulnerable to contact with humans. Legal sport hunting occurs in at least 18 range countries, with the largest numbers killed in Canada, Finland, Romania, Russia, Slovakia and the USA (Alaska only). CITES Appendix II, Red List LC.

Kodiak Island
form, male

**BROWN
BEAR**

Grizzly form,
female

Syrian form

Cubs

HB ♀ 180–247cm, ♂ 200–285cm; T 6–21cm; SH 120–170cm; W ♀ 150–450kg, ♂ 300–655kg (exceptionally to 800kg)
The world's largest terrestrial carnivore. The bear's white fur is actually unpigmented and transparent, its shifting hue depending on reflected light, e.g. golden at sunset/sunrise and bluish when it is cloudy or misty. Impurities such as oils from kills also stain the fur, contributing to a dirty cream or yellowish colour. Skin pink in young cubs, but completely black (as on nose) in adults; it was once thought to enhance the absorption of ultraviolet light, but the transparent hairs actually absorb UV before it reaches the skin. Hairs also hollow, increasing insulation (and the reason why captive animals sometimes appear greenish due to algal growth inside the hair). Closely related to Brown Bear; ranges of the two overlap minimally, chiefly in the western Canadian Arctic, where wild hybrids are rarely recorded; two were shot by hunters in Canada's Northwest Territory (Banks Island and Victoria Island) in 2006 and 2010.

Distribution and Habitat
Circumpolar in the Arctic; Canada, the USA (Alaska), Russia, Norway (Svalbard) and Greenland. Transient in northern Iceland, where it is usually shot, e.g. two in June 2008. Reliant on sea-ice habitats mostly within 300km of the coast, where marine productivity is the highest. Follows retreating ice northwards in summer where possible, e.g. northern Greenland where summer ice remains near the coast; otherwise forced inland, e.g. Hudson Bay, Canada. After the first few months of life, many individuals spend their entire lives on sea-ice.

Feeding Ecology
The most carnivorous bear species and profoundly dependent on seals and sea-ice for hunting; cannot hunt in open water. The most important prey species is Ringed Seal, followed by Bearded Seal, Harp Seal and Hooded Seal. Seals are caught by 'still-hunting', which involves patiently waiting at breathing holes, sometimes for hours; by carefully stalking basking seals either on ice or from water; or by 'pole-driving' into seal birth dens to catch newborn pups. Walruses are occasionally hunted, sometimes by stampeding colonies at haul-out sites, e.g. Wrangel Island, Russia, resulting in dozens of crushed animals, which are scavenged. Recorded killing Narwhals and Belugas, usually when the whales are trapped in ice. During lean periods, e.g. summer, when some populations are restricted to ice-free land, opportunistically eats berries, kelp, grasses, fish, seabirds, eggs, small mammals, hares and Reindeer/Caribou, but it cannot survive indefinitely on this diet. Humans are very rarely killed, e.g. six records in Canada and Alaska since 1968. Has a prodigious ability to fast when food is scarce, dependent on extensive fat reserves accumulated from seals. Fasting bears enter a similar physiological state to hibernation, but remain active and alert. This can occur during food shortages in any season (in contrast to other bears, which fast only during winter hibernation). Pregnant females can fast for 8 months from the time they leave the ice to den until the next ocean freeze-up after birth. Healthy males and non-pregnant females easily fast for 4–6 months, but apart from sheltering temporarily in severe weather, they do not overwinter in dens. Foraging mostly solitary, but congregations form around large carcasses, Walrus colonies and dumps. Readily scavenges, e.g. from beached whales and refuse.

Social and Spatial Behaviour
Solitary, but emerged cubs (≥3 months old) always accompany their mothers, and bears congregate amicably at food-rich areas and while waiting for the freeze-up, when unrelated adults often engage in highly social behaviour such as play-fights. Extremely mobile and has the largest home ranges of any carnivore; utilizes different areas in its range depending on ice dynamics, but does not wander randomly as once thought. Annual ranges are smaller in near-shore areas with stable ice, averaging 50,000km^2, compared with ranges on drift ice, which average 250,000km^2. Hudson Bay females have annual ranges of 8470–311,646km^2 (average 106,613km^2), while Beaufort Sea (Canada/Alaska) female ranges are 13,000–597,000km^2 (average 149,000km^2). Straight-line movements (i.e. minimum estimates) can exceed 50km per day and 6200km per year. Easily swims 25–40km and capable of extraordinary endurance in water: a monitored Beaufort Sea female swam continuously for 232 hours covering 687km, in 2–6°C water. Ranges and movements are thought to be similar for both sexes, but males are rarely radio collared as their neck circumference exceeds their head size.

Reproduction and Demography
Seasonal. Mates March–June. Oestrus duration poorly understood, but mating associations last 2–4 weeks; gestation 195–265 days, with delayed implantation. Cubs born mid-November–mid-January in traditional denning areas near the coast or in drift ice, where snowfall and topography allow excavation of snow dens. Litter size 1–3 (4 exceptionally recorded from captivity), most often 2. Weaning and independence occur together, coinciding with the mother's next breeding season, normally at 30 months, sometimes at 18 or 42 months, depending on food availability. Inter-litter interval averages 3.1–3.6 years. Both sexes are sexually mature at 3–3.5 years, but earliest breeding is at 4–6 (♀s) and 6–8 (♂s). Reproduction is possible until age 27 (♀s) and over 20 (♂s). MORTALITY Depending on resources, 25–65% of cubs die in their first year, most from starvation. Survival rates increase each year of life until prime adult natural mortality is 1–4%. Adults have no predators except other Polar Bears and humans; male bears are sometimes infanticidal, and Grey Wolves are (rarely) recorded killing young cubs. There are occasional records of Walruses killing bears when attacked. LIFESPAN 32 (♀s) and 29 (♂s) years in the wild, 42 in captivity.

Status and Threats
Total numbers estimated at 20,000–25,000. Protected by a five-nation treaty that restricts hunting to indigenous communities (Canada, Alaska and Greenland), or prohibits hunting (Svalbard and Russia, largely unenforced in latter). Annually, around 750 bears are legally hunted, most (~500) in Canada, which also permits trophy hunting under the quota to native people. Given the species' strict dependence on sea-ice, global warming is now recognized as the greatest long-term threat; southernmost populations, e.g. Hudson Bay, exposed to earlier than usual sea-ice thaws already display lowered condition and survival, and escalated conflict with people in their search for food. CITES Appendix II, Red List VU.

Pole-driving
for seal pups

Play-fighting

POLAR BEAR

Cubs

Waiting for ocean
freeze-up

Cozumel Raccoon

HB 35–43cm; T 22–35cm; W ♀ 2.9–3.5kg, ♂ 3–4.1kg

The smallest raccoon, with orange-tinged fur on the tail and along the spine, especially in males. Has been classified as a Northern Raccoon subspecies, but recent genetic analysis shows it is a distinct species. Three other insular Caribbean populations, formerly called Bahamas Raccoon *P. maynardi*, Barbados Raccoon *P. gloveralleni* (extinct) and Guadeloupe Raccoon *P. minor*, are actually introduced Northern Raccoons. **Distribution and Habitat** Endemic to Cozumel Island, Mexico. Relies heavily on mangroves, coastal wetland and (less so) adjacent tropical rainforest. **Feeding Ecology** Omnivorous, with over 50% of the diet made up of crabs; insects, fruits, and marine turtle eggs and nestlings make up most of the balance. Foraging solitary and mainly nocturnal. Individuals near settlements are heavier, suggesting they scavenge from food refuse. **Social and Spatial Behaviour** Poorly known. Thought to be essentially solitary, analogous to low-density unsubsidized Northern Raccoon populations. Limited telemetry data suggest a provisional range size of 0.7km². **Reproduction and Demography** Poorly known. Births believed to occur mainly November–January. MORTALITY No natural predators, but killed by feral dogs and possibly introduced Boa Constrictors. Frequent hurricane devastation of habitat causes significant mortality. LIFESPAN Unknown. **Status and Threats** Occurs only on Cozumel Is, where it numbers fewer than 150 individuals. Depends strongly on coastal habitats, which are under intense development pressure for tourism infrastructure and associated roads. Roadkills, predation and possibly disease from introduced predators constitute additional threats. Red List CE.

NORTHERN RACCOON *Procyon lotor*

Common Raccoon

HB 44–62cm; T 19.2–40.5cm; W ♀ 1.7–7.1kg (exceptionally to 10kg), ♂ 2.4–11kg (exceptionally to 28kg)

Largest and most familiar raccoon. Colour usually grizzled grey-brown, but varies from cinnamon to near black. All forms have the characteristic dark mask with pale eyebrows, banded tail and pale feet. Albinism occurs. Northern individuals, e.g. Idaho, can be twice as large and heavy as southern ones, e.g. S Florida. Tres Marias Raccoon (Tres Marais Island, Mexico) was considered a separate species, but is now classified as a Northern Raccoon subspecies. **Distribution and Habitat** S and C Canada, the USA, Mexico and C America over to the Panama Canal. Introduced in W Eurasia, Japan and on some Caribbean islands. There is almost no habitat it cannot occupy, but it prefers edge habitats and areas associated with water. Lives in close association with humans, including in urban areas and large cities. **Feeding Ecology** Extremely omnivorous and opportunistic, eating virtually all available edible items. Wilderness populations eat mostly fruits, hard mast, aquatic crustaceans, molluscs, amphibians, eggs, nestlings and small mammals. Many populations live largely or entirely on anthropogenic foods, including grains, crops, pet food, bird food and refuse. Sometimes kills domestic poultry. Foraging mainly nocturno-crepuscular and solitary; congregates at food patches such as dumps. Most food is found by its keen sense of smell, and captured or handled with its very dexterous front feet; often 'washes' food in water, though clean food is also submerged so the actual purpose is unclear. Does not hibernate, but northern populations may over-winter in dens for weeks or months, during which they live off accumulated fat. **Social and Spatial Behaviour** Sociality and spatial behaviour very flexible, depending on food availability. In natural settings, essentially solitary and defends ranges against same-sex individuals. Under high food availability, neighbouring females (usually related) have small, closely spaced ranges with high overlap, and males form coalitions of 3–4 that cooperate to defend a territory from other males. Average range size 0.05–0.8km² (urban), 0.5–3km² (rural) and to 25.6km² (wilderness). Density 0.5–6 raccoons/km² (wilderness), 1–27/km² (rural) and to 111/km² (urban). **Reproduction and Demography** Seasonal. Most populations mate February–March, with births April–June; southern populations are more variable. Gestation 54–70 days. Litter size 1–8, averaging 2–5. Kittens largely independent by 17–18 weeks, but often rejoin the mother for winter denning; the family finally breaks up the following spring. MORTALITY Highly variable, depending on latitude and harvest; 10–83% (juveniles), and 16–55% (yearlings and adults). Most mortality is from disease (canine distemper and rabies) and human harvest. LIFESPAN 12.5 years in the wild (typically <5), 17 in captivity. **Status and Threats** Widespread and extremely adaptable. Widely killed for fur in Canada and the USA, peaking at over five million/year in 1979–1980. Populations are very resilient to harvest, which is not considered a threat. Red List LC.

CRAB-EATING RACCOON *Procyon cancrivorus*

HB 54–76cm; T 25–38cm; W 3.1–7.7kg

Slender raccoon with a short dense coat. Fur greyish-brown with a buff, tawny or rufous shade, especially where the body colour transitions to the paler underparts. Legs and feet dark grey-brown, in contrast to Northern Raccoon, which always has pale feet. The two species overlap in S Costa Rica and W Panama. **Distribution and Habitat** S Costa Rica to N Argentina and Uruguay, and Trinidad. Inhabits a wide variety of habitats usually associated with water in forest, brush, wooded grassland, swamps and wetland, and on coastlines. Occurs in dry chaco scrubland, where water sources are mostly man-made. Tolerates rangeland and plantations; rarely occurs in urban areas. **Feeding Ecology** Omnivorous, eating mainly crabs, crayfish, snails, insects and fruits, especially of palms. Small lizards, snakes, birds, rodents and fish are also eaten. Apparently forages in coastal caves, either for dying bats or insects attracted to bat droppings on the cave floor. Sometimes kills poultry and causes damage in fruit plantations. Foraging mostly nocturnal, solitary and terrestrial. **Social and Spatial Behaviour** Poorly known. Solitary. Density estimated at 6.7 raccoons/km² (dry wooded rangeland, Paraguayan chaco). **Reproduction and Demography** Poorly known. Births apparently peak February–June (Suriname). Litter size 2–3. MORTALITY and LIFESPAN Unknown. **Status and Threats** Widespread and probably fairly common. Habitat loss is the main threat. Fairly widely persecuted as a pest and hunted for sport, but effects on populations are unknown. Frequently killed on roads in Brazil. Red List LC.

**PYGMY
RACCOON**

**NORTHERN
RACCOON**

**CRAB-EATING
RACCOON**

HB 40–54cm; T 19.2–30cm; W 1–1.5kg
Smallest coati, dark olive-brown with paler, rufous-tinged brown on the throat and chest. Face narrow and elongated, tapering to a sharp point at the nose, which is naked. Recent molecular and morphological analyses suggest sufficient differences to recognize two species, Western Mountain Coati *N. olivacea* (Colombia and Ecuador), and Eastern Mountain Coati *N. meridensis* (Venezuela), but this remains tentative. **Distribution and Habitat** Endemic to the Andes of Colombia, Ecuador and Venezuela. Inhabits cloud forest and paramo shrubland at 1300–4250m. Does well in reforested plantations of Andean Alder. **Feeding Ecology** Eats mainly adults and larvae of soil invertebrates, especially beetles, grasshoppers, locusts, millipedes, centipedes and ants. Frogs are the most common small vertebrate prey (based on limited data). Also eats fruits, leaves, grass roots and moss. Blamed for killing poultry and raiding potato crops, though neither is definite. Thought to forage diurnally in social groups. Leaves behind large areas of turned soil with many distinctive diggings made by the claws and probing nose; one moss bank of 35m² had over 5000 such holes. **Social and Spatial Behaviour** Poorly known. Observed in small groups and individually, suggesting sociality resembles that of other coati species. Group size typically 6–8, though 50–80 has been reported. The only range estimate is 0.11km² (single ♂ monitored for 3 months, Colombia). **Reproduction and Demography** Unknown. A litter of 4 is reported. **Status and Threats** Limited distribution with narrow habitat tolerances. Range is under significant pressure for agriculture, logging and pine plantations. Persecuted as a pest, and hunted in some regions for meat and fur. Red List DD.

SOUTH AMERICAN COATI *Nasua nasua*

Brown-nosed Coati, Ring-tailed Coati
HB 43–59cm; T 42–55cm; W 2–7.2kg
Large coati, typically pale to dark olive-brown, but colour varies from bright ginger to almost black. Most forms have a white lower jaw and buff or orange-buff throat. Dark muzzle compared with White-nosed Coati's; the two species do not overlap. **Distribution and Habitat** East of the Andes in Colombia and Venezuela to N Argentina and N Uruguay. Introduced to Isla Róbinson Crusoe, Chile. Occurs to 2500m in rainforest, cloud forest, riverine forest, dry chaco scrub, and wooded habitat in cerrado and Pantanal savanna. Tolerates disturbed habitats with cover, e.g. pasture-woodland mosaics. **Feeding Ecology** Omnivorous, eating mainly ground-litter invertebrates, especially beetles, ants, millipedes and arachnids, as well as fruits. Occasionally takes small vertebrates, and is an important predator of caiman nests. Foraging social, diurnal and mainly terrestrial, but also above ground among vine tangles and bromeliads. Most food is found by smell. Scavenges from carrion, human refuse and handouts. **Social and Spatial Behaviour** Social. Forms matrilineal bands of related females and their offspring, numbering 5–30 (exceptionally to 65). Adult males are usually solitary and temporarily join bands for breeding, though males sometimes accompany bands year-round, e.g. Iguaçu NP, Argentina. Range size estimates include 7.6–54.4km² for bands, and 2.2–9.6km² for solitary males. Density estimated at 6.2–13 coatis/km². **Reproduction and Demography** Seasonal. Mating August–October; births October–November (Iguaçu). Gestation 65–77 days. Litter size 1–7 (captivity). Group females synchronize births; they leave the band to give birth alone. MORTALITY Under low predation (Iguaçu), annual mortality 19.7–43.4% (juveniles), and 29.2% (band females). Known predators include Ocelot, Puma and Jaguar. LIFESPAN 17.7 years in captivity. **Status and Threats** Wide distribution and wide habitat tolerances, including the ability to occupy modified areas. Important meat species for some subsistence hunting communities. Red List LC.

WHITE-NOSED COATI *Nasua narica*

Coatimundi
HB 43–68cm; T 42–68cm; W 3.5–5.6kg
Similar to South American Coati, but usually dark or reddish-brown with pale 'frosting' on the forequarters, a white muzzle and less distinct bands on the tail (sometimes entirely absent). Coatis on Cozumel Island, Mexico are very small and sometimes treated as a separate species, Dwarf Coati *N. nelsoni*. **Distribution and Habitat** SW USA, Mexico, C America and Colombia west of the Andes. Inhabits wet and dry forests, woodland and scrub, and lives along vegetated watercourses in arid areas. Occurs close to humans in modified habitats with cover. **Feeding Ecology** Omnivorous. Eats mostly ground-litter invertebrates and fruits, which constitute 88–100% of the diet. Small vertebrates generally comprise <5% of the diet; an important predator of marine turtle nests. Sometimes raids cultivated fruits and gardens, and (rarely) kills poultry. Foraging social, diurnal and mainly terrestrial, but it is a capable climber that also forages in the subcanopy. Scavenges from carrion, human refuse and handouts. **Social and Spatial Behaviour** Social. Similar to South American Coati, with matrilineal bands of 5–22 (exceptionally to 30). Adult males are more solitary than in South American Coati, and rarely accompany bands except during breeding. Range size estimates include 0.19–0.41km² (tropical forest, Panama) to 7.9–22.4km² (semi-arid forest, Arizona) for bands; and 0.15–0.54km² (Panama), and 3.5–10.7km² (Arizona) for solitary males. Estimated density from 1.7 coatis/km² (Arizona) to 50–70/km² (Costa Rica, Panama). **Reproduction and Demography** Seasonal. Timing depends on location, with births April–May (Central America) and June (Arizona). Gestation 70–77 days. Litter size 1–6. Females synchronize births; they give birth alone and integrate the kittens with the group at 5–6 weeks. Males disperse to become solitary at 20–24 months. MORTALITY Most natural mortality is from predation, e.g. 76% of known deaths in SE Arizona are by Puma; Jaguar, Ocelot, large raptors and Boa Constrictor are also predators. Capuchins are significant predators of nest young, killing 100% of juveniles in some bands, e.g. Santa Rosa NP, Costa Rica. Canine distemper and rabies have produced die-offs in Arizona. LIFESPAN 9 years in the wild, 17 in captivity. **Status and Threats** Widespread, adaptable and secure in most of its range. Important for subsistence hunters, which can lead to local declines, e.g. N Mexico. Indiscriminate predator control, including by poisoning in the USA, led to large declines. The unusual Cozumel population numbers less than 150 and is Endangered. CITES Appendix III (Honduras), Red List LC.

MOUNTAIN
COATI

SOUTH
AMERICAN
COATI

WHITE-NOSED
COATI

Group
foraging

HB 30.5–42cm; T 31–44.1cm; W 0.8–1.1kg
Smallest procyonid in N America, with an unmistakable long bushy tail marked with black-and-white rings. Can be confused only with the closely related Cacomistle where the two species overlap in S Mexico. **Distribution and Habitat** USA from S Oregon through the SW, and Mexico to Oaxaca. Inhabits rocky desert areas, desert grassland, canyons, chaparral scrub and various forest types. Tolerates agricultural areas with cover, and human settlements including suburban areas. **Feeding Ecology** Eats arthropods, fruits and small vertebrates to the size of hares; mice and squirrels are typical mammal prey. Raids birds' nests for eggs and fledglings, and feeds on agave nectar. Rarely raids poultry. Foraging both terrestrial and arboreal, and exclusively nocturnal. Scavenges from carrion, human refuse and hummingbird feeders. **Social and Spatial Behaviour** Solitary. Ranges overlap extensively, with little evidence of territorial defence; male-female pairs sometimes associate loosely, but do not share dens (though captive animals are tolerant and den together). Range size 0.05–2.3km². Density usually 1–4 Ringtails/km², exceptionally reaching 10.5–20/km² (Central Valley, California). **Reproduction and Demography** Seasonal. Mates February–May; most births May–June. Gestation 51–54 days (the shortest among procyonids). Litter size 1–4, exceptionally 5 (captivity). MORTALITY Adult mortality (Trans-Pecos, Texas) 81%, mainly from predation by Great-horned Owl and, less so, Coyote, Northern Raccoon and Bobcat. LIFESPAN 16.5 years in captivity, much lower in the wild. **Status and Threats** Widespread and relatively common. Legally trapped in the SW USA, though fur is of poor quality; no information on the effects of harvesting. Red List LC.

CACOMISTLE *Bassariscus sumichrasti*

Central American Cacomistle
HB 38–47cm; T 39–53cm; W 0.7–1.2kg
Very similar to Ringtail, but larger and darker coloured, with muted facial markings, a darker body colour and dark lower legs. Tail less strikingly banded, with up to a third of the tip coloured dark brown or blackish. **Distribution and Habitat** S Mexico to W Panama. Occurs primarily in rainforest, evergreen forest, dry forest and dense scrub mostly associated with mountainous areas. Tolerates secondary forest, plantations and overgrown pasture. **Feeding Ecology** Diet similar to Ringtail's, but insects and fruits appear to be more important than vertebrates. Associated with bromeliads, which are rich sources of small prey as well as water. Foraging nocturno-crepuscular and mainly arboreal. **Social and Spatial Behaviour** Poorly known. Thought to be solitary. Extensive range overlap and mutual tolerance of adults in captivity suggests some social tendencies. Male ranges are slightly larger than those of females; ranges average 0.17–0.23km². **Reproduction and Demography** Poorly known. Gestation 63–65 days. Litter size 1–2. MORTALITY and LIFESPAN Unknown. **Status and Threats** Status poorly known. Locally common and able to occupy small forest fragments, but has disappeared from much of its former range due to deforestation. Hunted locally for meat by indigenous people. CITES Appendix III (Costa Rica; listed as EN), Red List LC.

OLINGOS *Bassaricyon* spp.

Northern Olingo HB 35–49cm; T 40–53cm; W 1–1.6kg. Panamanian or Western Lowland Olingo HB 29–37cm; T 37–42cm; W 1–1.2kg. Eastern Lowland Olingo HB 30–49cm; T 35–53cm; W 0.9–1.6kg. Andean Olingo HB 32–40cm; T33–43cm; W 0.75–1.1kg
Olingo taxonomy is disputed and the number of species is unclear. The most recent, ongoing revision of the genus indicates four species: Northern Olingo *B. gabbii*, Panamanian Olingo *B. medius*, Eastern Lowland Olingo *B. alleni*, and Andean Olingo *B. n.* sp. (Linnaean name still to be assigned). Olingos look similar, typically yellowish-brown to tan-brown with creamy or golden underparts; the head is often dark, either greyish or dark brown. Andean Olingo is the most distinctive, with shaggy fur that varies from black-tipped tan to rich russet-brown. Olingos resemble the closely related Kinkajou and are easily confused, especially as they sometimes feed in the same fruiting tree. They are smaller and have long tubular tails that lack Kinkajou's tapering prehensile tip. **Distribution and Habitat** Northern Olingo, Guatemala to C Panama; Panamanian Olingo, E Panama to W S America, west of the Andes; Eastern Lowland Olingo, S America, E of the Andes from Venezuela to Bolivia; Andean Olingo, endemic to the Colombian and Ecuadorean Andes. All olingos inhabit moist tropical forest; Andean Olingo is restricted to highland cloud forest. Olingos' ability to tolerate human-modified habitats is unclear; Northern Olingos apparently occur in secondary forest and plantations. **Feeding Ecology** Olingos eat mainly fruits, as well as nectar and flowers. Animal prey is rare, but Northern Olingos have been observed killing deer mice, and in one case a Variegated Squirrel; they may sometimes snatch hummingbirds from hummingbird feeders (which they also plunder for sugar-water). Foraging mostly nocturnal, exclusively arboreal and usually solitary, but olingos congregate in fruiting trees; observations of Eastern Lowland Olingo groups suggest communal foraging by small social groups, possibly family members. **Social and Spatial Behaviour** Olingos are usually observed alone and thought to be mainly solitary, but groups of 2–6 Eastern Lowland Olingos occur in Amazonia, Brazil. Only one range estimate exists, for a male Panamanian Olingo, 0.24km². Olingos are thought to occur in naturally low densities, though 20.4 Eastern Lowland Olingos/km² was estimated in northern Brazilian Amazonia; Panamanian Olingo is much less common than Kinkajou in Panamanian surveys. **Reproduction and Demography** Poorly known. Gestation in captive Northern Olingos 72–75 days, and 1 young is born. Wild births appear to peak in the dry season (Costa Rica), but births are recorded year-round in captive Northern Olingos. MORTALITY Unknown. LIFESPAN 25.3 years (a captive Northern Olingo), doubtless much less in the wild. **Status and Threats** Status of all species is poorly known. However, they are dependent on forest, occur in low densities and appear less tolerant of human disturbance than most procyonids. All have probably declined due to deforestation and habitat fragmentation, which are the chief threats. Indigenous people hunt olingos for meat, and they are sometimes captured for the pet trade. Current CITES and Red List classifications are obsolete given the taxonomic uncertainty.

RINGTAIL

CACOMISTLE

PANAMANIAN
OLINGO

ANDEAN
OLINGO

EASTERN
LOWLAND
OLINGO

NORTHERN
OLINGO

HB 40.5–76cm; T 37–57cm; W 1.4–4.6kg

Velvety, dense, uniformly coloured fur, ranging from honey-brown to dark wood-brown with lighter golden underparts. Rounded head has large eyes, small forwards-facing ears and a short muzzle. Together with Binturong, the only carnivore with a prehensile tail. Related to and resembles olingos, but is 2–3 times larger and has a tapering slender tail compared to olingos' bushy non-tapering tail. **Distribution and Habitat** C Mexico to C Bolivia and SE Brazil. Inhabits a variety of closed-canopy tropical forests, including cloud forest, rainforest and dry forest from sea level to 2500m. Occurs close to humans, including in forested urban parks and plantations with natural canopy. **Feeding Ecology** One of the most frugivorous mammal species, with ripe fruits comprising 90–99% of the diet. Balance is made up of nectar, flowers, and young leaves and buds. Eats fruits from over 100 tree species, most importantly figs which produce fruits year-round. Seeds are consumed, but most are passed intact. Substantial amounts of ants and insects are sometimes eaten, probably incidentally while eating fruits and nectar. Captives consume cereals, bread, meat, eggs and milk, though these are not recorded in wild individuals. Sometimes raids orchards; does not kill poultry. Foraging almost exclusively nocturnal and entirely arboreal; rarely comes to ground except when compelled. Forages alone, but members of the same social group congregate in fruiting trees. Can hang from its prehensile tail to use both forepaws for handling fruits, and has an extremely long tongue thought to aid it in consuming nectar from deep flowers, e.g. of Balsa. Does not scavenge except to visit hummingbird feeders and feeding stations at tourist lodges. **Social and Spatial Behaviour** Sociality is unusual. The only intensively studied population (Soberanía NP, Panama) lives in small social groups comprising an adult female, her juvenile offspring, a single subadult offspring and two adult males. Group members share a collective range, which is defended against other groups, but individuals mostly travel and forage alone. They congregate in diurnal dens where group members groom, sleep and play together. Some adult females live solitarily in small ranges, and are visited by group males for breeding. Groups occur elsewhere in the species' range, though it is unclear if sociality follows the same pattern. Individual range size 0.11–0.5km². Density varies widely depending on fruit abundance, estimated at 12.5–30 Kinkajous/km². **Reproduction and Demography** Aseasonal. Gestation 98–120 days. Litter size 1, rarely 2. Males do not assist in raising juveniles. Females apparently disperse (as subadults), while males remain in the group, possibly explaining the species' patrilineal group structure. MORTALITY Few predators given its nocturnal arboreal behaviour, but occasionally killed by Harpy Eagle, Black-and-chestnut Eagle, Ocelot and Jaguar. LIFESPAN 28.7 years in captivity (and one exceptional case of 40.5 years). **Status and Threats** Widespread and common in suitable habitats. Vulnerable to forest loss due to strict arboreality, and disappears wherever habitat is converted to pasture and agriculture. Hunted in some areas for meat and captured as pets, though harvest is unlikely to constitute a significant threat. Red List LC.

RED PANDA *Ailurus fulgens*

Lesser Panda

HB 51–73cm; T 28–53.3cm; W 3–6kg

Small robust species with a pale to dark chestnut-brown coat, and black legs and underparts. Thick bushy tail has alternating chestnut and buff rings. Individuals east of the Nujiang River, China, tend to be darker, with richer red colouration; western populations are paler, especially on the face and head, though there is high variability within all populations. Has formerly been classified with Giant Panda, procyonids and mustelids, but is currently treated as the only species in a unique family – the Ailuridae – related to all of these. **Distribution and Habitat** S China (most of the range), extreme N Myanmar, Bhutan, N Nepal and extreme NE India, including a disjunct low-altitude population on the Meghalaya Plateau, India. Records from N Laos are equivocal. Lives mainly in dense temperate forests with a bamboo-thicket understory at 1550–4800m. Meghalaya population lives in tropical forest at 700–1400m. **Feeding Ecology** Eats almost exclusively bamboo, primarily leaves as well as new bamboo shoots in spring, which collectively comprise up to 95% of the diet. Fruits including berries, and fungi are eaten in late summer–autumn. Captives are recorded eating small mammals, birds, insects, flowers and bark (also reported for wild animals, though authenticated observations are equivocal). Bamboo is handled with a 'false thumb', actually a highly modified sesamoid bone with its own pad (a feature shared with Giant Panda). Foraging cathemeral, with long feeding bouts interspersed with regular rest periods. Forages alone and mostly above ground, using fallen timber, low shrubs and tree stumps to reach bamboo. **Social and Spatial Behaviour** Solitary. Groups numbering up to five are reported, but there is no evidence of complicated sociality; groups probably comprise females with large cubs. Adults occupy relatively small and stable ranges that often overlap extensively. They demarcate ranges with regular scent marking and latrines. Range size 0.94–9.6km². **Reproduction and Demography** Seasonal. Mating January–mid-March; births June–July. Gestation 114–145 days (captivity). Litter size 1–4 (exceptionally 5 in captivity), averaging 2. MORTALITY In one study, 67% of cubs died before 6 months and 30% of adults died annually. Most deaths are human related; Leopard is a known predator. LIFESPAN 14 years in captivity. **Status and Threats** Restricted to suitable bamboo habitat, which is under severe pressure from forestry, cultivation and grazing. Thought to be declining in much of its range, especially in China, where a crude estimate of population decline is 40% since 1960. Poached for meat and fur in some parts of its range. CITES Appendix I, Red List VU.

KINKAJOU

RED PANDA

Western
form

Eastern form

Patagonian Skunk

HB 20–32cm; T 16.5–20.2cm; W 0.5–2.5kg

Smallest hog-nosed skunk, similarly marked to other hog-nosed skunks with two parallel white stripes, but distinct in often having dark brown, rufous-brown or creamy-brown fur on the back and base of the tail, extending variably down the flanks. Fur long and shaggy, especially in winter. **Distribution and Habitat** Endemic to Patagonia, i.e. S Chile and S Argentina. Has wide habitat tolerance, occupying rocky desert to forest to 700m, but apparently prefers open grassland and pastures for foraging. Inhabits ranchland and other moderately modified habitats, including in and around human dwellings. **Feeding Ecology** Diet similar to that of other hog-nosed skunks, comprising mainly invertebrates and small rodents: insects peak in the diet during summer, while rodents and carrion are more important in autumn–winter. Eats fruits and vegetables, sometimes raiding vegetable gardens and refuse dumps near settlements. Primarily nocturno-crepuscular, but becomes more diurnal during winter. **Social and Spatial Behaviour** Poorly known. Solitary. Home ranges overlap extensively, and vary from approximately 0.07–0.11km^2 (independent juveniles) to 0.16km^2 (adult female; adult male ranges unknown). Density estimates unknown, but apparently does not attain high numbers anywhere. **Reproduction and Demography** Poorly known; given the extreme winters of its range, likely to be seasonal, substantiated by anecdotal reports of juveniles, which occur mostly in spring. Gestation estimated at 42–60 days. Litter size 2–5. MORTALITY Poorly known. Puma is a confirmed predator. LIFESPAN Unknown. **Status and Threats** Status poorly known. Heavy fur harvests (mainly for blankets) led to its inclusion on CITES in 1983, banning exports from Chile and Argentina. No longer harvested commercially, but localized hunting and capture for the pet trade occurs. Due to its predilection for scavenging, it is often killed in poisoning campaigns targeting Culpeo in livestock areas. CITES Appendix II, Red List LC.

MOLINA'S HOG-NOSED SKUNK *Conepatus chinga*

Andean Hog-nosed Skunk

HB 30–49cm; T 13–29cm; W 1–3kg

Very similar to Striped Hog-nosed Skunk, and there is dispute over whether they comprise separate species. Molina's Hog-nosed Skunk is usually described as slightly smaller, typically with less white, though there is considerable overlap in all physical characteristics and the two species are often indistinguishable in appearance. **Distribution and Habitat** S Peru, C Chile, Bolivia, Paraguay, Uruguay, Argentina and SE Brazil. Inhabits a wide variety of grassland, steppe, bush, scrub and forested habitat from sea level to the Bolivian altiplano (>4000m). Apparently avoids moist dense rainforest, and tolerates secondary forest, agricultural areas and degraded livestock pastures with cover. **Feeding Ecology** Omnivorous, feeding mainly on invertebrates (especially ground beetles and their larvae), rodents, and small reptiles and their eggs. Scavenges from carrion, including the carcasses of hares, Guanacos and livestock. Nocturno-crepuscular. **Social and Spatial Behaviour** Poorly known. Solitary. Home ranges are considerably larger than those of other hog-nosed skunks, and male ranges are larger than female ranges. Range size (females and males combined) averages 1.09km^2 in protected pampas (Argentina), while a single individual of each sex on an Argentinean ranch covered 1.95km^2. Only one density estimate exists, of 5 skunks/km^2. **Reproduction and Demography** Poorly known, but thought to resemble other hog-nosed skunks in breeding seasonally, presumably with similar gestation length and litter size. MORTALITY Poorly known. Sometimes killed by domestic dogs on ranches; known natural predators include Puma and Grey Eagle-Buzzard. LIFESPAN 6.6 years in captivity. **Status and Threats** Status poorly known, but thought to be reasonably secure. Heavily harvested for fur in Argentina until the early 1980s, but there is limited commercial hunting presently. Has disappeared from habitat severely degraded by overgrazing and soil erosion from livestock and feral exotic ungulates, but is able to occupy well-managed ranchland. Vulnerable to indiscriminate poisoning for predators on farmland. Red List LC.

STRIPED HOG-NOSED SKUNK *Conepatus semistriatus*

HB 33–50cm; T 13.5–31cm; W 1.4–3.5kg

Largest of the South American hog-nosed skunks. Two parallel white stripes of very variable length and width run alongside the spine from the crown towards the base of the tail. Distal end of the tail is white, sometimes extending the length of the tail except for black fur at the tail's base. **Distribution and Habitat** S Mexico to Venezuela, Ecuador, Colombia and N Peru; separated by the Amazon Basin from disjunct populations in E and C Brazil. Inhabits deciduous forest, woodland savannah, shrubland and grassland with shrubby cover; appears absent from lowland rainforest. Tolerates disturbed habitats, including secondary forest, plantations, and pastures and clearings at forest edges. **Feeding Ecology** Omnivorous and thought to eat mainly invertebrates grubbed from the soil surface, supplemented with small mammals, birds, reptiles, amphibians and ripe fallen fruits. Chiefly nocturnal. **Social and Spatial Behaviour** Poorly known. Solitary. Only one range size estimate exists, from a radio-collared female in Venezuela that covered 0.18km^2 in the wet season, expanding to 0.53km^2 in the dry season. Thought to reach high densities, calculated at 6–12 skunks/km^2, though figures have not been validated by modern survey techniques. **Reproduction and Demography** Thought to be seasonal, though few data exist. Gestation approximately 60 days. Litter size 2–5. MORTALITY Poorly known. One of the most hunted bushmeat species by local people in NE Brazil. Known natural predators include Jaguar and Puma, and there is one record of a Crane Hawk killing a juvenile. LIFESPAN 6.5 years in captivity. **Status and Threats** Status poorly known, but it is thought to be reasonably secure. Though it is frequently hunted in some areas, it is widespread and tolerant of some habitat conversion. Red List LC.

HUMBOLDT'S
HOG-NOSED SKUNK

MOLINA'S
HOG-NOSED SKUNK

STRIPED
HOG-NOSED
SKUNK

HB 17–40cm; T 15–47cm; W ♀ 0.6–3.6kg, ♂ 0.7–5.5kg Domestic cat-sized skunk with highly variable white markings, usually variations of a V-shaped cape or double stripe along the back. Some individuals are almost entirely black; brown, reddish and albino phases occur. The most common skunk in N America, and the species typically encountered in urban areas. **Distribution and Habitat** S Canada, throughout the USA and N Mexico. Occurs in almost every habitat in its range, to 4200m, provided there is shelter and food. Thrives in disturbed habitats, and has probably benefited from forest conversion to pasture and fields. Common in urban areas. **Feeding Ecology** Omnivorous and highly opportunistic, feeding on all kinds of invertebrates, small mammals, birds, reptiles, amphibians, eggs, fruits, vegetables, grains and nuts. Eats carrion and readily scavenges from dumps, pet bowls and birdfeeders. Sometimes raids domestic beehives and (rarely) poultry coops. Not a true hibernator as it feeds throughout the winter, but it over-winters in dens in its northern range; in severe winters, subsists entirely on fat stores for as long as 3 months. Mainly nocturno-crepuscular. **Social and Spatial Behaviour** Solitary, but adults (especially females) sometimes den together. Ranges fluctuate geographically and seasonally, depending on food availability; estimates at 0.5–12km². Urban skunks typically have very small ranges, while the largest ranges are from the Canadian prairies. Density estimates vary extensively, depending on food availability and associated disease outbreaks, typically 1.8–4.5 skunks/km², but varying from 0.7 to 38/km², with the highest numbers recorded from urban parkland. **Reproduction and Demography** Seasonal. Mates February–early March; births April–June. Gestation 59–77 days; longer gestations probably involve a short period of delayed implantation in females that mate early. Litters average 5–9, exceptionally to 18. Independence and dispersal by late August–September. MORTALITY Populations naturally have high turnover rates, with frequent crashes (and rapid recovery) from disease cycles associated with starvation in harsh winters. Natural predators include large raptors (one Great-horned Owl nest contained the remains of 57 skunks), as well as cats and canids. LIFESPAN Averages 2–3.5 years in the wild, to 12 in captivity. **Status and Threats** Widespread, adaptable and common, and considered very secure in most of its range. Susceptible to rabies, though populations recover quickly. Red List LC.

AMERICAN HOG-NOSED SKUNK *Conepatus leuconotus*

Common Hog-nosed Skunk, White-backed Hog-nosed Skunk
HB 34–51cm; T 12–41cm; W 2–4.5kg
Largest of the hog-nosed skunks, with a white cape rather than two white dorsal stripes as in other species. Tail usually completely white, occasionally with a black underside at the base. Formerly classified as two species, Eastern (*C. leuconotus)* and Western (*C. mesoleucus)* Hog-nosed Skunks, but recent molecular data indicate that they comprise a single species. **Distribution and Habitat** SW USA, Mexico (absent from the Yucatan peninsula and Belize), and throughout C America to N Nicaragua; records into N Costa Rica are equivocal. Occurs in desert scrub, rocky areas, grassland, cacti/thorn brush, marshland, woodland and various forest types to 3000m. Apparently absent from evergreen forest and true desert. Inhabits pastures and ranchland, but is less tolerant of open agriculture and urban landscapes. **Feeding Ecology** Invertebrates, especially soil-living insects, grubs and larvae, comprise 50–90% of the diet. Also eats small mammals, reptiles, amphibians, carrion and a wide variety of fruits and vegetables. Drinks when water is available, but appears independent of water in dry habitats, e.g. Texas. Largely nocturnal. **Social and Spatial Behaviour** Solitary; surprisingly poorly known with no available radio-telemetry data. Spatial patterns assumed to be similar to those of other hog-nosed skunks; no range estimates. Does not reach high densities, estimated at 0.2–1.3 skunks/km². **Reproduction and Demography** Seasonal in the USA (probably true generally). Mates February–early March; births April–May. Gestation 60–70 days. Litter size 1–5. Independence and dispersal by late August at 12–16 weeks. MORTALITY Poorly known; observed mortality is largely anthropogenic, especially roadkill. LIFESPAN 16 years in captivity. **Status and Threats** Despite being considered a low conservation priority, it has declined sharply in the USA, thought to be from habitat loss to agriculture, pesticide use, roadkill, indiscriminate predator control and competition with feral pigs. Status poorly known in its Meso-American range, though subject to similar threats, especially in Mexico. Red List LC.

HOODED SKUNK *Mephitis macroura*

HB 27.8–36cm; T 27.5–43.5cm; W ♀ 0.7–1.2kg, ♂ 0.8–2.7kg
Small to medium-sized skunk with variable white markings, typically one or two narrow stripes along the flanks, sometimes with a white cape. Cape and especially the tail often appear greyish or silvery due to inter-mixing of black hairs. Smaller than Striped Skunk, with a more slender face and snout, and a proportionally longer, luxuriant tail. Individuals increase in size from south to north. **Distribution and Habitat** SW USA through Meso-America to NW Costa Rica. Occupies a wide variety of habitats to 3110m, including grassland, shrubland, arid lowland, dry and deciduous forests, rocky areas and riverine habitats. Occurs in agricultural areas and close to human settlements. **Feeding Ecology** Feeds mainly on invertebrates grubbed from the soil and leaf litter, as well as rodents, shrews, fruits, e.g. prickly pear, and eggs. Chiefly nocturnal. **Social and Spatial Behaviour** Solitary, but ranges overlap extensively, and adults congregate amicably at food patches, e.g. artificial feeding stations. Ranges in Mexico are 2.8–5km². Density estimates typically at 1.3–4 skunks/km²; densities to 25/km² in Mexico reported. **Reproduction and Demography** Thought to be seasonal; assumed to mate February–March, and give birth mid-April–June. Gestation approximately 60 days. Litter size 3–8 (typically 3–5). MORTALITY Annual adult mortality in Arizona 44–76%, mostly due to predation by Coyote, Bobcat and raptors. LIFESPAN 8 years in captivity. **Status and Threats** Secretive and inconspicuous, but probably more common than widely assumed. Can survive in human-modified habitats, and abundant in parts of its range, e.g. much of Mexico. Thought to have declined in Texas, but otherwise considered secure. Red List LC.

STRIPED SKUNK

AMERICAN
HOG-NOSED
SKUNK

HOODED SKUNK

Central American Spotted Skunk

HB 20–25cm; T 10–14.5cm; W 0.24–0.54kg

Formerly classified with Eastern Spotted Skunk, but recent molecular data indicate that they are separate species, though this is still disputed by some authorities and they are virtually indistinguishable by appearance. Southern species is most easily differentiated by range as the only spotted skunk in Central America. Overlaps the very similar Western Spotted Skunk in N Mexico, and much smaller Pygmy Spotted Skunk on Mexico's Pacific coast. **Distribution and Habitat** N Mexico to C Costa Rica. Inhabits bushland, grassland, scrub, thickets and various forest types to 2800m. Occurs in agricultural and settled areas with cover. **Feeding Ecology** Diet dominated by invertebrates, with around 50% of prey made up of insects and larvae from leaf litter and the soil surface. Balance of the diet is made up of small rodents, lizards, amphibians, birds' eggs and plant matter, including fruits and grains. Foraging almost entirely nocturnal. **Social and Spatial Behaviour** Poorly known, but assumed to be similar to that of other spotted skunks. Sightings are largely of single animals. Recorded denning in groups, the composition of which is unclear but most likely comprises mothers with large young. **Reproduction and Demography** Poorly known. Litter size 2–9. MORTALITY and LIFESPAN Unknown. **Status and Threats** Status has not been assessed, but it is widespread and relatively common in much of its range. Tolerant of some human activity and able to occupy farmland and settlements. Some populations test positively for exotic carnivore diseases, probably transmitted by domestic dogs and cats, though there is no evidence of population-level effects. Red List LC.

EASTERN SPOTTED SKUNK *Spilogale putorius*

HB 23–33cm; T 7–28cm; W ♀ 0.2–0.48kg,
♂ 0.28–0.9kg

The spotted skunks are very similar in appearance, with some controversy over their classification. The Eastern species generally has less white than the Western species, with discontinuous white stripes broken into discrete blotches, and less white to the tail-tip. The two species are separated by range except along the eastern Rocky Mountains though C Texas. When threatened, both employ a distinctive handstand display before spraying. **Distribution and Habitat** E USA to the Canadian border with Minnesota, and extreme NE Mexico. Inhabits forest, woodland, rocky habitat, brushland, vegetated dunes, and scrubby strips along canals and fences. Avoids very open habitats and wetland. Readily adapts to farmland, using outbuildings, haystacks and woodpiles for shelter. **Feeding Ecology** Feeds mainly on insects and rodents, the importance of which shifts seasonally; rodents become very important in winter when insect availability is low. Also eats birds, eggs, fruits, vegetables and carrion. Occasionally raids domestic poultry and grain stores. Does not hibernate, but winter activity diminishes in colder parts of its range. Almost entirely nocturnal. **Social and Spatial Behaviour** Solitary, but sometimes dens in small groups during winter. Occupies loosely defined home ranges that are not defended as territories. Male home ranges are 2.5–6.4 times larger than female ranges, and expand significantly during the spring breeding season. Range estimates include 0.3–1.9km² (♀ s, all seasons), 0.2–4km² (♂ s, non-breeding seasons) and 2.2–18.2km² (♂ s, spring, breeding). Density estimates range from 9 skunks/km² (Iowa farmland) to 40/km² (protected coastal habitat, Florida). **Reproduction and Demography** Seasonal. Mates March–April; births May–early June. In productive years, sometimes breeds again July–September, producing a second litter. Gestation 50–65 days, with 2 weeks of delayed implantation. Litter size 2–9, averaging 5–6. Females first breed at 10 months. MORTALITY Humans and domestic carnivores are the main causes in studied populations. Natural predators include Coyote, Red Fox, Bobcat and large owls. LIFESPAN 9.8 years in captivity. **Status and Threats** Despite reaching high abundances locally, the species has declined significantly in much of its range, particularly in the Midwestern USA. Causes are unclear, but may be due to a combination of fur over-harvests and intensification of farming, including widespread use of agro-pesticides. Red List LC.

WESTERN SPOTTED SKUNK *Spilogale gracilis*

HB ♀ 24–37cm; T 8.5–20.3cm; W ♀ 0.27–0.57kg,
♂ 0.26–1kg

Very similar to Eastern species in appearance, behaviour and ecology. Generally has broader, more continuous white stripes, with a larger white spot between the eyes and more white to the tail-tip. **Distribution and Habitat** W USA, N Mexico and extreme SW Canada. Inhabits a variety of broken or vegetated habitats, including forest, woodland, thickets, vegetated watercourses, rocky outcrops, dry valleys, cliffs and lava fields. Occurs on farmland and in other modified habitats in association with natural or anthropogenic cover including buildings. **Feeding Ecology** Diet similar to Eastern Spotted Skunk's. Insects and small mammals are the main prey items, supplemented with birds, small lizards, amphibians, eggs, fruits, vegetables and carrion. Does not hibernate, but is sometimes inactive and den-bound for weeks in areas with cold winters, e.g. British Columbia. Largely nocturnal. **Social and Spatial Behaviour** Solitary. Less studied than Eastern species; most information comes from island populations off the California coast, where it does not defend territories and ranges are small; seasonal ranges of males (average: 0.47km²) are larger than those of females (0.31km²). Density reaches 9–19 skunks/km² on Californian islands. **Reproduction and Demography** Seasonal. Mates September–October; births April–May. Gestation 210–230 days, with a prolonged period of delayed implantation, in contrast to Eastern species; this difference reproductively divides Western from Eastern. Litter size 2–6, averaging 3–4. MORTALITY Predators include domestic carnivores, Coyote, Red Fox, Bobcat and large raptors. LIFESPAN >10 years in captivity. **Status and Threats** Widely distributed, but occurs patchily across its range, and its status is poorly known. Often killed during predator-control programmes, by fur trappers and on roads. Island Spotted Skunk subspecies is considered to be 'of special concern' in California, but it has recently undergone significant increases, possibly due to declines in Island Foxes, combined with habitat recovery after the removal of feral livestock. Red List LC.

SOUTHERN
SPOTTED
SKUNK

EASTERN
SPOTTED
SKUNK

WESTERN
SPOTTED
SKUNK

Pygmy Skunk
HB 19–21cm; T 5–8.7cm; W ♀ 0.13–0.17kg,
♂ 0.15–0.23kg
Smallest skunk. Resembles other spotted skunks, but is considerably smaller, with creamy rather than crisp white stripes. Distinct from other spotted skunks, a continuous cream-white stripe runs across the forehead down the sides of the body, and stripes extend down the legs to the feet. **Distribution and Habitat** Endemic to Mexico's Pacific coast from Sinaloa to Oaxaca. Inhabits various coastal habitats including dry forest, desert scrub and vegetated dunes, usually below 350m but occasionally to 1630m. Occurs in agricultural areas with cover. **Feeding Ecology** Feeds mainly on invertebrates, especially insect larvae and adult beetles, ants, termites, millipedes, centipedes, spiders and scorpions. Eats vertebrates less often than other spotted skunks, but rodents (especially Spiny Pocket Mice) become important in the dry season, when insect abundance declines. Occasional food includes small lizards, birds, crustaceans, snails and vegetation, including fruits and seeds. Mainly nocturnal. **Social and Spatial Behaviour** Solitary. Occupies defined territories with male ranges overlapping one or more smaller female ranges. Exclusive core areas are maintained and males aggressively exclude other males, especially during the breeding season. Range size (averaged for both sexes) 0.2km². **Reproduction and Demography** Seasonal. Mates April–early August; births July–September. Gestation 43–51 days, possibly with a short period of delayed implantation. Litter size 1–6. MORTALITY Poorly known; Barn Owl and snakes are known predators. LIFESPAN Unknown. **Status and Threats** Locally abundant in some areas, but has a very restricted range that is under intense pressure from tourist resorts, towns and roads. Associated with this development, it is sometimes killed and stuffed as tourist souvenirs, and killed by domestic cats and dogs. Red List VU.

PALAWAN STINK BADGER *Mydaus marchei*

Pantot, Philippine Stink Badger
HB 32–49cm; T 1.5–4.5cm; W 0.85–2.5kg
Smaller stink badger species, with dark chocolate-brown fur and a very short, almost absent tail. White dorsal stripe rarely extends beyond the crown and shoulders, and is sometimes very faint or absent altogether. Well-developed anal scent glands, but secretion is apparently less noxious than Sunda Stink Badger's, described as pungent but inoffensive, smelling faintly of almonds and ants. Nevertheless, the secretion is defensive and stink badgers are left alone by domestic dogs and cats. **Distribution and Habitat** Endemic to three Philippine islands, Palawan, Busuanga and Calauit. Occurs in a wide range of lowland habitats, including primary and secondary forests, swamp forest, mangroves and shrub grassland. Tolerant of agricultural and settled areas, including rice paddies and cropland, provided cover is available. **Feeding Ecology** Grubs on the surface with its mobile snout, and uses its robust claws to excavate invertebrates such as worms, insects like mole crickets and beetles, and larvae. Apparently often forages near watercourses, where crabs and freshwater arthropods are eaten. Eats fallen ripened fruits, including mangoes and cheesewood fruits, possibly in part for the insects found feeding on them. Foraging mainly nocturnal. **Social and Spatial Behaviour** Poorly known, but most observations are of singletons, suggesting solitary behaviour. Adults deposit regular scent-marks on the soil surface while foraging, indicating that the anal glands are important for socio-spatial behaviour as well as for defence. **Reproduction and Demography** Poorly known. Litters thought to number 2–3, born in underground burrows. The only direct observations of wild young occur in November–March, though it is unclear if breeding is seasonal. MORTALITY and LIFESPAN Unknown. **Status and Threats** Extremely restricted range, but within it the species is widespread and appears to be common. Tolerant of agricultural areas, though probably requires some undisturbed habitat; disappears from areas of extensive rice cultivation that lack cover. Vulnerable to being killed on roads. Red List LC.

SUNDA STINK BADGER *Mydaus javanensis*

**Malayan Stink Badger, Teledu,
Indonesian Stink Badger**
HB 37.5–51cm; T 3.4–7.5cm; W 1.4–3.6kg
Larger of two stink badger species, with a longer snout and longer tail than Palawan Stink Badger. Fur dirty black with a dorsal white stripe that usually extends from a conspicuous crown patch to a white-tipped tail, but is reduced in some individuals. Possesses well-developed anal scent glands for defence, and can spray a pale green secretion up to 15m. Locals believe it can blind or asphyxiate dogs, and there are unconfirmed reports of people having been rendered unconscious by the secretion. **Distribution and Habitat** Java, Sumatra, Borneo and the Natuna Islands (Indonesia). Inhabits mainly forested habitat from sea level to 2100m. Also recorded in secondary forest, as well as fields and pastures adjacent to forested areas, suggesting some flexibility in habitat use. A prodigious digger, and dens in underground burrows that it excavates or which are dug by other species such as Sunda Porcupine, with which it sometimes shares burrows. **Feeding Ecology** Diet thought to be mainly worms and other invertebrates excavated from the soil surface, for which the large claws and elongated, probing snout are well adapted. Also recorded eating birds' eggs, some plant matter and carrion. Captive animals eat meat and eggs. Not recorded depredating poultry, but sometimes persecuted for digging up crops and plantation seedlings while foraging. Nocturnal. **Social and Spatial Behaviour** Unknown. Reported to occur in adult pairs or trios, though this remains to be confirmed. **Reproduction and Demography** Unknown. Litters thought to number 2–3, and are likely reared in underground burrows. **Status and Threats** Status poorly known. Likely to occur in many protected areas, and found sufficiently often in human-modified habitat to suggest it is not dependent on undisturbed forest. Does not appear to be widely hunted, though some ethnic groups consume the flesh and believe its parts have medicinal value. Red List LC.

PYGMY
SPOTTED SKUNK

PALAWAN
STINK BADGER

SUNDA
STINK BADGER

River Otter, Nearctic Otter, Canadian Otter

HB 58–73cm; T 31.7–47cm; W 3.4–15.5kg

The only otter in N America except for Sea Otter (coastal W USA and Canada). Uniform pale brown to dark brown, with a pale cream to greyish muzzle, throat and chest. **Distribution and Habitat** Most of Canada (excluding the NE), and the E and NW USA, including most of Alaska. Occurs in all types of non-polluted fresh and coastal water bodies with well-vegetated or rocky shorelines, including streams, rivers, ponds, lakes, estuaries, marshes, beaches and reservoirs. **Feeding Ecology** Eats mainly slow-moving, schooling and bottom-dwelling fish, especially carp, minnow, catfish, bowfin and suckers. Crustaceans, especially crayfish, are second most important in the diet. Fast-swimming fish such as salmon are taken infrequently except during spawning runs. Opportunistically eats frogs, reptiles, waterbirds and mammals, mainly Muskrats, Nutria (introduced) and rarely North American Beavers (probably juveniles). Occasionally catches small terrestrial mammals in deep snow, including Snowshoe Hares. Foraging mainly nocturno-crepuscular, with increased diurnalism where it is undisturbed and during winter; active through winter and forages under ice. Hunts alone or in social groups. Can swim to 11km/h and remain underwater for up to 4 minutes. Hunting success poorly known; solitary adult females catch prey on 38% (lake) to 62% (small inlet) of dives. Occasionally scavenges from carrion. **Social and Spatial Behaviour** Sociality complex and fluid. Adults are generally solitary, with the primary social unit comprising a mother and cubs, but sociality increases with food availability. Mother-cub families may include helpers, either yearlings from previous litters or unrelated young adults. Large clans of up to 18 (mainly males) form in coastal Alaska, probably linked to cooperative hunting of large pelagic fish; females sometimes join clans, but never while raising cubs. Members of social groups are gregarious and play together, e.g. 'snow-sliding'. Ranges vary from being largely exclusive to overlapping extensively. Average range sizes include 9.6km² (♀s) to 30.4km² (♂s) in SE Minnesota, and 70km² (♀s) to 231km² (♂s) in Alberta. **Reproduction and Demography** Seasonal. Mating December–April; births January–May (following year). Gestation 10–12 months, with delayed implantation (actual gestation 61–63 days). Litter size 1–6. Weaning at 12 weeks. Cubs can swim at 60 days. Sexual maturity at 12–18 months; first breeding rarely before 2–3 years. MORTALITY Most mortality is anthropogenic. The most important natural factor is starvation, mainly in the far north of the range. Adults are occasionally killed by predators ranging from Bald Eagle to Killer Whale. LIFESPAN 13 years in the wild, 25 in captivity. **Status and Threats** Extirpated from around 75% of its original range by the 1970s due to massive unregulated trapping and degradation of aquatic habitats. Rehabilitation of waterways and reintroduction programmes have restored it to much of its range; it is still extinct in about a third of its historic US distribution and in Mexico. Remains susceptible to pollutants and habitat loss. Legally trapped for fur where populations are considered secure, in 29 US states and 11 Canadian Provinces. CITES Appendix II, Red List LC.

SEA OTTER *Enhydra lutris*

HB 100–120cm; T 25–37cm; W ♀ 14.5–32.7kg, ♂ 21.8–45kg

Largest mustelid by weight. Rufous-brown to dark brown, typically with a cream or greyish head, the colour extending variably on the throat and chest. Young pups have woolly light brown fur. The only completely marine otter species; can live its entire life at sea. **Distribution and Habitat** Endemic to the N Pacific, in coastal NE Russia (Kamchatka Peninsula and associated islands), N Japan, British Columbia, Alaska, Washington and California. Inhabits coastal marine environments typically within 1km of shore. **Feeding Ecology** Diet overwhelmingly dominated by marine invertebrates such as sea urchins, clams, abalones, mussels and crabs. Squid and octopus are eaten especially during episodic abundances, e.g. California. Fish comprise only occasional prey, but are important to Aleutian Islands populations. Dives up to 100m (usually 25–40m) for an average of 60–120 seconds (up to 260 seconds), and finds prey mostly by sight with its strong underwater vision or by touch. Carries prey to the surface in a fold of skin in its armpit, and breaks open hard-shelled items by smashing them against a rock held on its belly (the rock is also transported in the armpit fold). Between 63% (subadults) and 82% (adults) of foraging dives are successful. Eats 20–33% of its body weight each day to compensate for heat loss to its cold environment. Able to drink sea water. **Social and Spatial Behaviour** Basically solitary, but with wide regional and temporal variation. Adults mostly occupy enduring ranges, and either space themselves out or overlap extensively depending on food availability. High-quality food patches such as squid flushes promote overlap, and otters sometimes aggregate in large 'rafts' of up to 2000 individuals to feed. During breeding and where females occupy small stable ranges, males establish exclusive territories and exclude other males. Rafts of non-breeding males form in some areas. Average range size 0.8–6.8km² (♀3) and 0.4–4.8km² (♂s) in California, and linear 38km (♀s) to 50km (♂s) in Washington. **Reproduction and Demography** Weakly seasonal; breeding occurs year-round, but births peak May–June (Washington and Alaska), and weakly January–March (California). Gestation 4–12 months (averaging 180–218 days), with delayed implantation. Litter size 1, exceptionally 2. Pups are born at sea and carried and suckled by the mother at rest; they can swim at 2 months. Weaning and independence coincide at 5–9 months. MORTALITY For pups varies from 17% (Kodiak Island, Alaska) to 53% (Amchitka Island, Alaska, with limited food). Adults generally have low mortality, e.g. 4–11% (Kodiak Is), mainly from starvation and predation. LIFESPAN 15 (♂s) to 20 (♀s) years in the wild, 30 in captivity. **Status and Threats** Previously very heavily exploited for its luxuriant fur, which reduced a total estimated population of 300,000 to <1000 by 1911. Strict protection and reintroductions have restored it to an estimated 95,000–107,000, though local declines continue to occur. The main threats are oil spills and other contaminants. Elevated Killer Whale predation in response to seal declines is implicated (though not proven) in declines in the Alaskan Aleutian Islands. Also killed by entanglement in fishing nets and by disease, including exotic disease introduced into waterways by humans. CITES Appendix I, Red List EN.

Snow-sliding

**NORTH AMERICAN
OTTER**

Group at sea

Nursing young

SEA OTTER

MARINE OTTER *Lontra felina*

Plate 69

Chungungo
HB 53–79cm; T 30–36.2cm; W 3.2–5.8kg
Smallest South American otter, uniformly glossy dark brown with a slightly paler muzzle, cheeks and underparts. Broad head has a short muzzle with heavy long whiskers (for which it is sometimes called the 'sea cat'), and the tail is relatively short. **Distribution and Habitat** Restricted to the Pacific coast of S America, from C Peru to Tierra del Fuego and Isla de los Estados, Argentina. Found along rugged rocky coasts rarely more than 100m offshore and 50m inland; occasionally travels up coastal freshwater inlets. Occurs in wharves and groynes near urban areas. **Feeding Ecology** Feeds primarily on marine crustaceans, fish and molluscs. Also takes freshwater shrimp in coastal rivers. Large prey, e.g. Southern King Crab, is taken ashore to eat, while small prey is consumed at sea; floats on its back to eat, but does not use rock anvils as do Sea Otters. Occasionally hunts on land, primarily chicks of colonially nesting seabirds; actively excavates the burrows of Peruvian Diving Petrel to reach chicks. Foraging cathemeral and mainly solitary; observed pairs and trios are mother-young groups. Adults congregate amicably at food patches such as fishery dumps. Sometimes scavenges from human fishing waste on shore. **Social and Spatial Behaviour** Solitary, with facultative territorial behaviour. Often found close together with entirely overlapping ranges, but otters do not move, forage or rest together. Adults, especially females, become more territorial and occupy exclusive core areas where food occurs in patches. Range size similar for the sexes, linearly 1.3–4.1km and <110m wide. **Reproduction and Demography** Poorly known. Pups recorded year-round, peaking September–November (C Chile), though seasonality may be more pronounced further south. Gestation 60–65 days. Litter size 1–4, averaging 2. MORTALITY Poorly known. Most documented deaths are anthropogenic; Killer Whale, sharks and large raptors (on pups) are predators. LIFESPAN Unknown. **Status and Threats** Reduced to <1000 adults in a series of discontinuous populations. Declining due to human over-fishing of prey, combined with illegal killing by fishermen as competitors and for fur. Also killed by entanglement in nets and by domestic dogs around docks and urban beaches. Almost extinct in Argentina. CITES Appendix I, Red List EN.

SOUTHERN RIVER OTTER *Lontra provocax*

Patagonian River Otter, Huillin
HB 57–80cm; T 35–43cm; W 8–14.5kg
Medium-sized otter, mid-brown to dark brown with a pale greyish muzzle, throat and upper chest, becoming beige on the underparts. **Distribution and Habitat** Patagonian Argentina and Chile. Occurs in freshwater lakes, rivers and wetland with dense bank vegetation, as well as marine rocky coasts, estuaries and marshes. Avoids disturbed waterways such as canals. **Feeding Ecology** Crustacean specialist, eating mainly large freshwater crabs and crayfish; small slow-moving fish are the second most important food category. Also eats molluscs and, less so, amphibians and birds. Foraging solitary and cathemeral. **Social and Spatial Behaviour** Solitary. Ranges are stable, with low overlap among males and greater overlap among females. Range size differs little between sexes, linearly 7.4–22.3km with very small core areas, usually <1.5km, centred around patches of thick vegetation for denning. Density averages 0.73 otters/km of marine coast (S Chile). **Reproduction and Demography** Thought to be seasonal, depending on location. Pups observed year-round in southern range, but mating occurs December–March, with births thought to occur July in C Chile. Gestation unknown; assumed to include delayed implantation. Litter size 1–4, averaging 2. MORTALITY and LIFESPAN Unknown. **Status and Threats** Now known only from seven discontinuous populations that occupy a fraction of its former range. Serious threats include widespread habitat degradation by water pollution, wetland draining, dredging, canalization and clearing of bank vegetation, as well as illegal hunting for fur, which is intense in some areas. CITES Appendix I, Red List EN.

NEOTROPICAL OTTER *Lontra longicaudis*

Neotropical River Otter
HB 46–66; T 37–84cm; W 5–12kg (exceptionally to 15kg)
Very similar to Southern River Otter, but generally larger and more heavily built with paler fur around the cheeks and muzzle. The two species are separated by range. **Distribution and Habitat** N Mexico to N Argentina. Occurs mostly in clear, flowing streams, rivers and lakes in a variety of forest, woodland and scrub habitats from sea level to 3000m. In coastal areas, occurs near fresh water such as river outlets and rain-fed lagoons. Avoids silted and sluggish water bodies. Occurs in suitable habitat near human settlements. **Feeding Ecology** Eats mainly fish, especially catfish, characins and cichlids, as well as crustaceans, and large aquatic insects and their larvae. Molluscs, amphibians, reptiles, birds and mammals (including rats and Nutria) are opportunistically taken. Fruit is recorded in the diet, but is possibly consumed while eating frugivorous fish. Small prey is eaten in the water, while large prey is carried to shore or tree snags to consume. Foraging solitary and mainly diurno-crepuscular; becomes nocturnal in areas of human disturbance. Scavenging is suggested by the presence of Capybara fur in scats. **Social and Spatial Behaviour** Poorly known. Solitary. Adults very actively scent-mark ('spraint') prominent features such as logs, root systems, sand bars and rocks, but ranging and territorial patterns are poorly studied. Density estimates include 0.8–2.8 otters/km of shore. **Reproduction and Demography** Semi-seasonal. Mating mainly February–May, but year-round in some areas. Gestation 56–86 days; it is unclear if a short period of delayed implantation accounts for the wide variation. Litter size 1–5, averaging 2–3. Pups start swimming at around 2.5 months. MORTALITY Predation known by anacondas, caimans, piranhas and Jaguar. LIFESPAN Unknown. **Status and Threats** Species has a wide range, but its status is poorly known. Heavily hunted for fur before 1970, leading to localized extinctions in much of its range; continued illegal hunting remains a threat. More importantly, habitat conversion to agriculture and ranching, combined with wetland drainage, damming and water pollution, cause declines. Sometimes killed intentionally and accidentally by fishermen. CITES Appendix I, Red List DD.

MARINE OTTER

SOUTHERN RIVER OTTER

NEOTROPICAL OTTER

HB 86.4–130cm; T 45–75cm; W ♀ 22–26kg,
♂ 26–34kg

The largest Latin American mustelid and the world's second largest (by weight, after Sea Otter). Strongly built with a very robust, powerful skull and a long, muscular paddle-like tail. Uniformly dark chocolate-brown with creamy-white markings on the chin and throat that vary from light stippling to a solid patch that extends onto the chest; markings are unique to individuals. **Distribution and Habitat** Endemic to S America, from Venezuela to E Paraguay and S Brazil; likely extinct in Argentina and Uruguay. Requires clear, slow-moving waterways or lakes in intact forest, woodland and wetland. Occasionally occupies agricultural canals and artificial reservoirs, but does not tolerate waterways in open and highly disturbed habitats. **Feeding Ecology** Primarily piscivorous; adults eat around 3kg of fish a day, mainly characins (piranha family), catfish, perch and cichlids. Most fish captured measure 0.1–0.4m, but Giant Otter is capable of taking catfish measuring over 1m. Other prey types infrequently recorded include reptiles (including anacondas to 3m, caimans to 1.5m and large turtles), small mammals, birds, crustaceans and molluscs. Hunts in family groups, but with little evidence of cooperation except in subduing large prey; group members provision cubs and large prey is sometimes shared, otherwise individuals mostly eat their own catches. Exclusively diurnal. Small prey is eaten in the water and large prey is taken to the bank or fallen trees to consume. Adults catch up to 3.2 fish/hr. **Social and Spatial Behaviour** Social and territorial. Lives in highly cohesive family groups comprising a monogamous adult pair and its offspring from one or more generations. Groups usually number 4–13; largest recorded groups, numbering to 20, are probably temporary associations of two or more families. Groups defend their territories from neighbours, largely by scent-marking and mutual avoidance, but clashes are aggressive and occasionally fatal. Group territories are linearly shaped and spaced along riverbanks or lake edges. Group territory size (linearly) 5.2–32km, averaging 9.3–11.4km. **Reproduction and Demography** Weakly seasonal. Only the alpha pair breeds in the group. Litters are born in dens in low sloping banks with dense cover. Births occur mainly in the dry season in areas with marked seasonality, e.g. August–October (Suriname) and April–September (Brazilian Pantanal). Gestation 52–70 days. Litter size 1–5, averaging 2–3. Cubs can swim at 6 weeks and accompany the group as it forages from 3–4 months; weaning at 8–9 months. Dispersal first occurs at 10 months, but usually at 2–3 years; males thought to disperse more often than females. MORTALITY Poorly known. Adults are sometimes killed in intraspecific fights. Black and Spectacled Caimans are occasional predators, largely on cubs. LIFESPAN 12.8 years (minimum estimate) in captivity. **Status and Threats** Formerly heavily hunted for fur, leading to extirpation or declines across much of its range. Hunting is rare today, but its narrow ecological needs and slow reproductive rates make it vulnerable to habitat modification, pollution of waterways by mining contaminants, e.g. mercury, agricultural run-off and over-fishing. Persecuted by fishermen as perceived competitors for fish, and caught accidentally in fishing nets. CITES Appendix I, Red List EN.

EURASIAN OTTER *Lutra lutra*

European Otter, Common Otter

HB ♀ 59–70cm, ♂ 60–90cm; T 35–47cm;
W ♀ 6–12kg, ♂ 6–17kg

Medium-sized otter, uniformly dull mid-brown to dark brown with pale cream to white underparts. Has been considered a separate species in Japan, the Japanese Otter *L. nippon* (now probably extinct), though the evidence is equivocal. **Distribution and Habitat** Eurasia, from W and N Europe to the Russian Far East and S China; scattered populations across C and SE Asia, with isolated populations in S India, Sri Lanka, Sumatra and N Africa. Inhabits a wide variety of aquatic habitats, always with bank vegetation, rocks or debris, including rivers, streams, lakes, wetland, swamp forest, mangroves and coastal areas from sea level to 4120m. Occupies rural and urban water bodies only if they are sufficiently intact with bank cover. **Feeding Ecology** Eats principally fish, freshwater amphibians and invertebrates, especially crabs, crayfish and aquatic insects. Occasional prey items consist of birds, including seabirds such as fulmars and guillemots in coastal populations, water voles, rats, rabbits (which are killed in their burrows) and reptiles. Plant matter sometimes appears in the diet, but is probably taken incidentally while consuming prey. Rarely takes carrion. Foraging largely nocturno-crepuscular and solitary, except for mothers with attendant cubs. Adults eat approximately 1–1.5kg of fish a day, representing around 12% (♂s) to 28% (♀s with cubs) of their body weight a day. **Social and Spatial Behaviour** Solitary and territorial. Males maintain large territories encompassing numerous female territories. Females are essentially solitary, but 2–3 adult females (probably related) may share and jointly defend a group territory, though each female maintains an exclusive core area and rarely interacts with the others. Territories are small and mostly exclusive in high densities, with size and overlap increasing as density declines. Average territory size (linearly) 7–18.7km (♀s), and 15–38.8km (♂s); individual males are recorded using linear areas up to 84km. Density estimates 1 otter/5km (riverbank) to 1/2–3km (lakeshore). **Reproduction and Demography** Aseasonal or seasonal, depending on seasonality of food availability. Mating in W–N Europe and Russia January–April; births peak April–May. Gestation 60–63 days. Litter size 1–4, exceptionally 5 (captivity). Cubs can swim at 2 months, are weaned at around 4 months and stay with their mother until 10–12 months. Sexual maturity at 18 (♂s) to 24 (♀s) months. MORTALITY Rates poorly known, but the most important factors are roadkills, starvation and intraspecific aggression. LIFESPAN 16 years in the wild (but rarely older than 4), 22 in captivity. **Status and Threats** Has undergone significant declines in much of its range, due to a combination of overfishing, habitat loss and hunting. Highly vulnerable to canalization of rivers, clearing of riverine vegetation, dam construction, draining of wetlands and aquaculture, and pollution associated with development. Roadkill is the main mortality factor in most of Europe. Hunted intensively, mainly for traditional medicinal use, in Asia. CITES Appendix I, Red List NT.

Family group

GIANT OTTER

EURASIAN OTTER

Oriental Small-clawed Otter

HB 36–47cm; T 22.5–27.5cm; W 2.4–3.8kg

Smallest otter, uniformly dark brown, sometimes with an ashy-grey tinge. Greyish-white lower face, throat and upper chest. Claws are vestigial on all the feet. Most closely related to African clawless otters. **Distribution and Habitat** S China through Indochina, Myanmar, Nepal to N India (with an isolated population in S India), the Philippines (Palawan), Borneo, Sumatra and Java. Inhabits freshwater bodies from sea level to 2000m, including rivers, mountain streams, lakes, peat swamps, mangroves and coastal wetland. Occupies anthropogenic habitats provided there is cover, e.g. rice paddies. **Feeding Ecology** Specializes in freshwater crabs, followed by small fish such as mudskippers, catfish and gouramis. Also eats snails, aquatic insects, shellfish, amphibians, snakes and small mammals. Captives leave shellfish in the sun until the heat opens them, but it is unknown if wild animal do likewise. Foraging social and mainly diurno-crepuscular, but nocturnal when disturbed. Hunts by sight and by searching in mud and bank debris with its very dexterous forepaws.

Social and Spatial Behaviour Intensely social, living in extended family groups of 2–15, thought to comprise a breeding pair and its offspring of one or more generations. Group members inhabit a stable range and are always together, moving, foraging and resting as one. Range sizes and densities unknown. **Reproduction and Demography** Monogamous, with breeding apparently restricted to the alpha pair, but reproduction is poorly known. Gestation 60–86 days. Litter size 2–7, averaging 4–5. Captive males help raise pups, and other group members are thought to act as helpers in the wild. MORTALITY Most documented deaths are anthropogenic. LIFESPAN 15 years in captivity. **Status and Threats** Threatened by habitat degradation, especially from damming, draining of wetlands and peat swamps, water pollution from agricultural run-off, and aquaculture. This is exacerbated by loss of prey from overfishing and pollution, combined with direct persecution; throughout the whole of Asia, all otters are killed for fur, meat and traditional medicinal beliefs, and as perceived pests by fishermen and shrimp farmers. CITES Appendix II, Red List VU.

HAIRY-NOSED OTTER *Lutra sumatrana*

HB 57.5–82.6cm; T 35–51cm; W 5–8kg

Medium-sized otter, very dark brown with white or yellowish upper lips, chin and throat patch. Very distinctive bulbous muzzle and broad nose covered entirely with fur. The rarest and least known of Asian otters. **Distribution and Habitat** Endemic to extreme SE Asia, with discontinuous populations known from S Myanmar, S Vietnam, S Cambodia, extreme S Thailand, S Malaysia, Borneo and Sumatra. Recorded mainly from peat swamps, flooded forest, wetland and mangroves, as well as rivers, lakes and mountain streams. **Feeding Ecology** Thought to eat mainly fish such as gouramis, climbing perch, walking catfish and snakeheads. Water snakes are frequent prey in Thailand. Also eats small numbers of crabs, insects, frogs, lizards, birds and small mammals. Camera-trap records show that it is cathemeral. Reportedly a very fast and agile

swimmer that weaves in and out of mangrove roots and bank debris to capture fish. **Social and Spatial Behaviour** Virtually unknown. Usually reported as solitary, but reliable records exist of small groups numbering to five, including an observation of two adults with a pup. **Reproduction and Demography** Poorly known. Limited records of pups cluster in November–February, and litters in the wild number 1–2 (based on very few observations). MORTALITY and LIFESPAN Unknown. **Status and Threats** Very rare, to the point where it was considered extinct until recent 'rediscoveries'. Now confirmed from a handful of sites, where it is seriously threatened by development and degradation of habitat, combined with widespread poaching for the illegal fur trade to China and, to a lesser degree, for meat and traditional medicinal beliefs. CITES Appendix II, Red List EN.

SMOOTH-COATED OTTER *Lutrogale perspicillata*

Indian Smooth-coated Otter, Smooth Otter

HB 59–75cm; T 37–43.2cm; W 7–11kg

Largest Asian otter. Glossy, short, very smooth fur typically chestnut-brown to dark brown. Creamy-white or yellowish throat patch runs from the upper lips and chin to between the front legs. **Distribution and Habitat** S Asia, from SE Pakistan through India, Indochina, S China, Borneo, Sumatra and Java; isolated population in SW Iraq. Inhabits lowland and plains water bodies, including rivers, lakes, wetland, peat swamps and mangroves. Found in rice paddies and artificial ponds. **Feeding Ecology** Eats mainly fish, often taking introduced pest species such as tilapia and European Carp. Also eats crustaceans (especially in mangroves and rice paddies), snails, insects, frogs, and occasional meals of birds, and rodents such as Ricefield Rat. Foraging cathemeral, with elevated diurnalism during winter. Forages in family groups, with some evidence of cooperative driving of fish into reeds, where they are caught. In parts of India and Bangladesh it is tamed and used by fishermen to catch fish and herd them into nets. **Social and Spatial Behaviour** Highly social, living in

groups of 2–11, thought to centre around a female and her offspring from multiple generations. Adult males are sometimes observed with groups, but it is unclear if they are permanent members. Range size estimated at 7–12km of river to each family group. Density estimates (linearly) include 1–1.3 otters/km of shore. **Reproduction and Demography** Seasonal. Mates August–September; births October–February (India and Nepal); some evidence that it breeds year-round under high food availability. Gestation 60–63 days. Litter size 1–5. MORTALITY Poorly known, but groups mount a formidable defence against predators; a credible record exists of a family group fatally injuring an Indian fisherman who had captured a cub in his net. LIFESPAN 20 years in captivity. **Status and Threats** Fairly widely distributed, but estimated to have declined by >30% over the past 30 years. Vulnerable to the same threats as other Asian otters; being essentially a lowland species, it is particularly vulnerable to the intense development pressure on wetlands throughout S Asia. CITES Appendix II, Red List VU.

Foraging in
shallows

**ASIAN SMALL-CLAWED
OTTER**

HAIRY-NOSED OTTER

Family
group

SMOOTH-COATED OTTER

HB 57.5–76cm; T 38.5–44cm; W ♀ 3.5–4.7kg,
♂ 4.5–6kg (exceptionally to 9kg)
Smallest African otter, chocolate-brown, sometimes with a reddish cast. Throat, chest and sometimes belly are marked with white to creamy-white blotches, unique to individuals. Chin and upper lip are white. **Distribution and Habitat** Sub-Saharan Africa, from Guinea to Ethiopia, south to S Africa. Inhabits unsilted and unpolluted freshwater habitats with dense bank vegetation, including lakes, streams, rivers and large swamps. Rarely found >10m from water. **Feeding Ecology** Eats mostly small fish (<20cm), especially cichlids, catfish, barbell, tilapia and introduced trout. Also eats crabs and frogs, especially in the relatively fish-poor waters of southern Africa; diet almost entirely fish in the richer great lakes of C and E Africa. Opportunistically eats aquatic insects, their larvae and occasionally waterbirds. Foraging solitary or in small social groups, and mostly diurnal with elevated nocturnalism during moonlight. Almost all fishing occurs within 10m from shore. Around 50% of dives produce a catch, averaging one capture every 66 seconds.

Scavenging on land not recorded, but takes dead and live fish from fishing nets. **Social and Spatial Behaviour** Sociality flexible. Adults are often solitary, and small family groups (up to five) usually comprise a female and kittens, but non-breeding adults (especially males) often associate in larger groups numbering up to 21. Individuals and groups appear to be non-territorial, with high overlap between ranges. Range size averages 5.8km² (♀s) and 16.2km² (♂s) in KwaZulu-Natal, S Africa. **Reproduction and Demography** Apparently seasonal (Lake Victoria); mating in July and births in September, though litters are recorded at other times elsewhere. Gestation approximately 60 days. Litter size 1–3. MORTALITY Nile Crocodile is considered the main predator, but there are no details. LIFESPAN Unknown. **Status and Threats** Widespread and numerous in good habitat, especially C and E Africa's great lakes. However, gradually declining in most of its range due to siltation and pollution. Also drowned in fishing nets, killed by fishermen and valued for traditional medicinal beliefs in many areas. CITES Appendix I, Red List LC.

CAPE CLAWLESS OTTER *Aonyx capensis*

African Clawless Otter
HB ♀ 73–73.6cm, ♂ 76.2–88cm; T 46.5–51.5cm;
W ♀ 10.6–16.3kg, ♂ 10–21kg
Large and powerful otter, uniformly mid-brown to dark brown with a white to pale grey muzzle, throat and upper chest. Feet are partially webbed; the hindpaws have small grooming claws, while the forepaws lack claws except for vestigial 'fingernails'. **Distribution and Habitat** Most of sub-Saharan Africa; absent from the Congo Basin and arid SW and NE Africa. Inhabits lakes, rivers, streams, ponds, estuaries, mangroves and marine habitats close to fresh water, which is required for drinking. Relatively tolerant of pollution and siltation, and occupies urban streams, canals and reservoirs. **Feeding Ecology** Crustacean specialist, eating mainly crabs, crayfish and lobsters. Also eats fish, octopus, frogs and molluscs. Cape populations increase fish consumption during the cold winter, when fish are sluggish and easily caught. Aquatic insects, birds and small mammals (mostly riparian rodents and shrews) are occasionally consumed. Sporadically kills domestic ducks and geese on farmland. Foraging mainly nocturno-crepuscular, with increased diurnalism where undisturbed, and solitary or in small social groups. Uses its very dexterous forepaws to find and capture prey, mostly at

shallow depths under 1.5m. Locates and pursues fish mainly by sight. **Social and Spatial Behaviour** Females are mainly solitary, while males are either solitary or live in small groups numbering up to five that occupy a shared range; group members generally travel alone, but regularly meet in a 'fission-fusion' pattern, sometimes sharing large prey. Adult female ranges have relatively little overlap, while males vary from little (especially for groups) to extensive overlap. Average territory size (linearly) 17km (♀s) and 42km (♂s) in fresh water, S Africa. Density estimates 1 otter/1.4–5km. **Reproduction and Demography** Thought to be aseasonal, with birth peaks in the rainy season. Gestation 60–63 days. Litter size 1–3. Kittens born with pale smoky-grey, woolly fur. They are weaned at around 45–60 days and remain with the mother for up to 12 months. MORTALITY Most known mortality is anthropogenic; Nile Crocodile and African Fish Eagle occasionally kill otters. LIFESPAN 13 years in captivity. **Status and Threats** Widespread and tolerant of some habitat modification, but threatened by severe degradation with associated pollution, siltation and eutrophication. Persecuted as a perceived problem on fish farms, and valued for traditional medicinal beliefs. CITES Appendix I (Cameroon, Nigeria), Appendix II elsewhere, Red List LC.

CONGO CLAWLESS OTTER *Aonyx congicus*

Swamp Otter
HB 79–97cm; T 41–56cm; W 14–25kg
Very similar to Cape Clawless Otter and sometimes regarded as the same species, but limited genetic and morphological data indicate they are distinct. They overlap at the edges of their respective distributions in Uganda and Rwanda (and possibly elsewhere). **Distribution and Habitat** Endemic to the Congo Basin; exact range limits unknown, given confusion with Cape Clawless Otter. Inhabits rivers, wetland and swamps in undisturbed rainforest. Not known from marine habitats, but may occur in coastal lagoons and mangroves. **Feeding Ecology** Diet includes earthworms, frogs, freshwater crabs, fish and aquatic insects. Earthworms are a key component of the diet based on observations from Mbeli Bai, Rep Congo, where otters grub in the mud for

earthworms with their dexterous forepaws, consuming up to three a minute. Diurno-crepuscular where undisturbed. **Social and Spatial Behaviour** Poorly known. Most sightings are of individuals or mothers with young. Small groups of up to four are reported. Forages close to Spotted-necked Otter without conflict (Mbeli Bai). **Reproduction and Demography** Poorly known. Probably breeds year-round given its equatorial range. Litter size 1–3. Kittens born with white fur that darkens to adult colouration by 2 months. MORTALITY and LIFESPAN Unknown. **Status and Threats** Status poorly known. Likely occurs in suitable habitat throughout the Congo Basin, which is well preserved over vast areas. Forest loss and over-hunting for bushmeat are primary threats, thought to prompt local declines near settlements. CITES Appendix I, Red List LC.

Foraging pair

**SPOTTED-NECKED
OTTER**

Male
coalition

**CAPE
CLAWLESS
OTTER**

CONGO CLAWLESS OTTER

Chinese Ferret-badger

HB 31.5–42cm; T 13–21.1cm; W 0.8–1.6kg

Ferret-badgers are very similar in appearance, with slight external differences between the described species; their taxonomy is in need of review. Small-toothed Ferret-badger co-occurs with Large-toothed Ferret-badger in much of its range; they are virtually indistinguishable except that the Small-toothed is slightly smaller, generally has a shorter white dorsal stripe and has markedly smaller premolars. **Distribution and Habitat** S and C China, Taiwan, N Myanmar, N India and N Indochina. Inhabits forest, woodland, scrub and dense grassland. Tolerates cultivated areas with cover and occurs near human settlements. **Feeding Ecology** Feeds chiefly on soil-living invertebrates, especially earthworms and insects, as well as fruits and seeds. Less important food includes small mammals, herptiles, carrion and eggs. Not known to prey on poultry. Foraging almost exclusively nocturnal. **Social and Spatial**

Behaviour Foraging solitary, but ranges overlap extensively and mixed-sex groups of up to four adults share setts, suggesting some sociality, including the possible maintenance of group ranges as in Eurasian Badger. Range size does not differ for sexes and averages 1.3km² (range 0.51–4.7km²). **Reproduction and Demography** Poorly known. Thought to be seasonal, at least weakly. Mating supposed to be in March (China), but births occur through May–December (Taiwan). Gestation 53–80 days (captivity). Litter size 1–4. MORTALITY Unknown; rabies, canine distemper and SARS (rarely) confirmed, but there is no evidence of population impacts. LIFESPAN 10.5 years in captivity. **Status and Threats** Status poorly known, but the species is widespread and considered common in much of its range. It is used for meat and traditional medicinal practices in Indochina, and heavily hunted in S China; despite this, it apparently remains relatively widespread. Red List LC.

LARGE-TOOTHED FERRET-BADGER *Melogale personata*

Burmese Ferret-badger

HB 33–43cm; T 14.5–23cm; W 1.5–3kg

Largest of the ferret-badgers, although the differences between them are slight. Dentition relatively massive and well developed. Usually described as having more white colouration than other ferret-badgers. White dorsal stripe typically extends to at least the mid-point of the spine, the distal half of the tail is white and there is extensive white 'frosting' on the body fur. Sometimes considered the same species as Bornean and Javan Ferret-badgers. **Distribution and Habitat** NE India, Myanmar, Thailand, Indochina and S Yunnan, China. Occurs in similar habitats (including anthropogenic habitats) to Small-toothed Ferret-badger; optimum habitat is unknown. **Feeding Ecology** Diet less well known than that of Small-toothed Ferret-badger, but assumed to be similar. Considerably more massive dentition

suggests it is more predatory on small vertebrates, but this remains to be confirmed by field studies. Foraging almost exclusively nocturnal. **Social and Spatial Behaviour** Unknown. Most records are of single adults or females with young; it is unknown if it is social or semi-social, as recently revealed for Small-toothed Ferret-badger. **Reproduction and Demography** Poorly known. Reputed to be seasonal, with captives giving birth mainly May–June. Litters in captivity 1–3, but may not be representative. MORTALITY and LIFESPAN Unknown. **Status and Threats** Status poorly known, but hunting pressure is very high in most of its range and it rarely appears in camera-trap surveys, e.g. it was not photographed in 8499 camera-trap days in 2003–2006 in Nam Et-Phou Louey National Protected Area, Laos. Occurs in a number of protected areas in Thailand. Red List DD.

BORNEAN FERRET-BADGER *Melogale everetti*

Kinabalu Ferret-badger, Everett's Ferret-badger

HB 33–45cm; T 14.5–17cm; W 1–2kg

Usually described as mainly brown (rather than greyish), with typical ferret-badger markings that are less extensive than in the mainland species. The only ferret-badger in its range. **Distribution and Habitat** Endemic to Borneo. Most definite records are from the Kinabalu massif, N Sabah, in montane broadleaved forest habitat at 900–3700m, but there is one certain record from lowland forest in E Sabah, and it possibly occurs more widely in Borneo. It is unknown if it tolerates agricultural habitats.

Feeding Ecology Poorly known. Reportedly eats soil-living invertebrates (especially earthworms), lizards, small birds, rodents and fruits. Nocturnal. One record exists of scavenging in roadside refuse dumps. **Social and Spatial Behaviour** Unknown. Most records are of single adults or females with young. **Reproduction and Demography** Unknown. **Status and Threats** Status very poorly known. Very restricted known range, much of which is undergoing conversion to agriculture, and remaining habitat is exposed to hunting. High potential that the species is threatened. Red List DD.

JAVAN FERRET-BADGER *Melogale orientalis*

HB 35–40cm; T 14.5–17cm; W 1–2kg

Physically indistinguishable from Bornean Ferret-badger and sometimes considered the same species (and both are sometimes treated as a subspecies of Large-toothed Ferret-badger). The only ferret-badger in its range. **Distribution and Habitat** Endemic to Java and Bali. Confirmed only from three isolated hill and montane areas, where it occurs in primary and secondary forests. It is unknown if it occupies the intervening lowland areas. Recorded from modified habitats, including rubber plantations near

human settlements. **Feeding Ecology** Few records exist, but the diet is assumed to be similar to that of other ferret-badgers. Anecdotally reported in association with tourist refuse in Gunung Gede NP, West Java, but it is unclear if it scavenges from dumps. Nocturnal. **Social and Spatial Behaviour** Unknown. **Reproduction and Demography** Unknown. **Status and Threats** Status very poorly known. Restricted known range that is exposed to a high rate of forest loss and hunting. High potential that the species is threatened. Red List DD.

SMALL-TOOTHED
FERRET-BADGER

LARGE-TOOTHED
FERRET-BADGER

BORNEAN
FERRET-BADGER

JAVAN FERRET-BADGER

SUMATRAN HOG-BADGER *Arctonyx hoevenii*

Plate 74

HB 51–71cm; T 8–18cm; W c. 4–8kg

Hog-badgers were considered to be a single species, *A. collaris*, until 2008, when a long-overdue review of the genus recognized three species. Sumatran Hog-badger is the smallest and darkest species, with sparse dirty-black body fur and a creamy-white tail, throat, chin and blaze. **Distribution and Habitat** Endemic to Sumatra. Restricted to 700–3780m in the forested foothills and slopes of the Barisan mountain chain. Occurs in montane and mossy forests as well as subalpine meadows; camera-trap surveys indicate it is more common in higher montane forest than in lower foothill forest. **Feeding Ecology** Thought to have a diet consisting almost entirely of soil-living invertebrates, primarily earthworms, beetle larvae and ants. Captive animals refuse raw meat, though one animal was trapped using a squirrel carcass as bait. Foraging thought to be mainly by a well-developed sense of smell. 'Grubs' in soft soil with its muzzle to locate prey, which is excavated and lapped up with its long cylindrical tongue, leaving characteristic funnel-shaped depressions. Limited observations indicate it is cathemeral. **Social and Spatial Behaviour** Unknown. Early accounts suggest adults form pair bonds, but most sightings and camera-trap images are of single animals. **Reproduction and Demography** Unknown. Females have three pairs of mammae, suggesting litters numbering up to 6, but there are no records. MORTALITY Unknown. Reported as aggressive when threatened, and like all hog-badgers uses pungent secretions from well-developed anal scent-glands to deter predation. LIFESPAN Unknown. **Status and Threats** Status poorly known. Camera-trap photographs and frequency of diggings suggest it is common in intact forest at 800–2600m. Considered a local delicacy in some areas, and snared intentionally and accidentally, but impacts are unknown. Red List NE.

NORTHERN HOG-BADGER *Arctonyx ablogularis*

Chinese Hog-badger

HB 54.6–70cm; T 11.4–22cm; W c. 5–10kg

Medium-sized hog-badger with a shaggy coat that is long and soft in winter. Colour blackish with interspersed white hairs on the hindquarters, mid-back and sides, becoming near-white in some individuals. More white on the face and throat than in Sumatran Hog-badger. **Distribution and Habitat** E–S China, NE India and probably sub-Himalayan Bhutan, Nepal and N Bangladesh. Considered the most generalist of the hog-badgers, with a wide habitat tolerance. Occurs from sea level to 4300m (China) in temperate forest, scrubland and montane meadows. Lives in agricultural habitats and close to villages. **Feeding Ecology** Omnivorous and opportunistic, with a high proportion of small rodents and snails in the diet, as well as herptiles, birds, earthworms, beetles, larvae, roots, acorns and leaves. Earthworms are especially important in the diet from late spring to autumn. Foraging solitary and primarily nocturnal. Unlike other hog-badgers, hibernates over winter (November–March), known from its northern range (C China); it is not clear if it hibernates in milder areas of its range. **Social and Spatial Behaviour** Unknown. Most records are of single adults; pairs live tolerantly in captivity. **Reproduction and Demography** Poorly known, but thought to be seasonal (C China). Mating apparently April–May; births February–March, indicating a long period of delayed implantation (estimated at 5–9.5 months for a captive female). Litter size 1–6. Captive juveniles first eat solid food at 85 days; weaning and independence co-occur at 4 months. MORTALITY Unknown; allegedly preyed on by Dhole, Leopard, Grey Wolf and Asiatic Black Bear in C China. LIFESPAN 13.9 years in captivity. **Status and Threats** Status poorly known. The most widespread hog-badger species and common in some areas, but it is heavily hunted for human consumption and killed as 'by-catch' in snares. Severely threatened in SE China. Red List NE.

GREATER HOG-BADGER *Arctonyx collaris*

HB 65–104cm; T 19–29cm; W 7–15kg

Largest hog-badger, robust with a massively built skull, and described as resembling a small bear in the field. The lightest coloured species, pale grizzled grey to yellowish-grey, with more white on the face, head and neck than in Northern Hog-badger (with which it overlaps in S China, and possibly N Myanmar and NE India); some individuals have almost entirely white or creamy-white heads. Lower limbs are black, extending variably over the shoulders and neck. **Distribution and Habitat** SE Asia, from E India, extreme S China, Myanmar and Thailand to Indochina. Records from Malaysia are equivocal. Occurs at 500–1500m primarily in undisturbed lowland and hill forests as well as bamboo stands. Reported close to villages, but rarely in modified habitat; occurs in rubber plantations close to forests. **Feeding Ecology** Poorly known; thought to be omnivorous and possibly specializing partially in earthworms (which captive animals relish) like Sumatran Hog-badger, but not to the same extent. Captive animals consume meat, reptiles, fish, bread, milk and fruits, especially plantains. Cathemeral. **Social and Spatial Behaviour** Unknown. Assumed to be solitary. **Reproduction and Demography** Very poorly known. Litter size thought to be 2–3. MORTALITY Important prey of Leopard in Thailand and also killed by Tiger. LIFESPAN 7 years (minimum estimate) in captivity. **Status and Threats** Relatively widespread, but its large size and apparent lack of wariness of humans and dogs make it a common target of hunters. Intensively hunted in much of its range and over-hunted to extinction in much of Indochina. Severely threatened in Laos, Vietnam and perhaps Myanmar, where it now occurs only patchily. Hunting intensity is lower in Thailand, where the species is considered relatively secure. Red List NT.

SUMATRAN HOG-BADGER

NORTHERN HOG-BADGER

GREATER HOG-BADGER

AMERICAN BADGER *Taxidea taxus*

Plate 75

HB 42–72cm; T 10–15.5cm; W 4–12kg
Squat low-slung badger with a broad wedge-shaped head. Coat grizzled yellow-grey with buff underparts and dark limbs. Face has a black mask, black cheek stripes and a distinctive white blaze from the nose to the nape, sometimes extending along the spine. **Distribution and Habitat** SW Canada, most of the USA, to C Mexico. Inhabits mainly open habitats from sea level to 3600m, including prairie, grassland, open woodland, scrubland and desert. Occurs on farmland but cannot tolerate intensive agriculture. **Feeding Ecology** Omnivorous, but strongly reliant on small burrowing mammals such as ground squirrels, marmots, prairie dogs, pocket gophers, voles, mice and lagomorphs. Other prey includes arthropods, birds, eggs and carrion, and occasionally reptiles, amphibians, fish and molluscs. Consumes grains, seeds and grass, but most vegetation is probably taken incidentally while eating prey. Does not kill livestock except for very rare kills of newborn lambs and poultry. Foraging solitary and usually nocturnal. Very well adapted to excavate prey from burrows; uses its keen sense of smell to locate burrows and plug their entrances with soil and sod, or sometimes snow, rocks or other objects, to block escape routes. Coyotes often form hunting 'partnerships' with badgers, which is supposedly advantageous to both species, but there is little evidence that badgers benefit. Caches surplus food; scavenges from carcasses and occasionally from human refuse. **Social and Spatial Behaviour** Solitary. Adults occupy stable ranges; male ranges are usually, but not always, 2–4 times larger than female ranges. Territorial behaviour is limited; there is high range overlap, adults actively avoid each another and individuals often use the same dens at different times. Average range estimates include 2.4 (♀s) to 5.8km² (♂s) in Utah to 13 (♀s) to 44 km² (♂s) in Illinois, with very large ranges at the distribution's extreme (SW Canada) of 9–87.3km² (♀s) and 51–450km² (♂s). Density estimated at 40–500 badgers/100km², but as low as 0.4 badgers/100km² in SW Canada. **Reproduction and Demography** Seasonal. Mating late July–August; births late-March–early April. Gestation 210–240 days, with delayed implantation. Litter size 1–5, averaging 2. Weaning at 6 weeks, dispersal at 3–4 months. Some females breed at 4–5 months, but most breeding follows the first winter. MORTALITY Natural mortality is mostly from starvation during prey crashes and predation; adult badgers are occasionally killed by bears, Puma, Grey Wolf and Coyote. LIFESPAN 14 years in the wild, 26 in captivity. **Status and Threats** Generally common and widespread. Main threats are agricultural intensification on grassland, associated prey declines and retributive killing for badgers' excavations, which can damage crops and equipment, and (rarely) cause injury to livestock. Endangered in British Columbia and Ontario. Red List LC.

EURASIAN BADGER *Meles meles*

Anakuma (Japan)
HB 50–90cm; T 11.5–20.5cm; W 3.5–17kg
Largest badger species, typically brindled silvery-grey with black legs, and alternating black-and-white facial stripes. Colouration in Asian populations varies extensively, from yellowish-grey with facial stripes reduced to dark 'spectacles', to entirely black with pale cheeks; some authorities regard them as separate species, Asian Badger *M. leucurus* (east of the Volga River/Ural Mountains) and Japanese Badger *M. anakuma* (Japan). Albino, melanistic and erythristic individuals occur. **Distribution and Habitat** Most of Europe and N–E Asia, including the UK, W Europe, SW Asia to N Afghanistan, C Asia, Russia, China, Korea and Japan. Inhabits forest, scrubland, grassland, steppes and semi-desert with scrub cover. Readily occupies farmland, fields and urban habitats. **Feeding Ecology** Omnivorous, feeding mainly on soil-living invertebrates, especially earthworms (a major component of the diet in most populations) and insects, followed by wild and cultivated fruits including berries, hard mast, grains, tubers and mushrooms. Small mammals such as mice, voles and shrews are important prey; hedgehogs, rabbits, small birds, herptiles and eggs are opportunistically consumed. Exceptionally kills very young lambs and poultry. Foraging mainly solitary, but it congregates, sometimes in large groups >20, at food-rich patches including artificial feeding sites. Nocturno-crepuscular, with increased diurnalism where undisturbed. Undergoes partial hibernation with opportunistic foraging, but can remain underground for months in severe winters, living entirely off fat reserves. Scavenges from carrion, birdfeeders, pet bowls and human refuse. **Social and Spatial Behaviour** Sociality flexible and complex. Most gregarious at high densities, e.g. lowland UK, living in large communal clans numbering up to 29, averaging 5–8 adults. Clan adults are mostly inter-related, with a minority of unrelated immigrants. Clans share territory with extensive burrow systems called setts, where members gather to interact before setting off to forage alone. Clans actively repel strangers; neighbouring ranges may overlap, but core areas with main setts are defended, sometimes violently. Clans tend to be smaller outside the UK; mated pairs with shared ranges are typical in much of mainland Europe, and badgers are mainly solitary in Japan, with large male ranges overlapping numerous female ranges. Territory size 0.1–0.4km² (Japan) and 0.3–1.5km² (UK), to 4–24.4km² (Poland). Density estimates include 0.2–0.25 badgers/km² (Finland), 0.9/km² (Ireland), 1.6–2.6/km² (Poland), 4/km² (Japan, suburbs) and 4.7–25.3/km² (UK). **Reproduction and Demography** In clans, multiple adults of both sexes breed, and females often mate with males from neighbouring clans. Breeding occurs year-round, but mating peaks February–May (UK) and April–August (Japan), with most births December–April. Gestation includes a very variable period of delayed implantation with a range of 90–300 days. Litter size 1–5, averaging 2–3. Weaning begins at 12 weeks, but extends to 6 months under low food availability. MORTALITY Annual mortality rates estimated at approximately 50% of cubs and 30% of adults. Natural mortality occurs mainly from starvation (especially cubs) and predation; important prey of Amur Tiger (Russia). LIFESPAN 14 years in the wild, 16 in captivity. **Status and Threats** Widespread, common and found in many protected areas throughout its range. Main threat in Europe is roadkill, which claims 50,000 individuals in Britain and 10–15% of Danish and Dutch populations. Persecuted as a pest and for illegal 'baiting' with terriers. Controversially culled as a carrier of bovine tuberculosis (bTB) in the UK, despite strong evidence that culls do not reduce incidence of the disease. Red List LC.

AMERICAN
BADGER

East Asian
forms

At sett
entrance

EURASIAN
BADGER

Ratel

HB 74–96cm; T 14.3–26cm; W ♀ 6.2–13.6kg,
♂ 7.7–14.5kg

Powerfully built and conspicuously bicoloured, with black sides and underparts contrasting with a silver-grey to dark grey cape. Melanistic individuals occur in African rainforest. Skin very thick and loose, providing protection against snakebites, bee stings and predators. A formidable tenacious animal that sometime stands up to Lions and Leopards, but there are few carnivores surrounded by as many falsehoods. **Distribution and Habitat** Sub-Saharan Africa, the Arabian Peninsula and C–S Asia from Kazakhstan to India. Occupies all kinds of wet and dry forests, woodland, grassland, alpine heath (to 4050m), steppes, scrub, wetland, semi-desert and true desert. Occurs in agricultural areas with cover. **Feeding Ecology** Omnivorous and highly opportunistic. The most important prey is small mammals to the size of Springhare (2kg), and reptiles including monitors, African Rock Python and highly venomous snakes, e.g. Cape Cobra, Puff Adder; has a high tolerance to venom. Occasionally kills larger vertebrates, e.g. Aardwolf, but reports of it killing large ungulates by castration are specious. Also consumes invertebrates, birds, nestlings, eggs, carrion, fruits including berries, and seeds. Bees, and their honeycomb and honey, are readily eaten, but there is no evidence that Honeyguide (*Indicator indicator*) directs badgers to hives. Raids domestic beehives and poultry coops. Foraging mainly nocturno-crepuscular, with increased diurnalism during cold winters. Forages alone, but congregates at food-rich patches.

Foraging is largely by its excellent sense of smell, and most prey is excavated after following scent trails. Climbs strongly and raids nests for eggs and nestlings; raptor chicks are important seasonal prey in the Kalahari. Scavenges, including from other carnivores' kills, and at camp grounds and dumps. **Social and Spatial Behaviour** Solitary. Male ranges are massive, marked by high overlap and avoidance rather than active territorial defence; each male range overlaps as many as 13 female ranges (Kalahari). Female ranges are more exclusive, but also with little evidence of territorial defence. Moves constantly while foraging, covering up to 40km/day, averaging 14km (♂s) and 8km (♀s) between rest periods. Kalahari range size 85–194km², averaging 126km² (♀s), and 229–776km², averaging 541km² (♂s). **Reproduction and Demography** Aseasonal. Gestation 50–70 days. Litter size 1, rarely 2. Weaning at 2–3 months, but cubs are entirely dependent on their mothers to 10–12 months, gradually diminishing to independence at 16–22 months. MORTALITY Annual mortality under protection estimated at 46% (cubs, to independence) and 34% (adults, Kalahari), mainly by starvation, predation by large cats and hyaenas, and infanticide. LIFESPAN 7 years in the wild, 28 in captivity. **Status and Threats** Widespread, adaptable and has a broad habitat tolerance. However, it has an unusually low reproductive rate and is vulnerable to local extinction under elevated threats, especially persecution by apiarists and livestock owners. Also killed for traditional medicinal uses and superstitious beliefs, such as claims that it excavates gravesites. Red List LC.

WOLVERINE *Gulo gulo*

Glutton, Skunk-bear

HB 65–105cm; T 17–26cm; SH 36.5–43.2cm;
W ♀ 6.6–14.8kg, ♂ 11.3–18.2kg

Largest terrestrial mustelid, heavily built with a bear-like head, short powerful limbs and a bushy tail. Colouration typically dark brown with a blond to rusty-brown fringe running from the shoulders along the sides to the base of the tail. Forehead often grizzled grey to blond. Cream to white markings on the chest are common; they extend to the front legs and feet in some individuals. **Distribution and Habitat** Circumpolar, mostly N of 50°N from Fenno-Scandinavia through Russia, N Mongolia, N China, Canada, Alaska and the W USA in the Rockies. Inhabits coniferous and deciduous forests, open rocky terrain and Arctic tundra. Strongly associated with deep snow and dead timber for denning. Actively avoids areas of human disturbance, including agricultural land, roads, skiing fields and recent logging. **Feeding Ecology** Capable of killing adult Moose and Reindeer in deep snow, and relies heavily on ungulates (particularly during winter), but most are consumed as winter-killed or predator-killed carrion; scavenges whale and seal carcasses in coastal Alaska. Actively hunts marmots, porcupines, beavers, ground squirrels, lagomorphs and small rodents. Also eats birds such as Ptarmigan and grouse, eggs, invertebrates, fruits including berries, and fungi. Known to raid trap-lines for captured fur-bearers, and sometimes kills livestock (mainly lambs) and domestic Reindeer (Fenno-Scandinavia). Foraging cathermal, solitary and mostly terrestrial, though it is a strong climber and swimmer known to forage in trees and water. Hoards surplus food under rock piles, ice or snow,

sometimes creating large caches, e.g. 20 Red Foxes and 100 Ptarmigan in one cache (Russia). **Social and Spatial Behaviour** Solitary, with extensive stable ranges. Male ranges are larger and overlap multiple female ranges. Adults exclude same-sex conspecifics, but range overlap can be considerable. Covers large daily distances to 35km, driven in part by the search for carrion; summer movements tend to be larger than winter ones. Territory size estimates include 31–560km² (♀s) and 133–1131km² (♂s) in N Sweden; 53–232km² (♀s) and 488–917km² (♂s) in NW Alaska, and 175–692km² (♀s) and 845–2127km² (♂s) in Idaho. Density poorly known, but estimated at 4.7 Wolverines/1000km² (SC Alaska) to 15.3/1000km² (Montana). **Reproduction and Demography** Seasonal. Mating May–August; births January–April. Gestation 215–272 days (captivity), with delayed implantation. Litter size 1–5, averaging 2–3. Weaning at 7–8 weeks. Independence at around 8–10 months, dispersal at 12–13 months. Females breed on average at 3.4 years. MORTALITY Annual adult and subadult mortality respectively is 26% and 43% (trapped), and 12% and 7% (untrapped). Most natural mortality is due to starvation and predation by Grey Wolf, Puma and other Wolverines. LIFESPAN 13 years in the wild, 18 in captivity. **Status and Threats** Wide distribution and numerous large populations, but the species occurs in naturally low densities, and it is sensitive to persecution and disturbance. It has declined in large areas of the southern USA and southern Europe. Threatened by over-trapping, predator-control programmes, illegal killing for livestock depredation and habitat conversion. Red List LC (VU in Europe).

HONEY BADGER

WOLVERINE

Scanning

Huroncito
HB 30–35cm; T 6–9cm
Very small, with a pale grizzled grey body and chocolate-brown to black underparts. Wide, wedge-shaped white crown covers the head, distinguishing it from the considerably larger Lesser Grison, which has a narrow white brow. **Distribution and Habitat** Endemic to W Argentina and a narrow band of C–E Chile. Inhabits cold, arid and semi-arid shrubland, steppes and open scrubby woodland from sea level to 2000m. **Feeding Ecology** Poorly known.

Chief prey thought to be small burrowing rodents such as Tuco-tucos and Mountain Cavies; one record of predation on Elegant Crested Tinamou. An adult living under the ranger station at Cabo Dos Bahias scavenged handouts. **Social and Spatial Behaviour** Unknown; assumed to be solitary like other small weasels. **Reproduction and Demography** Unknown. MORTALITY Black-chested Buzzard-eagle is a known predator. **Status and Threats** Status essentially unknown. Rarely observed or encountered during wildlife surveys, suggesting it is naturally rare. Red List DD.

LESSER GRISON *Galictis cuja*

HB 27.3–52cm; T 12–19cm; W 1–2.5kg
Grizzled yellow-grey to brownish-grey upperparts. Black face and underparts bounded by a narrow white or creamy band across the brow running variably down the neck and shoulders. **Distribution and Habitat** SE Peru, S Bolivia, S Brazil through Paraguay, Uruguay, Argentina, and C Chile. Occurs in desert, steppes, grassland savannah, shrubland, marshland, woodland and forest from sea level to 4200m. Tolerates agricultural and pastoral habitats. **Feeding Ecology** Diet dominated by small mammals, especially mice and cavies, and introduced European Hare and European Rabbit; focuses almost entirely on hares/rabbits under high availability. Also eats birds, small reptiles, frogs, eggs and invertebrates. Fruits are consumed, but appear rare in the diet. Blamed for killing poultry, but depredation is poorly

quantified. Cathemeral. **Social and Spatial Behaviour** Poorly known. Adults usually observed alone but also seen in small groups, including adult pairs and their juveniles, suggesting monogamous pair bonds. Up to a dozen animals recorded playing together. **Reproduction and Demography** Gestation 39 days. Litter size 2–5. Juveniles recorded March–October, suggesting weak seasonality. Mated pairs apparently cooperate to raise kittens; extent of monogamy unknown. MORTALITY Ocelot and Black-chested Buzzard-eagle are known predators. LIFESPAN Unknown. **Status and Threats** Widespread. Broad habitat tolerance and considered secure, though there is no detailed information on status. Persecuted for killing poultry, often killed on roads, e.g. C Brazil, and occurs often in the diet of urban feral dogs, e.g. University of Sao Paulo, Piracicaba campus, SE Brazil. Red List LC.

GREATER GRISON *Galictis vittata*

HB 45–60cm; T 13.5–19.5cm; W 1.4–4kg
Larger than Lesser Grison, with a proportionally shorter tail and paler grizzled grey fur with a smoother, short-haired appearance. The two species may overlap only in SE Bolivia and S Brazil. **Distribution and Habitat** From S Brazil through N South America to SE Mexico. Occurs in low and mid-elevation forest woodlands, palm savannah, grassland and wetland to 1500m. Tolerates disturbed forest, plantations, open fields and agricultural land with cover. **Feeding Ecology** Carnivorous, eating rodents to the size of agoutis, marsupials including Southern Opossum, reptiles, amphibians, fish, invertebrates and eggs. Captives eat fruits and some plant matter. Sometimes raids domestic poultry. Foraging cathemeral. Hunting terrestrial, but readily pursues prey into trees and deep water. Hunts alone, in

adult pairs or in small family groups. **Social and Spatial Behaviour** Assumed to be largely solitary; adult pairs and family groups occur, but sociality is poorly understood. Only range estimate is for one female (Venezuela) over 2 months, 4.15km². **Reproduction and Demography** Gestation 39–40 days. Litter size 1–4. Juveniles recorded March–October, suggesting weak seasonality. Males often recorded with mothers and kittens, but it is unknown if they assist in raising juveniles. MORTALITY Unknown. Carries Chagas trypanosomes, but impacts are unknown. LIFESPAN 10.5 years in captivity. **Status and Threats** Secure over much of its range, but threatened at the extremes, e.g. Mexico and Costa Rica. Tolerant of some disturbance, but hunting pressure and habitat conversion to open agriculture leads to local declines. Red List LC.

TAYRA *Eira barbara*

Grey-headed Tayra, Eira
HB 55.9–71.2cm; T 36.5–47cm; W 2.7–7kg
Largest mustelid in Latin America, excluding otters. Typically dark smoky-brown to black, often with a creamy-yellow or white throat patch and a pale head varying from slightly grey to blond. Completely pale individuals occur, including a golden form in N South America. **Distribution and Habitat** C Mexico, through C America to N Argentina, Paraguay and S Brazil. Occurs in various dry and wet forests and forest woodland to 2400m, and in meadows, grassland and savannah in forested mosaics. Tolerates agriculture, plantations and pasture in association with forest. **Feeding Ecology** Omnivorous, eating mammals to the size of agoutis, Southern Opossum, juvenile primates and neonate sloths, as well as iguanas, birds, eggs, invertebrates (adults, eggs and pupae), fruits, honey and carrion. Chases large prey,

including Brocket Deer and large adult primates, but successful hunts are unknown. Usually forages alone, though adult pairs and family groups are known. Cathemeral, and equally at home on the ground or in trees. **Social and Spatial Behaviour** Poorly known. Assumed to be mainly solitary; small social groups occur, but are poorly understood. Limited data indicates that adult ranges overlap extensively. Range estimates 5.3–16km² (♀s) and 24.4km² (1 ♂) in Belize. **Reproduction and Demography** Thought to breed year-round. Gestation 63–70 days. Litter size 1–3. Weaning at around 2–3.5 months. Males are occasionally recorded in family groups, but apparently do not help raise juveniles. MORTALITY Unknown. LIFESPAN 18 years in captivity. **Status and Threats** Widespread and often common in much of its range. Main threat is loss of forested habitat, leading to local endangerment, e.g. in Mexico. Red List LC.

PATAGONIAN
WEASEL

LESSER
GRISON

GREATER
GRISON

TAYRA

African Striped Weasel, White-napped Weasel
HB ♀ 24–35cm, ♂ 27–33cm; T 13.8–21.5cm;
W ♀ 21–29kg, ♂ 28.3–38kg
Small weasel with a long sinuous body, very short limbs and a long tail. Fur black with a yellowish-white dorsal stripe starting at the crown; stripe splits into paired stripes that run along each side. Tail is white. **Distribution and Habitat** Sub-equatorial Africa, from S Kenya and S Uganda to coastal DR Congo, and south to S Africa. Inhabits woodland savannah, grassland, scrubland, forest (its range stops at the limits of the forested Congo Basin) and vegetated semi-arid desert, e.g. the Kalahari. Occurs in plantation, agricultural and pastoral habitats. **Feeding Ecology** Rodent specialist, hunting mainly small mice, rats and mole rats to its own size; an adult may kill 3–4 rodents a night. Also eats small reptiles, insects and eggs. Foraging mainly nocturnal, terrestrial and solitary. Forages chiefly by scent, and is well suited to entering small rodent burrows; a powerful burrower, but has not been observed excavating prey. Rodents are killed with a nape bite and vigorous kicking by the hind legs, which may dislocate the neck; large prey is sometimes killed by a throat bite. Caches surplus kills in burrows. **Social and Spatial Behaviour** Poorly known. Assumed to be solitary; most sightings are of adult individuals or females with pups. **Reproduction and Demography** Possibly seasonal. Breeding September–April (southern Africa), with births from November. Gestation 30–33 days. Litter size 1–3. Weaning at 11 weeks (captivity). Sexual maturity at 8 months. MORTALITY Poorly known. Occasionally killed by domestic dogs and large owls. Rabies is recorded. LIFESPAN 6 years in captivity. **Status and Threats** Considered uncommon to rare, but it is inconspicuous and elusive, and there is little accurate information on its status. Killed on roads in rural areas and sought after for traditional medicinal use in S Africa. Red List LC.

LIBYAN WEASEL *Ictonyx libyca*

Saharan Striped Polecat, North African Striped Weasel
HB 20.7–26cm; T 11.4–18cm; W 0.2–0.6kg
Small and compact weasel with a black face, limbs and underparts. White stripes interleaved with variable black inter-stripes cover the body. Tail long and white with interspersed black hairs, and sometimes with a black tip. Fur longish with a silky appearance. Unbroken white band encircles the face, running from the forehead behind the eyes to the base of the throat; this helps distinguish it from the similar Zorilla. It has well-developed anal glands and secretes a pungent fluid when threatened. **Distribution and Habitat** N Africa, on the edges of the Sahara in the coastal band of Mediterranean N Africa from Egypt to Mauritania, and through the Sahel from Mali to Sudan. Scattered records exist across the Sahara itself, but it is unclear if it occurs throughout. Occupies mainly sub-desert habitats such as stony desert, massifs, steppes, oases and sparsely vegetated dunes. Found close to settlements in cultivated areas. **Feeding Ecology** Poorly known. Thought to feed mainly on small desert rodents, birds, reptiles, eggs and invertebrates. Nocturnal. **Social and Spatial Behaviour** Unknown. Most records are of single adults; assumed to be solitary. **Reproduction and Demography** Poorly known. Thought to be seasonal; all records of young occur January–March. Litter size 1–3. MORTALITY Unknown. LIFESPAN 5.5 years in captivity. **Status and Threats** Status poorly known. Widely distributed and locally abundant in some coastal dune areas. Hunted in Libya and Tunisia in the belief that its body parts increase human male fertility. Red List LC.

ZORILLA *Ictonyx striatus*

Striped Polecat
HB 28–38cm; T 16.5–28cm; W ♀ 0.4–1.4kg,
♂ 0.7–1.5kg
Larger than similar species (Libyan and Striped Weasels), jet-black with four white stripes that unite on the crown and run the length of the body to the tail, which is white interspersed with black hairs. Face distinctively marked with a cluster of three white blotches on the forehead and on each temple. Overlaps Libyan Weasel in the Sahel. Ejects a noxious anal secretion when threatened. **Distribution and Habitat** Throughout sub-Saharan Africa, except the Sahara and Congo Basin. Occurs in a wide variety of habitats from sea level to 4000m, including wet and dry woodland savannahs, grassland, forest, dunes, wetland, montane heath, semi-desert and desert. Absent from equatorial forest and desert interiors. Readily inhabits agricultural and cultivated habitats. **Feeding Ecology** Eats mainly small rodents and insects. Also eats herptiles, birds, chicks, eggs, arachnids and other invertebrates. Largest prey includes Springhare, ground squirrels and large snakes, including venomous ones such as cobras. Occasionally kills domestic poultry. Nocturnal and terrestrial. Hunting is solitary, but juveniles sometimes help the mother in subduing large prey such as snakes. Prey is hunted by sight and smell, with rodents and insects often killed in burrows or excavated. **Social and Spatial Behaviour** Poorly known. Adults are largely solitary. Captive males are intolerant of each other, but females with juveniles tolerate other mother-kitten families in captivity. **Reproduction and Demography** Poorly known. Reported to give birth mainly November–February in southern Africa, but lactating females are recorded February–October in E Africa. Gestation 36 days. Litter size 1–3, exceptionally to 5 (captivity, the maximum reared being 3). Weaning at around 8 weeks. Females first breed at 10 months (captivity). MORTALITY Poorly known. Large raptors, especially owls, are confirmed predators, and it is frequently killed by domestic dogs in rural areas. LIFESPAN 13.3 years in captivity. **Status and Threats** Widespread habitat generalist and common to abundant in suitable protected habitat. Roadkills, domestic dogs and persecution for poultry depredation kill significant numbers in rural areas, but probably constitute only a localized threat. Valued in traditional medicinal beliefs in some areas. Red List LC.

STRIPED
WEASEL

LIBYAN
WEASEL

ZORILLA

HB 28.8–47.7cm; T 14.5–20.1cm; W 0.3–0.72kg
Very dark chocolate-brown with a striking buff-yellow cape dappled with red-brown blotches extending from the nape over the back and sides. Bushy tail grizzled yellow-white, usually with a dark tip. A conspicuous white stripe encircles the face, the tops of the ears are white, and the muzzle and chin are creamy-white. Assumes a distinctive arching posture when threatened, which is followed by ejecting a noxious anal-gland secretion if unheeded. **Distribution and Habitat** N China, Mongolia, C Asia, the Middle East and SE Europe. Inhabits temperate and arid steppes, grassland, scrubland, rocky upland, salt marshes, semi-desert and open desert habitats. Occurs in cultivated areas, orchards and vegetable gardens near settlements, in urban parkland and on university campuses. **Feeding Ecology** Diet dominated by small mammals, especially ground squirrels, jirds, hamsters, voles, rats, mice and rabbits. Also consumes insects (especially during spring–summer flushes), birds, herptiles and snails, and fruits including berries. Sometimes kills domestic poultry and rabbits. Foraging mainly nocturno-crepuscular, solitary and by scent; reputedly has poor eyesight. Eats carrion and scavenges from buildings, including raiding larders for smoked meat and cheese. Caches surplus food in burrows. **Social and Spatial Behaviour** Solitary. Limited data indicates it occupies small stable ranges with moderate overlap. Outside breeding, adults may be aggressive to each other and sometimes fight furiously during encounters. Only known range estimates (Israel) 0.5–0.6km², with little difference between sexes. **Reproduction and Demography** Seasonal. Mating March–June; births February–May (following year). Gestation 243–327 days, with delayed implantation. Litter size 1–8, averaging 4–5. Weaning at 50–54 days and dispersal at 61–68 days. Females sexually mature at 3 months; males typically breed after their first year. MORTALITY Poorly known. Most documented mortality is anthropogenic. LIFESPAN Almost 9 years in captivity. **Status and Threats** Nowhere common, and threatened by conversion of steppe habitats to cultivation, combined with large-scale poisoning of rodents, e.g. China and Mongolia. Killed in small numbers for fur and persecuted for killing poultry. Red List VU.

YELLOW-THROATED MARTEN *Martes flavigula*

Himalayan Yellow-throated Marten
HB 45–65cm; T 37–45cm; W 1.3–3kg
Large marten with a long tail up to 70% the length of its body. Head, nape, hindquarters and tail normally dark brown with a highly variable tawny-brown cape covering the rest of the upper body; it is entirely absent in some animals, especially in Peninsular Malaysia, Borneo and Sumatra. Throat and chest always golden lemon-yellow, sometimes extending down the forelimbs. Chin and cheeks are white. **Distribution and Habitat** From the Russian Far East through S China, the northern Indian subcontinent, N Afghanistan, N Pakistan, Indochina, Sumatra, Java and Borneo. Inhabits temperate and tropical forests at 200–3000m. Occurs in secondary forest and plantations, but avoids open anthropogenic habitats. **Feeding Ecology** Poorly studied. Omnivorous. Eats small mammals, birds (including large ones such as pheasants), reptiles, amphibians, invertebrates, eggs, fruits including berries, flowers and nectar. Diet shifts to take advantage of seasonal foods, e.g. focusing on fruits and flowers during spring and summer. Has been observed pursuing large mammals including Himalayan Tahr, Himalayan Musk Deer and gorals; seven martens were observed feeding on a Chinese Goral, and tahr has been found in scats, but it is extremely unlikely to kill such large prey and presumably takes carrion. Primarily diurnal, and forages both terrestrially and arboreally. Scavenges boiled rice at guardposts in Khao Yai NP, Thailand. **Social and Spatial Behaviour** Poorly known. More often seen in pairs and trios than alone. Although pair composition is unclear, the frequency of occurrence suggests greater sociality than occurs in other marten species. Also found in larger groups, e.g. at carcasses, which are probably temporary aggregations. Only range estimates (Phu Khieo WS, Thailand) are 1.7–11.8km² (♂s) and 8.8km² (1 ♀). **Reproduction and Demography** Poorly known. Thought to be seasonal. Mating occurs June–August; births March–June (following year). Gestation 220–290 days, with delayed implantation. Litter size 2–5. MORTALITY Unknown. LIFESPAN 14 years in captivity. **Status and Threats** Widespread and considered secure. Presumably undergoes declines with forest loss and fragmentation. Remains relatively common even in areas with high hunting pressure, e.g. Indochina, perhaps because few communities eat it. CITES Appendix III (India, Tibet), Red List LC.

NILGIRI MARTEN *Martes gwatkinsii*

HB 50–70cm; T 35–50cm; W 1–3kg
Very similar to Yellow-throated Marten and considered by some authorities to be the same species. Lacks the latter's yellow cape, but the shoulders and torso sometimes tend to pale rufous-brown. Separated as a distinct species due mainly to its isolated discontinuous distribution. **Distribution and Habitat** Endemic to the W Ghats, India, where it is restricted to six disjunct populations. Occurs in forest patches in undeveloped montane and hilly areas at 120–1400m. Sometimes found in coffee, cardamom and wattle plantations. **Feeding Ecology** Poorly known. Assumed to be similar to that of Yellow-throated Marten. Has been observed pursuing Indian Giant Squirrel, Indian Spotted Chevrotain and Bengal Monitor, and eating nectar of cultivated Kapok trees. Raids honey from domestic beehives. **Social and Spatial Behaviour** Poorly known. Has been sighted singly and in pairs, consistent with the little that is known of Yellow-throated Marten sociality. **Reproduction and Demography** Unknown. **Status and Threats** Range very restricted and fragmented, and thought to be naturally rare based on the frequency of encounters. The Western Ghats are under intense anthropogenic pressure, and further habitat loss and fragmentation are the main threats. Also killed by beekeepers as a perceived pest, and illegally hunted by some communities for meat. CITES Appendix III, Red List VU.

MARBLED
POLECAT

YELLOW-
THROATED
MARTEN

NILGIRI
MARTEN

Beech Marten

HB 40–54cm; T 22–30cm; W 1.1–2.3kg

Typically rich dark brown with a slightly paler, greyish head, and light tawny underfur especially on the sides and underparts. Throat has a distinctive white or cream patch that often extends down the front legs. **Distribution and Habitat** Mainland W and C Europe through C Asia, Bhutan, N India, Nepal, N Myanmar to Mongolia and China. Introduced to Wisconsin, USA. Inhabits forest, shrubland, forest edges, hedgerows and rocky hillsides to 4200m. Occurs near humans, including in densely populated urban areas (W and C Europe). **Feeding Ecology** Diet varies seasonally and regionally with fluctuating proportions of two main food groups: small mammals (especially voles, mice and rabbits), and fruits including berries. Also eats insects, birds, herptiles, eggs, seeds and other plant items. Urban populations frequently eat commensal birds like pigeons. Sometimes kills domestic poultry. Foraging solitary and nocturno-crepuscular. Urban martens readily scavenge from refuse, birdfeeders, pet bowls and handouts. **Social and Spatial Behaviour** Solitary and territorial, with male ranges overlapping multiple female ranges. Ranges tend to be smallest in urban areas, intermediate in rural areas and largest in forested habitat. Range size 9.5–880ha, averaging 37–49ha (♀s) and 111–113ha (♂s) in rural/village areas. **Reproduction and Demography** Seasonal. Mating July–August; births March–mid-April (following year). Gestation 236–275 days, with delayed implantation (embryonic development ~30 days). Litter size 1–8, averaging 3–4. Weaning at around 6–8 weeks in late May–early June, and dispersal from 6 months. MORTALITY Most known mortality is anthropogenic. LIFESPAN 18.1 years in captivity, much lower in the wild. **Status and Threats** Widespread and adaptable, reaching high densities in urban habitats. Considered a nuisance in C Europe due to its habit of sheltering in car-engine spaces and chewing leads and hoses, e.g. 160,000 damaged cars in Germany in 2000, for which it is sometimes legally and illegally killed. Hunted for fur in India and Russia. CITES Appendix III (India), Red List LC.

PINE MARTEN *Martes martes*

European Pine Marten, Eurasian Pine Marten

HB 45–58cm; T 16–28cm; W 0.8–1.8kg

Similar to Stone Marten, with which it overlaps in most of W and C Europe, but distinguished by a yellow throat patch. **Distribution and Habitat** UK, mainland W and C Europe to Fenno-Scandinavia, W Siberia (Russia), Turkey, N Iraq and N Iran. Occurs mainly in mature intact forest, woodland and scrubland with dense understory. Inhabits coastal shrubland, pasture and grassland with cover, but avoids open areas. Occurs near settlements, but does not colonize urban areas as do Stone Martens. **Feeding Ecology** Preys predominantly on small mammals, especially Field Vole, Red-backed Vole, field mice, Wood Lemming, squirrels and lagomorphs. Other important prey includes invertebrates, birds, herptiles, fruits including berries, eggs and carrion (mainly wild and domestic ungulate carcasses). Sometimes kills domestic poultry. Foraging primarily nocturnal and solitary. Scavenges, mainly from carrion and rarely from urban sources. Caches excess food. **Social and Spatial Behaviour** Solitary and territorial, with male ranges overlapping multiple female ranges. Range size correlates with forest cover and rodent density. Smallest ranges are in mature forest with high rodent abundance, e.g. Poland and Germany; largest known ranges are in open habitat in Finland and Scotland. Average range size 1.4–9.8km² (♀s) to 2.3–28.6km² (♂s). **Reproduction and Demography** Seasonal. Mating July–August; births March–April (following year). Gestation 230–274 days, with delayed implantation (embryonic development ~30 days). Litter size 2–8, averaging 3–5. Weaning at around 6–8 weeks; dispersal from 6 months throughout winter. MORTALITY Annual mortality in protected forest (Poland) 38.4% (adults and subadults, sexes combined), from canine distemper, winter starvation, predation by Eurasian Lynx, Red Fox and raptors, and poaching. LIFESPAN 5 years in the wild, 17 in captivity. **Status and Threats** Formerly very heavily hunted for fur, resulting in declines and local extinctions, especially in Russia and Fenno-Scandinavia. Stricter controls have led to recovery and it is now fairly widespread, but still harvested at questionable levels in some areas. Also illegally persecuted as a pest, the main reason it disappeared from much of the UK. Red List LC.

JAPANESE MARTEN *Martes melampus*

Yellow Marten, Tsushima Island Marten

HB 47–54.5cm; T 17–22.3cm, W 0.7–1.7kg

Small slender marten, rich dark brown with a large rich-yellow throat patch; populations on Kyushu and northern Honshu moult in winter to a vivid orange-yellow with a white to pale grey head. Tail sometimes has a white tip. **Distribution and Habitat** Endemic to Japan (introduced to Hokkaido and Sado for fur, native to other islands). Records from the Korean peninsula are equivocal. Occurs mainly in broadleaved forest, woodland and subalpine shrubland. Inhabits rural and urban areas with natural forest patches. **Feeding Ecology** Feeds mainly on small mammals, insects, centipedes, earthworms, spiders, snails, fruits including berries, and seeds. Considered an important seed disperser in subalpine areas due to the amount of fruits it consumes. Also eats birds, frogs and various plant items. Occasionally takes domestic poultry and is easily trapped with chicks. Foraging nocturnal and solitary. **Social and Spatial Behaviour** Adults are solitary and probably territorial; they deposit scats at range borders, and ranges overlap little within the same sex. Range size 0.5–1km², similar for females (average 0.63km²) and males (average 0.7km²; Tsushima Island). **Reproduction and Demography** Seasonal. Mating late July–mid-August; births mid-April–early May (following year). Gestation 230–250 days, with delayed implantation (embryonic development 28–30 days). All known litters number 2 (based on few observations). MORTALITY Rates unknown; main factors on Tsushima Is are roadkills (72%) and feral dogs (9%). LIFESPAN Unknown. **Status and Threats** Relatively widespread and common within its limited range. Main threats are habitat conversion (including to forestry monocultures), roadkills and predation by feral dogs. Legally trapped for fur (not on Tsushima Is) in December–January. Red List LC.

STONE
MARTEN

PINE
MARTEN

Winter form,
Kyushu and
northern Honshu

JAPANESE MARTEN

HB 45–65cm; T 25.3–50cm; W ♀ 1.3–3.2kg,
♂ 3.5–5.5kg (exceptionally to 9kg)
Largest marten. Colour varies from pale grey-brown to silver-tipped black, but it is typically chocolate-brown with a pale brown or greyish-brown head and shoulders. Some individuals have white or cream markings on the throat, chest and groin. Recent genetic analyses suggest Fisher is sufficiently distinctive to be classified in its own genus, but this remains tentative. **Distribution and Habitat** S Canada, extreme SW Alaska, the NW USA, N Minnesota and New England. Strongly prefers intact low–mid-elevation (generally <1250m) forest with a dense canopy. Uses logged forest and forest-clearing mosaics, but avoids large open areas, clearcuts and human disturbance. **Feeding Ecology** Preys predominantly on North American Porcupine and Snowshoe Hare. Also eats squirrels, voles, rats, mice, birds, herptiles, invertebrates, fungi and carrion (mainly ungulate carcasses in winter). Rarely kills domestic poultry. Foraging solitary, nocturno-crepuscular and mainly terrestrial, but Fishers are extremely capable climbers; they chase porcupines to the ground, where they are killed by biting repeatedly at the face. Fishers return to large carcasses for 2–3 days, and cache small surplus kills. **Social and Spatial Behaviour** Solitary and territorial. Male ranges overlap multiple female ranges. Average range size 2.1–29.9km² (♀s) and 9.2–38.7km² (♂s). Rapidly covers large distances, up to 90km in 3 days. Density estimates include 12 Fishers/100km² (Maine) to 27–32.7/100km² (Quebec and eastern Ontario). **Reproduction and Demography** Seasonal. Mating March–May; births January–early May (following year). Gestation 236–275 days, with delayed implantation (embryonic development ~50 days). Litter size 1–6, averaging 2–3. Weaning at 2–3 months, and dispersal from 5–6 months. MORTALITY Most known mortality is anthropogenic; predators include Cougar, Canada Lynx, Coyote, Wolverine and Golden Eagle. LIFESPAN 7.5 years in the wild, 10 in captivity. **Status and Threats** Disappeared from most of its historic distribution due to intensive over-trapping for furs; it has now recovered or been reintroduced into much of its former range. Adapts poorly to modified forest, and habitat loss and fragmentation are the key threats. Legally trapped, which risks impacting populations in low-quality habitat. Red List LC.

SABLE *Martes zibellina*

Japanese Sable
HB 35–56cm; T 11.5–19cm; W ♀ 0.7–1.6kg, ♂ 0.8–1.8kg
Honey-brown to very dark brown, usually with a paler head and a small white or cream throat patch (often absent). Japanese Sables are often rich yellow or tawny-brown with a light grey head, similar to some winter forms of Japanese Marten. **Distribution and Habitat** Russia (about 95% of range), N Mongolia, NE China, N Korea and Japan (Hokkaido). Closely tied to temperate debris-rich, dense-canopy forest. Avoids open areas and disturbed habitat. **Feeding Ecology** Eats mainly red-backed voles, mice, Siberian Chipmunk, pikas and Mountain Hare, as well as seeds, berries, nuts, invertebrates and birds; occasionally eats fish and freshwater crustaceans. Capable of killing adult Musk Deer in deep snow. Occasionally kills domestic poultry. Foraging solitary, cathemeral and mostly terrestrial, though Sables are very agile climbers. Scavenges, mainly from winter-killed ungulate carcasses, and caches surplus food. **Social and Spatial Behaviour** Solitary and thought to be territorial, but often with high range overlap. Average range size 7.2km² (♀s) and 13.1km² (♂s) with little intrasexual overlap in open larch-taiga in NE China, and only 1.12km² (both sexes) with high intrasexual overlap in high-quality forest, Japan. **Reproduction and Demography** Seasonal. Mating June–August; births April–May (following year). Gestation 236–315 days, with delayed implantation (embryonic development 25–40 days). Litter size 1–5, averaging 2–3. Weaning at around 7–8 weeks. MORTALITY Trapping accounts for most mortality (especially in Russia); Red Fox is a confirmed predator in Japan. LIFESPAN 5.5 years in the wild, 15 in captivity. **Status and Threats** Sable fur is luxuriant and highly sought-after; historical over-harvests caused widespread declines and local extinctions. Hunting bans and reintroductions (especially in Russia) have led to recovery in much of Sable's former range. It continues to be commercially hunted, mainly in Russia (250,000 in 1984–1989), and is farmed for fur. Forest loss is a significant threat. Red List LC.

AMERICAN MARTEN *Martes americana*

HB ♀ 32–40cm, ♂ 36–45cm; T 13.5–23cm;
W ♀ 0.3–0.85kg, ♂ 0.47–1.3kg
Small slender marten ranging from tawny-beige with dark limbs to uniformly dark chocolate-brown. Head usually paler, buff-brown or greyish. All forms have a cream to yellow throat and chest patch. **Distribution and Habitat** Canada, Alaska and the W USA, marginally into Minnesota, New York and Maine. Prefers mature temperate forest and woodland with a closed canopy and dense understory. Uses forest edges and alpine shrubland, but avoids open or disturbed habitat including clearcuts and recently logged areas. **Feeding Ecology** Eats voles, mice, rats, chipmunks, Red Squirrel and Douglas's Squirrel; preys heavily on Snowshoe Hare during cyclical abundances. Also eats invertebrates, herptiles, fruits and seeds; coastal populations consume high proportions of fish and birds. Rarely takes domestic poultry. Foraging solitary and mainly nocturno-crepuscular. Hunts by constant searching on the ground and in trees; forages under snow during winter. Scavenges from carrion and caches excess food. **Social and Spatial Behaviour** Solitary and territorial, with large male ranges overlapping multiple smaller female ranges. Average range size 2.3–27.6km² (♀s) and 4.3–45km² (♂s). Density estimates 0.4–1.5 martens/km². **Reproduction and Demography** Seasonal. Mating July–August; births late March–April (following year). Gestation 220–275 days, with delayed implantation (embryonic development 27–28 days). Litter size 1–5, averaging 2–3. Weaning at 6–7 weeks; kittens kill small prey at 2.5 months. MORTALITY Adult mortality before trapping 7% (both sexes) increasing to 51% (♀s) to 74% (♂s) during trapping (Maine). Mortality in unharvested populations 13–44%, mainly from predation by Bobcat, raptors and other martens. LIFESPAN 14.5 years in the wild (typically <5), 15 in captivity. **Status and Threats** Fur over-harvest and forest loss have reduced populations, mainly in New England and the coastal W USA, where it remains fragmented and rare. Has recovered and is considered common in much of its remaining range. Vulnerable to over-harvests when combined with logging. Red List LC.

FISHER

SABLE

AMERICAN
MARTEN

HB 38–50cm; T 11.4–15cm; W ♀ 0.76–0.85kg,
♂ 0.96–1.1kg
The only ferret native to N America. Yellowish-buff on the body, darkening to dark brown on the back. Head creamy-white with a brownish-black mask and chocolate-brown crown. Limbs chocolate-brown to black, and the tail has a black tip. **Distribution and Habitat** Great Plains of the USA and N Mexico. Extinct in the wild by 1987, and all present populations result from reintroductions. Restricted to short–mid grass plains and prairies in obligate association with prairie-dog colonies. **Feeding Ecology** Dependent on prairie dogs, which comprise around 90% of the diet. The most important species is Black-tailed Prairie Dog, as well as White-tailed Prairie Dog and Gunnison's Prairie Dog (Arizona only). Occasional prey includes small rodents such as deer mice, voles and ground squirrels, as well as Cottontail and White-tailed Jackrabbit. Hunting mainly nocturno-crepuscular and underground; pursues prairie dogs into their burrows, where most kills occur. Above-ground hunts are less successful; adult prairie dogs often mount an effective defence on the surface. Does not hibernate and hunts hibernating prairie dogs throughout winter. Sometimes caches surplus kills in burrows. **Social and Spatial Behaviour** Solitary. Adults establish enduring ranges, closely tied to active prairie-dog colonies. Same-sex adults avoid each other, but ranges overlap by as much as 42% in areas of high prairie-dog density. Male ranges are about twice the size of female ranges. Range sizes 23–188ha, averaging 56–65ha (♀s) and 128–132ha (♂s). **Reproduction and Demography** Seasonal. Mating March–April; births May–June. Gestation 42–45 days. Litter size 1–6, averaging 3–4. Weaning at 6 weeks, and kits venture above ground at 60 days. Dispersal in late autumn at around 5–6 months. MORTALITY Rates of disappearance (including some emigration) are 53–86% annually. Main factors are disease and predation, especially by Coyote and large raptors. LIFESPAN 12 years in captivity, but much lower in the wild. **Status and Threats** The species was decimated by exotic disease (canine distemper and sylvatic plague), and the massive anthropogenic decline of prairie dogs. Eighteen reintroduction efforts have re-established wild ferrets, but only three populations are self-sustaining, in South Dakota and Wyoming. Approximately 500 adults in the wild. Red List EN.

SIBERIAN WEASEL *Mustela sibirica*

Siberian Polecat, Kolinsky, Himalayan Weasel
HB ♀ 25–30.5cm, ♂ 28–39cm; T 13.5–23cm;
W ♀ 0.36–0.45kg, ♂ 0.65–0.82kg
Uniformly rich orange-brown with slightly paler underparts. Summer coat tends to be darker brown than winter coat. Face has a dark brown to black mask with a white muzzle and chin. Sometimes regarded as a separate species in Japan, Japanese Weasel *M. itatsi*. **Distribution and Habitat** Temperate Asia; C to Far-eastern Russia, N Mongolia, E and S China, Japan, N and S Korea, and marginally into Nepal, Bhutan and N Indochina. Inhabits forest, forest-steppe, dense grassland, vegetated scrubland and wetland. Occurs in cultivated areas, plantations and urban areas, but avoids open anthropogenic habitat. **Feeding Ecology** Small rodents and shrews are the most important prey. Also eats insects, earthworms, crustaceans, herptiles, birds, fledglings, eggs, and fruits including berries. Locally a significant nest predator of colonially nesting birds, e.g. Little Tern (Nakdong Estuary, S Korea), and occasionally raids domestic poultry. In urban habitats, e.g. Japan, scavenges from refuse, handouts (cakes, bread, etc) and fish remains from dock areas. Foraging mainly nocturno-crepuscular and terrestrial, though it swims well. **Social and Spatial Behaviour** Solitary. Adults live in stable ranges that overlap with those of other adults, but avoid each other except male-female pairs when breeding. Range size estimates only known from Japan: 1.3–1.7ha (♀s) and 1.4–4.4ha (♂s). **Reproduction and Demography** Seasonal. Mating late February–March; births early April–June. Litter size 2–12, averaging 5–6. Nests in burrows and tree and rock cavities, and under buildings and haystacks in urban-rural areas. MORTALITY Unknown. LIFESPAN 8.8 years in captivity. **Status and Threats** Widespread and common in many areas. Hunted for fur (legally in Russia) and meat, but hunting is probably a threat only at the range limits, e.g. Laos and Tibet. Red List LC.

STEPPE POLECAT *Mustela eversmanii*

Steppe Weasel
HB 29–56.2cm; T 7–18.3cm; W ♀ 0.4–0.8kg,
♂ 0.75–1.2kg
Formerly classified with the closely related European Polecat, which it strongly resembles, but tends to be lighter with a paler face and fewer dark guard hairs over the body. The two species overlap in E Europe and W Russia. Domestic ferret is thought to descend from both Steppe and European Polecats. **Distribution and Habitat** E Europe through S Russia, NC Asia, Mongolia, and N and C China. Mainly inhabits steppe, grassland and vegetated semi-desert to 2600m. Mostly avoids densely vegetated areas, including most forest types. Occurs in pastures and on farmland, including areas with crops such as corn and cereals. **Feeding Ecology** Small rodents and lagomorphs are the mainstay, particularly voles, mice, gerbils, hamsters, sousliks, marmots, zokors and pikas. Also eats grassland birds (especially in spring and autumn), including grouse, Ptarmigan and pheasants, plus reptiles, fish, eggs and invertebrates. Hunting mainly nocturno-crepuscular and terrestrial, with most prey captured and killed in burrows. Occasionally scavenges, including from carrion, and caches surplus kills, e.g. up to 50 sousliks in Russian cases. **Social and Spatial Behaviour** Poorly known. Adults are solitary, with large male ranges overlapping numerous female ranges. The only range estimates (NW Hungary) average 127ha (♀s) and 354ha (♂s). **Reproduction and Demography** Seasonal. Mating February–April; births April–June. Gestation 36–41 days. Litter size 4–14, averaging 8–9. Weaning at around 6 weeks; dispersal at 3–4 months. Females first breed at 9–10 months. MORTALITY and LIFESPAN Unknown. **Status and Threats** Has a wide range and is common in many areas; regarded as secure. Hunting, conversion of grassland habitats, and widespread hunting and poisoning of prey are potential threats, especially in E Europe, Mongolia and China. Legally hunted for fur in Russia. Red List LC.

**BLACK-FOOTED
FERRET**

**SIBERIAN
WEASEL**

STEPPE POLECAT

EUROPEAN MINK *Mustela lutreola* Plate 83

HB ♀ 32–40cm, ♂ 28–43cm; T 12–19cm;
W ♀ 0.4–0.6kg, ♂ 0.6–1.1kg
Uniformly glossy chestnut-brown to dark coffee-brown with slightly paler underparts. Upper lips, chin and tip of the muzzle are white, sometimes extending down the throat and chest. **Distribution and Habitat** W Russia, and isolated populations in Belarus, Estonia, Latvia, Ukraine, coastal Romania and the French–Spanish border. Rarely more than 100m from fresh water; preferred habitats are slow-flowing streams, lake shores and wetlands that do not completely freeze in winter. **Feeding Ecology** Semi-aquatic and eats small prey along waterways, especially small rodents such as Water Vole, Bank Vole, Wood Mouse and introduced Muskrat. Other prey includes small fish, frogs, crustaceans, aquatic insects, and occasionally birds and reptiles. Foraging solitary, nocturno-crepuscular, and both terrestrial and aquatic; an excellent swimmer, pursuing prey underwater in dives of 5–20 seconds. Caches surplus kills. **Social and Spatial Behaviour** Solitary. Large male ranges overlap one or more smaller female ranges, with low to moderate overlap within sexes (especially males). Ranges expand in autumn–winter to locate unfrozen water. Linear range size 0.3–5.1km (♀s) and 2.9–11.4km (♂s). Density estimates 2–12 minks/10km of waterway. **Reproduction and Demography** Seasonal. Mating February–April (early May in captivity); births April–June. Gestation 35–48 days, without delayed implantation (71–76 days reported from fur farms). Litter size 2–7, averaging 4–5. Weaning at 7–9 weeks, at which juveniles can capture small prey. MORTALITY Most recorded mortality is anthropogenic, mainly trapping. LIFESPAN 10 years in captivity. **Status and Threats** Over-hunting for fur and habitat destruction (water pollution, damming and draining) has extirpated the species from >80% of its historic range and at least 10 countries. Remaining populations are discontinuous and threatened by ongoing habitat degradation, illegal and accidental trapping, persecution and roadkills. Introduction of the larger, competitively aggressive American Mink is thought to be an additional factor. Red List EN.

AMERICAN MINK *Neovison vison*

HB ♀ 30–40cm, ♂ 33–43cm; T 12.8–23cm;
W ♀ 0.45–1.1kg, ♂ 0.6–2.3kg
Very similar to European Mink but larger, and the white markings on the chin rarely extend to the upper lip. More closely related to American weasels than to European Mink, and classified in its own genus with the now-extinct Sea Mink *N. macrodon*. **Distribution and Habitat** Most of Canada and the USA, including Alaska; absent from the southern USA. Introduced in Eurasia, Chile and Argentina. Inhabits densely vegetated waterways, marshes, wetlands, swamp forest and coastal beaches. **Feeding Ecology** Eats small mammals, birds, slow-swimming fish, herptiles, eggs and aquatic invertebrates. In N America, preys heavily on Muskrat, and population fluctuations of the two species are closely linked. Important nest predator of waterfowl and colonially nesting birds such as gulls and terns. Readily preys on domestic poultry. Foraging solitary, nocturno-crepuscular, and both terrestrial and aquatic. Dives to 6m deep and swims underwater for up to 35m. Caches surplus food. **Social and Spatial Behaviour** Solitary. Male ranges overlap one or more smaller female ranges. There can be high intrasexual overlap of ranges. Linear range size 1–4.2km (♀s) and 1.5–11.1km (♂s). Density estimates from North American wetlands vary between 1.6–5.4 minks/km² (Wisconsin) and 25–42/km² (Louisiana cypress-tupelo swamp). **Reproduction and Demography** Seasonal. Mating February–April (to early May in Alaska); births April–June. Gestation 39–79 days, with a brief period of delayed implantation (embryonic development 30–32 days). Litter size 2–8, averaging 4–5. Weaning at 7–9 weeks. MORTALITY In N America mainly from trapping. LIFESPAN Rarely >3 years in the wild, 8 in captivity. **Status and Threats** Widespread and common. The most important American fur bearer and widely trapped, with 400,000–700,000 wild mink harvested each year. Threatened in S Florida by wetland modification and degradation. Red List LC.

EUROPEAN POLECAT *Mustela putorius*

Western Polecat, Common Polecat, Ferret
HB ♀ 20.5–38.5cm, ♂ 29.5–46cm; T 7–14cm;
W ♀ 0.4–0.92kg, ♂ 0.5–1.7kg
Dark chocolate-brown to near black, with buff-yellow underfur that is obvious on the sides and neck. Face buff to silvery-white with a dark mask. The progenitor of the domestic ferret (probably with interbreeding from the closely related Steppe Polecat), and they hybridize with fertile offspring. **Distribution and Habitat** W Russia, N Scandinavia throughout W and C Europe, including the UK (absent from Ireland), and N Morocco. Inhabits lowland forest, wooded steppes, vegetated dunes, marshes, meadows and river valleys in open habitat. Occurs in agricultural areas and close to human settlements. **Feeding Ecology** Eats mainly small rodents, shrews, lagomorphs, frogs and toads. Other prey includes birds, reptiles, fish, eels, invertebrates and eggs; fruits and other plant items are eaten mainly by young animals. Takes domestic poultry, occasionally becoming a serious pest. Foraging solitary, mainly nocturno-crepuscular and terrestrial. Scavenges from carrion and occasionally from human refuse, birdfeeders and food scraps, especially during winter food shortages. Caches excess food, e.g. 40–120 frogs (W Russia). **Social and Spatial Behaviour** Solitary. Male ranges overlap one or more smaller female ranges. Female ranges tend to be exclusive; male ranges have variable overlap and are most exclusive during the breeding season. Range size 0.65–1.65km (♀s) and 1–3.05km (♂s), linear ranges along riverbanks, Poland, and 0.42–1.21km² (♀s) and 1.07–2.8km² (♂s) in fragmented forest-agricultural mosaic, Luxembourg. **Reproduction and Demography** Seasonal. Mating March–June; births April–early August, peaking mid-July. Gestation 40–43 days, without delayed implantation. Litter size 2–13, averaging 4–6; large litters may be due to hybridization with ferrets. Weaning at 5–7 weeks. MORTALITY Mostly anthropogenic; starvation important during severe winters, especially for newly independent subadults. LIFESPAN <5 years in the wild, 14 in captivity. **Status and Threats** Widespread and relatively common. Persecuted heavily in the past, leading to declines, e.g. the UK, but now protected and recovering in many areas. Localized threats include roadkills, prey declines, trapping for fur and persecution as a pest. Hybridization with feral ferrets erodes genetic purity in some area, e.g. the S UK. Red List LC.

EUROPEAN
MINK

AMERICAN
MINK

EUROPEAN
POLECAT

Domestic ferret

Common Weasel

HB 11.4–26cm; T 7–9cm; W 0.025–0.3kg

The world's smallest carnivore, tiny with a short tail that always lacks a black tip. Except in southern populations, the brown upper fur moults to pure white in winter; transitional winter forms occur in temperate areas. Sometimes considered a separate species in Egypt, Egyptian Weasel *M. subpalamta*. **Distribution and Habitat** Global, from approximately 35–40°N to the Arctic, encompassing Canada, Alaska, the NE USA, Eurasia from the UK to Japan, C Asia and N Africa. Introduced in New Zealand, Malta, Crete, the Azores and São Tomé. Occurs in virtually all habitats with cover and rodents, from forest to semi-desert, at sea level to 4000m. Inhabits agricultural, pastoral and urban areas. **Feeding Ecology** Small rodent specialist, hunting mainly mice, rats, Meadow Vole, Field Vole, Water Vole, lemmings and cotton rats. Larger mammals, including lagomorphs, moles and squirrels, are also killed, especially by male weasels. Other food includes birds, fledglings, eggs, herptiles, fish and invertebrates (mainly beetles and earthworms). Takes poultry in rare cases. Foraging solitary and cathemeral; constantly active to maintain a very fast metabolism, making 5–10 kills/24hr. Hunting mostly terrestrial; its tiny size and tubular body enables it to enter burrows and rodent snow tunnels. Also pursues prey into trees and water. Mostly takes live prey, but occasionally scavenges from carrion during winter. Caches surplus kills in burrows. **Social and Spatial Behaviour** Solitary and territorial. Territories are aggressively defended from same-sex adults, but males often abandon their territories during breeding in search of females. Ranges 0.2–7ha (♀s) and 0.6–26ha (♂s). Densities fluctuate extensively depending on rodent population cycles, e.g. 2.4–13 weasels/km² in Finland depending on vole numbers. **Reproduction and Demography** Aseasonal, with a peak in spring to late summer. Gestation 34–37 days, without delayed implantation. Litter size 1–19, averaging 4–10. Weaning at 4–7 weeks, and juveniles can kill at 6–7 weeks. Sexual maturity (both sexes) 3–4 months. MORTALITY Populations turn over very rapidly, with annual adult mortality of 75–97%, mainly from food shortages and predation. LIFESPAN Rarely >2 years in the wild, 10 in captivity. **Status and Threats** Widely distributed with a broad habitat tolerance. Apparently naturally rare in N America and has undergone local declines in some areas, e.g. the UK. Vulnerable to the use of rodenticides, and persecuted intensely in some areas as a predator of game birds. Red List LC.

STOAT *Mustela erminea*

Ermine, Short-tailed Weasel

HB 17–34cm; T 4.2–12cm; W 0.06–0.37kg

Small weasel with a rusty-brown to chocolate-brown summer coat that moults to white in winter, except in southern populations. Tail has a black tip. In N America (called the Short-tailed Weasel), it is smaller with a proportionally shorter tail than the sympatric Long-tailed Weasel. **Distribution and Habitat** Global, from approximately 35°N to the Arctic, throughout Eurasia, N America and Greenland. Introduced in New Zealand. Occurs in a very wide range of habitats, from Arctic tundra to semi-desert, at sea level to 3000m. Occurs on farmland and pasture. **Feeding Ecology** Very similar diet to Least Weasel's, but kills greater proportions of larger prey, especially rabbits and squirrels. Where lagomorphs are absent, small rodents are the mainstay. Prodigious nest predator of eggs and fledglings, and also eats adult birds, herptiles, invertebrates and fruits. Takes domestic poultry. Foraging mostly nocturnal and terrestrial, but driven by high energetic requirements that necessitate flexible foraging patterns. Caches surplus prey and occasionally scavenges from carrion and human refuse. **Social and Spatial Behaviour** Solitary and territorial. Large male ranges overlap one or more smaller female ranges, and adults repel same-sex intruders. Males abandon their territories during breeding in search of females. Range size 2–135ha (♀s) and 8–313ha (♂s). Density estimates 3–10 weasels/km², exceptionally reaching 22/km² during rodent irruptions. **Reproduction and Demography** Seasonal. Mating June–August; births April–May (following year). Gestation 223–378 days, with delayed implantation (embryonic development 28–30 days). Litter size 4–18, averaging 6–8. Weaning 4–12 weeks. Females sexually mature at 4–6 weeks, and may be pregnant before they are weaned. MORTALITY Annual mortality 40–54% (3–6-month-old subadults) to 78–83% (2.25–2.5-year-old adults), mainly from prey shortages and predation. LIFESPAN Rarely >3 years in the wild, 10 in captivity. **Status and Threats** Widely distributed and relatively common to abundant. Legally trapped in much of its range, and persecuted as a predator of game birds and poultry, but neither apparently represents a serious threat. Red List LC.

ALTAI WEASEL *Mustela altaica*

Altai Mountain Weasel, Alpine Weasel

HB 21.7–28.7cm; T 9–14.5cm; W ♀ 0.12–0.22kg, ♂ 0.22–0.35kg

Pale buff-brown upperparts, darkening slightly in summer, with pale cream to creamy-yellow underparts. Feet conspicuously creamy-white, and no black tip to tail. **Distribution and Habitat** C and N Asia, in the Himalayan, Pamir, Altai and Tien Shan mountains, as well as most of S Russia, China and Mongolia; uncertain in N Korea. Inhabits alpine meadows, grassland, steppes and rocky areas at 1500–5200m. Occurs around and in remote human habitation in sheds, barns and cellars, but avoids degraded agricultural landscapes. **Feeding Ecology** Poorly known. Small rodents and lagomorphs, especially pikas, zokors, hamsters, souslisks and voles are probably the main prey. Also eats birds (including Ptarmigan), eggs, herptiles, invertebrates and fruits. Caches surplus kills. **Social and Spatial Behaviour** Poorly known. Assumed to be similar to the solitary territorial patterns of most *Mustela* weasels. **Reproduction and Demography** Poorly known; likely seasonal. Mating February–March (Kazakhstan). Gestation 35–40 days, apparently without delayed implantation. Litter size 2–8, exceptionally to 13. MORTALITY Poorly known. Populations appear to undergo large fluctuations, as for other weasels. LIFESPAN Unknown. **Status and Threats** Declining across much of its range due to habitat degradation from over-grazing by livestock, and agricultural poisoning campaigns of prey. CITES Appendix III (China, India), Red List NT.

Summer

LEAST WEASEL

Winter

Summer

STOAT

Winter

ALTAI WEASEL

Felipe's Weasel

HB 21.7–22.5cm; T 11.1–12.2cm; W c. 0.12–0.15kg
Very small weasel, uniformly very dark brown including the tail, with pale buff-orange underparts and sometimes a small dark patch on the chest. Easily confused with Long-tailed Weasel, but lacks its black-tipped tail. Natural history and ecology are almost entirely unknown. **Distribution and Habitat** Restricted to the Andes of Colombia and N Ecuador, where it is known from only six physical records and five sightings. All records are from high Andean forest at 1123–2700m, usually in riparian areas and river valleys (though this is probably a sampling artifact). **Feeding**

Ecology Unknown. Assumed to be a rodent specialist, as other weasels. **Social and Spatial Behaviour** Unknown. All records are of single animals. **Reproduction and Demography** Unknown. **Status and Threats** Occurs in a very restricted range and is naturally rare based on the relative rates it appears in surveys and museum records, e.g. compared to Long-tailed Weasel. Only known from mid-high elevation forest habitat, which is under intense pressure in the Andes for forestry and conversion to agriculture or pasture. Local people kill weasels for depredation on poultry and domestic guinea pigs, though it is unclear whether Colombian Weasels are involved. Red List VU.

LONG-TAILED WEASEL *Mustela frenata*

HB ♀ 20.3–22.8cm; ♂ 22.8–26cm; T 7.6–15.2cm;
W ♀ 0.8–0.25kg, ♂ 0.16–0.45kg
Long-bodied weasel with a proportionally long tail measuring half to two-thirds the body length. Summer coat rich rusty-brown to chocolate-brown, with creamy-white to rich yellow underparts. Northern populations moult to pure white in winter; tail always has a black tip, regardless of region and season. Southern populations do not moult to white. They have distinctive white or yellow facial markings in C America, Mexico and the southern USA. **Distribution and Habitat** S Canada, the USA, Mexico, C America, Venezuela, Columbia, Ecuador, Peru, extreme NW Brazil and N Bolivia. Occurs in virtually all habitats, from Arctic-alpine to tropical, and absent only from true desert interiors. Most abundant in open woodland, brushland, riparian grassland, meadows and marshes. Occupies modified habitats including logged areas, fence rows, fields, cropland and pastures. **Feeding Ecology** Opportunistic generalist with a wide diet, but rodents, lagomorphs and insectivores are the mainstay. Common prey includes voles, deer mice, cotton rats, wood rats, chipmunks, shrews, moles, pikas, cottontails, Volcano Rabbit (Mexico) and Snowshoe Hare (mainly juveniles). Also takes birds, reptiles (including kingsnakes and Bullsnakes), insects and eggs; an important nest predator in some areas. Occasionally preys on small carnivores, mainly other small weasels. Takes domestic poultry, and there is a credible record of predation on 3-day-old piglets. Foraging mainly nocturno-crepuscular but flexible; diurnalism increases during winter. Does not hibernate. Hunts mainly by sight and smell, ceaselessly investigating burrows, crevices, nests and log hollows for quarry. Prey is subdued by a distinctive hold in which it tightly entwines the victim with its body and administers a killing bite to the nape (small prey) or a suffocating bite to the throat. Adult weasels make 3–4 kills a day; a Colorado

population estimated at 8000 weasels killed an estimated 11,000,000 prey items (mostly rodents) annually. Caches surplus prey, and scavenges from carrion and human refuse. **Social and Spatial Behaviour** Solitary, with enduring ranges that overlap, but with exclusive core areas. Adults mostly avoid each other except during the breeding season, when males seek out females and aggressively repel other males. Mixed-sex pairs occasionally associate temporarily outside the breeding season, e.g. sharing burrows. Male ranges encompass one or more female ranges, and males attempt to increase their range size during breeding to encounter more females. Average range size 0.52km² (♀s) and 1.8km² (♂s) in fragmented agricultural habitat, Indiana, but 0.1–0.24km² (sexes combined) in better quality habitat. Density estimates 0.4–0.8 weasels/km² (agricultural habitat) to 19–38/km² (high-quality oak forest and marshland). **Reproduction and Demography** Seasonal. Mating July–August (N America); births April–May (following year). Gestation 205–337 days, with delayed implantation (embryonic development 21–28 days). Litter size 4–9, averaging 6. Kittens weaned at around 35–36 days, independence at 3 months. Females sexually mature at 9–12 weeks; males typically breed in their second year. MORTALITY Rates poorly known, but weasel populations turn over rapidly, with high mortality balanced by high reproduction. Natural mortality is mainly from predation, especially by raptors, foxes and Coyote. LIFESPAN Rarely over 3 years in the wild. **Status and Threats** The most widespread mustelid in the western hemisphere, and it has the greatest habitat tolerance of American weasels. Populations fluctuate significantly and rapidly, depending on prey availability; abundance declines under agricultural intensification, prompting local extinctions when combined with low prey numbers. Legally trapped in its N American range, but no longer sought after for commercial trade. Red List LC.

AMAZON WEASEL *Mustela africana*

Tropical Weasel

HB 24–38cm; T 16–21cm; W c. 0.1–0.3kg
Small weasel, uniformly red-brown to dark brown including the tail. Underparts pale buff to cream, bisected with a dark brown stripe running from the lower belly to the chest or throat; stripe is discontinuous in 2–3 segments in some individuals. Natural history and ecology are very poorly known. **Distribution and Habitat** Restricted to the Amazon Basin of Bolivia, Brazil, S Colombia, E Ecuador and E Peru. All records come from tropical lowland forest. **Feeding Ecology** Unknown. Assumed to

be a rodent specialist, as other weasels. Foraging possibly diurnal; one sighting was at 10.00h. **Social and Spatial Behaviour** Unknown. Most records are of single animals, with one sighting of four animals, probably a mother-offspring group. **Reproduction and Demography** Unknown. **Status and Threats** Status largely unknown. Occurs over a wide range where much of the habitat is well preserved, and is recorded from a number of protected areas. Habitat tolerances are unknown, but deforestation and conversion to agriculture and pasture are considered the primary threats. Red List LC.

COLOMBIAN
WEASEL

Winter

LONG-TAILED
WEASEL

Summer

AMAZON
WEASEL

YELLOW-BELLIED WEASEL *Mustela kathiah*

Plate 86

HB 20–29cm; T 12.5–18cm; W c. 0.15–0.3kg
Small weasel, russet-brown to dark maroon-brown, with rich buff-orange or yellowish underparts and a conspicuously white lower face. Feet sometimes have white flecking or small patches. Natural history and ecology are very poorly known. **Distribution and Habitat** SE China (about 80% of its range), N and E Indochina, N Myanmar, and the Himalayas from Bhutan to Kashmir. Inhabits mainly temperate forest, steppes and desert at 1000–5200m. Occurs in highly degraded forest patches and scrub mosaics, e.g. in Hong Kong at sea level to 200m, and in N Laos. **Feeding Ecology** Unknown. Assumed to prey mainly on rodents, other small vertebrates and invertebrates, similar to other weasels. **Social and Spatial Behaviour** Unknown. All records are of single animals. **Reproduction and Demography** Unknown. **Status and Threats** Status poorly known, but the species occurs in highly degraded habitat and in areas that are under intense hunting pressure, e.g. near Phongsaly town, Laos, suggesting that it is tolerant of habitat loss and resilient to anthropogenic pressure. CITES Appendix III (India), Red List LC.

STRIPE-BACKED WEASEL *Mustela strigidorsa*

Back-striped Weasel
HB 25–32.5cm; T 13–20.5cm; W 0.7–2kg
Small weasel, dark mahogany-brown to rich russet-brown, with a fine silvery-white or creamy dorsal stripe from the crown to the base of the tail. Cheeks and throat creamy-white to rich yellow-cream, narrowing to a thin yellow line running along the belly to the groin. Natural history and ecology are very poorly known. **Distribution and Habitat** NE India, N and C Myanmar, S China, N Thailand, N and C Laos and Vietnam. Most records are from evergreen hill and montane forests to 2500m, and associated secondary forest, bamboo stands, scrub and grassland. **Feeding Ecology** Poorly known. One was observed to attack a bandicoot rat estimated at three times its size, and another was seen carrying a suspected White-bellied Rat. Local people report that it takes domestic poultry; two animals for sale in a Laotian wildlife market were killed apparently while raiding a chicken coop, and the only individual captured for science (NE Thailand) was baited with a live chicken. **Social and Spatial Behaviour** Unknown. All records are of single animals. Very rarely appears in wildlife surveys, including by camera trapping, suggesting it is naturally rare, but this remains to be confirmed. **Reproduction and Demography** Unknown. **Status and Threats** Status very poorly known. Does not have especially high economic value and appears in wildlife markets moderately frequently, mainly in China, where 3000–4000 animals were harvested annually in the 1970s. Sometimes killed as a perceived poultry pest. Red List LC.

MALAY WEASEL *Mustela nudipes*

HB 30–36cm; T 24–26cm; W c. 1kg
Unmistakable weasel, bright orange to reddish-brown, usually with a paler orange neck and a creamy-white to pale orange head. Distal half of the tail is pale orange to white. Natural history and ecology are very poorly known. **Distribution and Habitat** Sumatra, Borneo, Peninsular Malaysia and S Thailand; reports from Java are erroneous. Recorded from lowland and hill tropical forests, heath forest, swamp forest, montane forest and montane scrub from sea level to 1700m. Sometimes occurs in degraded habitat, including exotic plantations and peri-urban areas. **Feeding Ecology** Unknown. Diet assumed to be similar to that of other small *Mustela* weasels, i.e. small rodents, birds, eggs and herptiles. **Social and Spatial Behaviour** Unknown. Most records are of single animals, with one sighting of two animals, always on the ground or on ground-level objects such as rocks and logs. **Reproduction and Demography** Unknown. **Status and Threats** Status very poorly known. Apparently widespread within its range, but rarely recorded and difficult to see. Has been recorded in degraded habitats, including urban areas, suggesting some tolerance for anthropogenic pressures; its presence in higher altitude forest buffers it to some extent from the extreme pressures on lowland forest in the Sundaic region. Red List LC.

INDONESIAN MOUNTAIN WEASEL *Mustela lutreolina*

HB 29.7–32.1cm; T 13.6–17cm; W 0.3–0.34kg
Small dark weasel, uniformly dark brown with slightly paler underparts, sometimes with inconspicuous creamy-white patches on the chin, throat, upper chest and groin. Otherwise unmarked. One of the least known carnivore species; known only from 15 museum records and a single sighting in the wild. Natural history and ecology are very poorly known. **Distribution and Habitat** Java and S Sumatra (uncertain elsewhere in Sumatra), where it is restricted to highlands at 1000–3000m. All records are from hill and montane forests, with a single sighting taking place in alpine scrub above the treeline at 3000m (Mt Kerinci, Sumatra). **Feeding Ecology** Unknown. Diet assumed to be similar to that of other small *Mustela* weasels, i.e. small rodents, birds, eggs and herptiles. **Social and Spatial Behaviour** Unknown. All records are of single animals, except for the single confirmed wild sighting, which was of four animals moving in a 'train' fashion, almost certainly a female and three large juveniles. Very rarely appears in wildlife surveys, including by camera trapping, suggesting it is naturally rare, but this remains to be confirmed. **Reproduction and Demography** Unknown. **Status and Threats** Status very poorly known. Its montane forest habitat is under significant pressure from illegal forestry and conversion, e.g. to coffee plantations, especially in Java. However, the species' habitat tolerances and its dependence on undisturbed forest are unclear. Not especially valued for trade, and hunting is less intense in montane habitat than in comparable lowland areas. Red List DD.

YELLOW-BELLIED
WEASEL

STRIPE-BACKED
WEASEL

MALAY WEASEL

INDONESIAN MOUNTAIN WEASEL

SKULLS

Finding a carnivore skull in the field is a rare event, but for the very fortunate or those who spend a lot of time outdoors, the following section provides a useful guide for identification.

The Carnivora are distinguished by their unique cheek teeth. Early in carnivoran evolution, the upper last premolar and lower first molar evolved into a pair of flattened cutting shears called the carnassials. The carnassials are an unmistakable signature of the true carnivores. In most mammals that evolved carnivory, the posterior molars became the flattened carnassial form for slicing meat. In true carnivores, however, the carnassials are located in the middle of the tooth row, in front of the posterior molars. This arrangement freed up the molars in the Carnivora for different purposes. Over time, in different species, they became more robust, with reduced shearing edges and greater crushing surfaces, enabling bones and insects, and seeds, fruits and other plant matter to be eaten.

The resulting dual-purpose dentition incorporates teeth that both shear (the carnassials) and crush (the molars). It is typified in dogs, and in fact most modern carnivores have elements of both tooth types. From this type of dentition, a transition to an even more herbivorous diet is possible – the carnassials themselves gradually become less blade-like and more molar-like to deal with tough vegetation. In the Giant Panda and Red Panda, the carnassial shear has disappeared entirely and both species are completely herbivorous. In an alternative evolutionary route, the rear molars gradually shrink until there is little or no crushing ability and the dentition is all carnassial, with teeth that can only slice meat. This hyper-carnivorous dentition type is most advanced in the cats, and modern cats have lost all their rear molars or retain them only as residual, useless pegs.

The range of carnassial-molar combinations in carnivores furnished them with unique evolutionary flexibility to rise to dominance over other early mammal groups that used their rear molars for carnivory and lacked that plasticity; most of those groups are now extinct.

Although not as uniquely carnivoran as the carnassials, the canines are the most distinctive and emblematic teeth in carnivore skulls. The canines are used primarily for securing and killing prey, and are also deployed in defensive bites, for instance during territorial fights. The smallest and foremost teeth in the skull are the incisors, used mainly for grooming fur as well as for some food handling, for example for plucking fur or feathers or stripping peel from fruits.

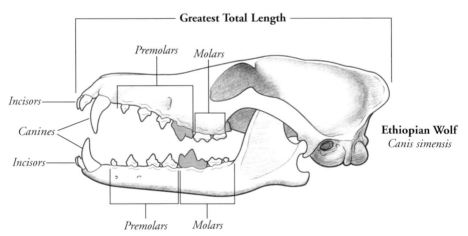

Ethiopian Wolf
Canis simensis

Skull
An Ethiopian Wolf skull, showing the different teeth categories present in carnivores. In the Carnivora, the unique carnassials (shaded) are made up of the upper last premolar and the lower first molar. Measurements with the illustrations that follow give skull length for each species (in scientific nomenclature known as the Greatest Total Length, referring to the length between the two most distant points on the skull).

FELIDAE

Plate 1

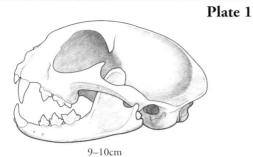

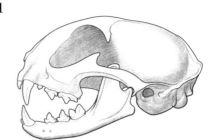

9–10cm

Chinese Mountain Cat
Felis bieti

8–11.2cm

Wildcat
Felis silvestris

Plate 2

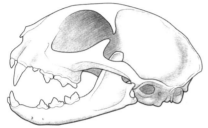

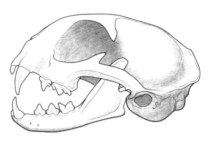

7.2–8.7cm

Black-footed Cat
Felis nigripes

8–9.5cm

Sand Cat
Felis margarita

9.8–14cm

Jungle Cat
Felis chaus

Plate 3

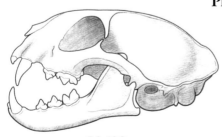

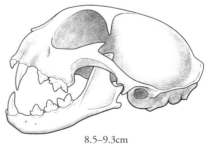

7.9–10.2cm

Leopard Cat
Prionailurus bengalensis

8.5–9.3cm

Pallas's Cat
Otocolobus manul

Plate 4

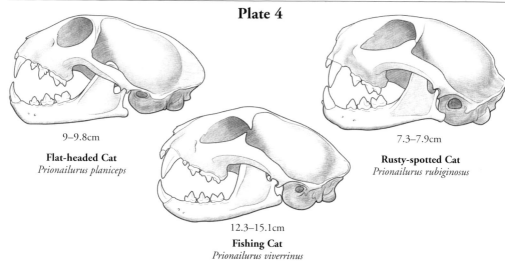

9–9.8cm

Flat-headed Cat
Prionailurus planiceps

7.3–7.9cm

Rusty-spotted Cat
Prionailurus rubiginosus

12.3–15.1cm

Fishing Cat
Prionailurus viverrinus

Plate 5

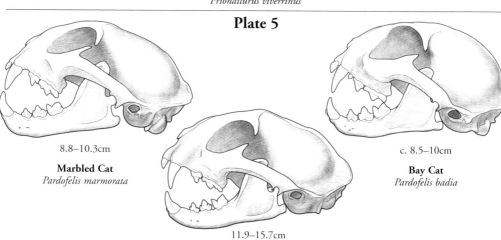

8.8–10.3cm

Marbled Cat
Pardofelis marmorata

c. 8.5–10cm

Bay Cat
Pardofelis badia

11.9–15.7cm

Asiatic Golden Cat
Pardofelis temminckii

Plate 6

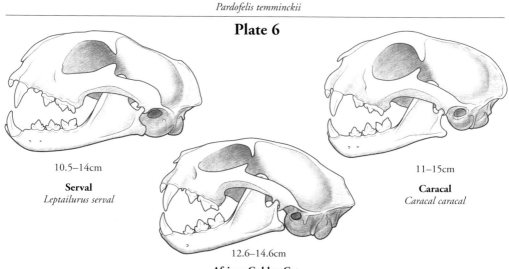

10.5–14cm

Serval
Leptailurus serval

11–15cm

Caracal
Caracal caracal

12.6–14.6cm

African Golden Cat
Profelis aurata

Plate 7

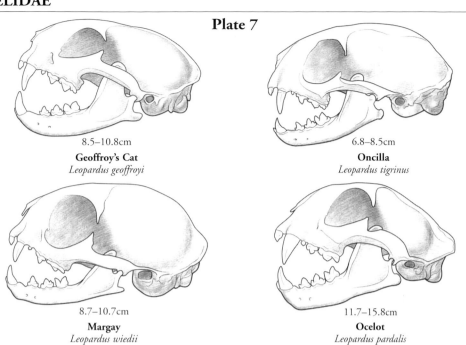

8.5–10.8cm

Geoffroy's Cat
Leopardus geoffroyi

6.8–8.5cm

Oncilla
Leopardus tigrinus

8.7–10.7cm

Margay
Leopardus wiedii

11.7–15.8cm

Ocelot
Leopardus pardalis

Plate 8

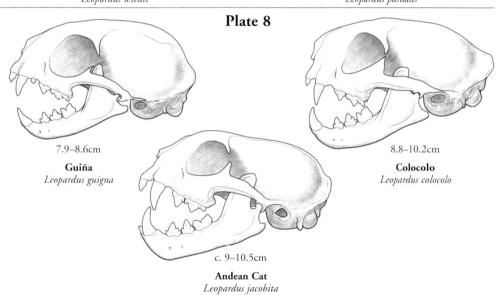

7.9–8.6cm

Guiña
Leopardus guigna

8.8–10.2cm

Colocolo
Leopardus colocolo

c. 9–10.5cm

Andean Cat
Leopardus jacobita

Plate 9

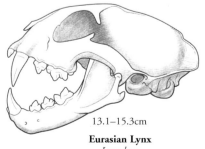

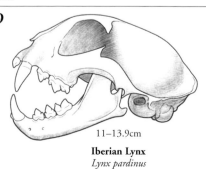

13.1–15.3cm

Eurasian Lynx
Lynx lynx

11–13.9cm

Iberian Lynx
Lynx pardinus

Plate 10

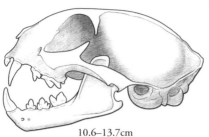

10.6–13.7cm
Bobcat
Lynx rufus

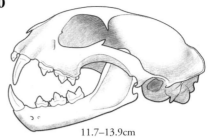

11.7–13.9cm
Canada Lynx
Lynx canadensis

Plate 11

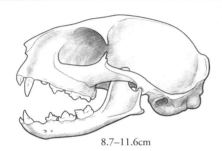

8.7–11.6cm
Jaguarundi
Puma yagouaroundi

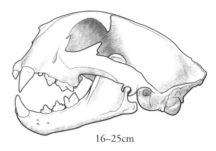

16–25cm
Puma
Puma concolor

Plate 12

15–19.3cm
Cheetah
Acinonyx jubatus

Plate 13

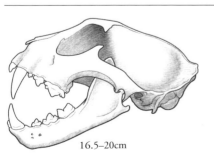

16.5–20cm
Snow Leopard
Panthera uncia

14.3–18cm
Clouded Leopards
Neofelis nebulosa, N. diardi

FELIDAE

Plate 14

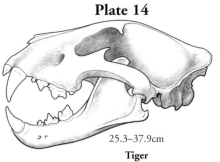

25.3–37.9cm

Tiger
Panthera tigris

Plate 15

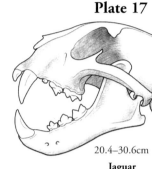

26.7–42cm

Lion
Panthera leo

Plate 16

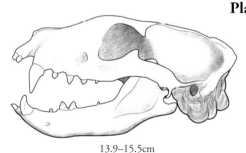

17–28.2cm

Leopard
Panthera pardus

Plate 17

20.4–30.6cm

Jaguar
Panthera onca

HYAENIDAE

Plate 18

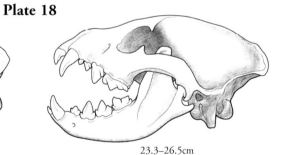

13.9–15.5cm

Aardwolf
Proteles cristata

23.3–26.5cm

Striped Hyaena
Hyaena hyaena

Plate 19

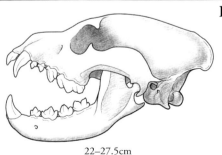

22–27.5cm

Brown Hyaena
Parahyaena brunnea

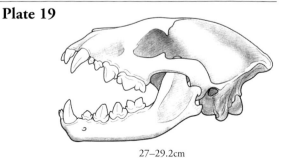

27–29.2cm

Spotted Hyaena
Crocuta crocuta

Plate 20

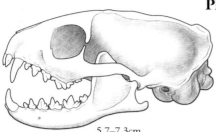

5.7–7.3cm

Small Indian Mongoose
Herpestes auropunctatus

6.8–8.4cm

Small Asian Mongoose
Herpestes javanicus

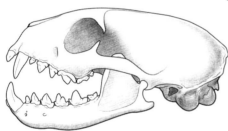

8.3–10cm

Short-tailed Mongoose
Herpestes brachyurus

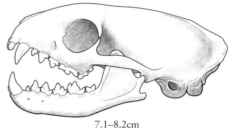

7.1–8.2cm

Indian Grey Mongoose
Herpestes edwardsii

Plate 21

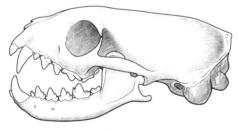

c. 8–9cm

Collared Mongoose
Herpestes semitorquatus

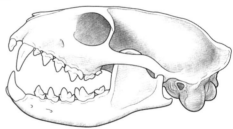

8.5–9.8cm

Crab-eating Mongoose
Herpestes urva

8–9cm

Ruddy Mongoose
Herpestes smithii

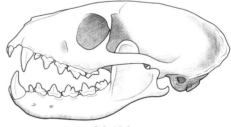

9.8–10.3cm

Striped-necked Mongoose
Herpestes vitticollis

HERPESTIDAE

Plate 22

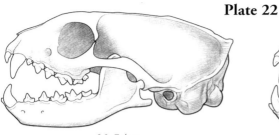

6.2–7.4cm

Cape Grey Mongoose
Herpestes pulverulentus

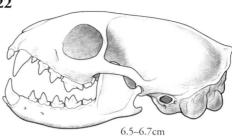

6.5–6.7cm

Common Slender Mongoose
Herpestes sanguineus

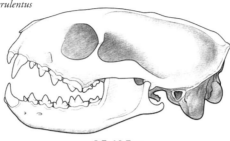

8.7–10.7cm

Egyptian Mongoose
Herpestes ichneumon

Plate 23

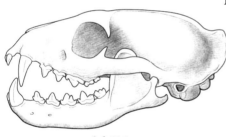

9.6–11.1cm

Marsh Mongoose
Atilax paludinosus

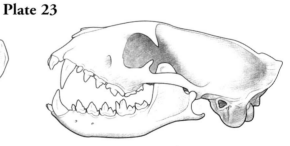

10.5–11.4cm

Long-nosed Mongoose
Herpestes naso

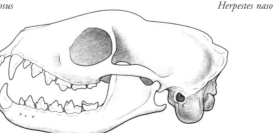

9.6–11.6cm

White-tailed Mongoose
Ichneumia albicauda

Plate 24

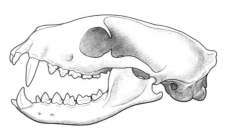

9.3–9.7cm

Bushy-tailed Mongoose
Bdeogale crassicauda

c. 9–11cm

Jackson's Mongoose
Bdeogale jacksoni

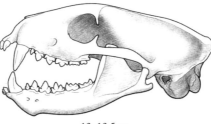

12–12.5cm

Black-legged Mongoose
Bdeogale nigripes

Plate 25

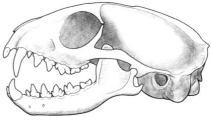

5.8–6.5cm

Meerkat
Suricata suricatta

5.8–7.8cm

Yellow Mongoose
Cynictis penicillata

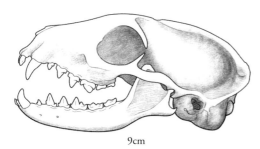

9cm

Selous's Mongoose
Paracynictis selousi

9cm

Meller's Mongoose
Rhynchogale melleri

Plate 26

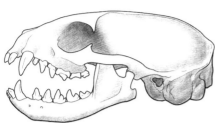

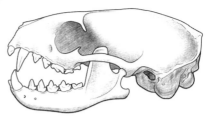

c. 6cm

Pousargues's Mongoose
Dologale dybowskii

5–6cm

Somali Dwarf Mongoose
Helogale hirtula

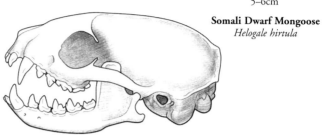

5–6cm

Common Dwarf Mongoose
Helogale parvula

Plate 27

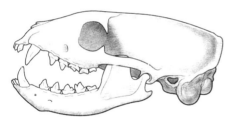

6.7–7.3cm

Gambian Mongoose
Mungos gambianus

6.2–7.9cm

Banded Mongoose
Mungos mungo

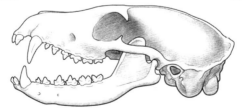

9.5–10cm

Liberian Mongoose
Liberiictis kuhni

HERPESTIDAE

Plate 28

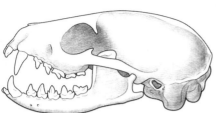

6–6.7cm

Angolan Cusimanse
Crossarchus ansorgei

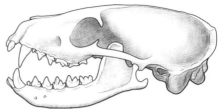

6.5–7cm

Flat-headed Cusimanse
Crossarchus platycephalus

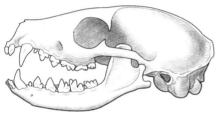

6.4–7.4cm

Common Cusimanse
Crossarchus obscurus

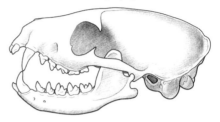

6.5–8cm

Alexander's Cusimanse
Crossarchus alexandri

EUPLERIDAE

Plate 29

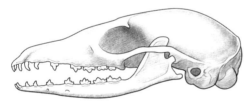

8.4–9.7cm

Falanouc
Eupleres goudotii

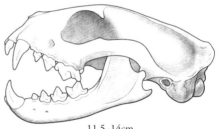

11.5–14cm

Fosa
Cryptoprocta ferox

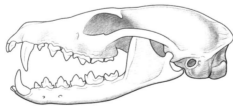

10.4–11cm

Fanaloka
Fossa fossana

EUPLERIDAE

Plate 30

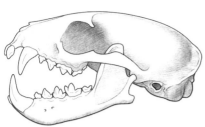

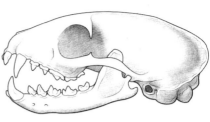

7.4cm

Broad-striped Vontsira
Galidictis fasciata

5.8–6cm

Narrow-striped Boky
Mungotictis decemlineata

6.5–7.2cm

Ring-tailed Vontsira
Galidia elegans

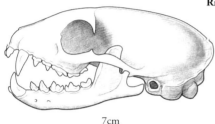

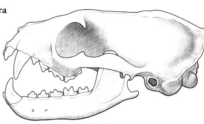

7cm

Brown-tailed Vontsira
Salanoia concolor

c. 7–8.5cm

Grandidier's Vontsira
Galidictis grandidieri

PRIONODONTIDAE & VIVERRIDAE

Plate 31

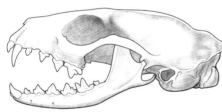

7.2–7.9 cm

Banded Linsang
Prionodon linsang

6.4–7.3cm

Spotted Linsang
Prionodon pardicolor

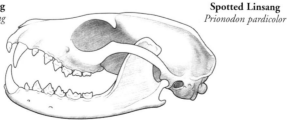

9.2–12cm

Small-toothed Palm Civet
Arctogalidia trivirgata

Plate 32

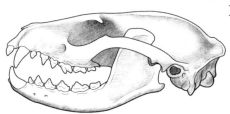

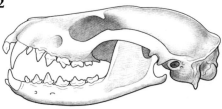

10–10.6cm

Golden Palm Civet
Paradoxurus zeylonensis

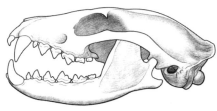

10.9–11.5cm

Brown Palm Civet
Paradoxurus jerdoni

c. 11–14.5cm

Sulawesi Palm Civet
Macrogalidia musschenbroekii

Plate 33

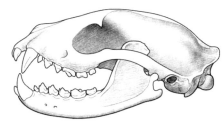

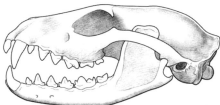

11.4–13.7cm

Masked Palm Civet
Paguma larvata

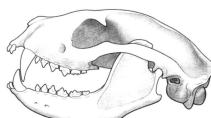

9.5–12cm

Common Palm Civet
Paradoxurus hermaphroditus

13–15.6cm

Binturong
Arctictis binturong

Plate 34

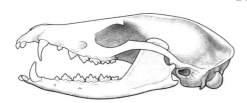

c. 9.5–11.3cm

Owston's Civet
Chrotogale owstoni

9.6–10.7cm

Banded Civet
Hemigalus derbyanus

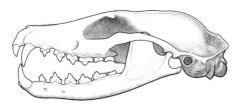

c. 9.5–10.5cm

Hose's Civet
Diplogale hosei

10.9–12.5cm

Otter Civet
Cynogale bennettii

Plate 35

c. 14.5cm

Malabar Civet
Viverra civettina

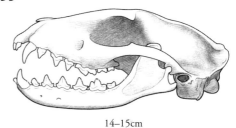

14–15cm

Large-spotted Civet
Viverra megaspila

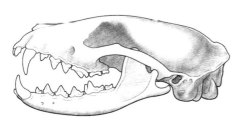

9.2–10.2cm

Malay Civet
Viverra tangalunga

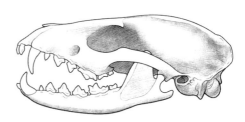

12.7–14.4cm

Large Indian Civet
Viverra zibetha

Plate 36

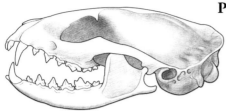

8.8–10.7cm

Small Indian Civet
Viverricula indica

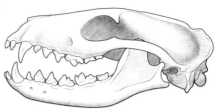

11.5–15.5cm

African Civet
Civettictis civetta

Plate 37

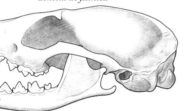

7.7–9cm

Abyssinian Genet
Genetta abyssinica

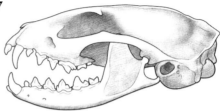

7.8–9.3cm

Hausa Genet
Genetta thierryi

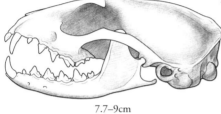

7.7–9.8cm

Servaline Genet
Genetta servalina

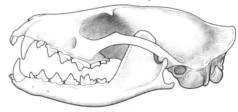

10.7–12.1cm

Giant Genet
Genetta victoriae

Plate 38

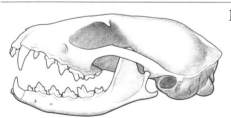

8.4–9.2cm

Small-spotted Genet
Genetta genetta

8.7–9.4cm

Miombo Genet
Genetta angolensis

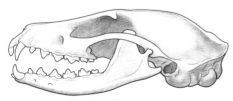

9.4–10cm

Johnston's Genet
Genetta johnstoni

9.7–11.4cm

Aquatic Genet
Genetta piscivora

Plate 39

8.5–9.1cm

Rusty-spotted Genet
Genetta maculata

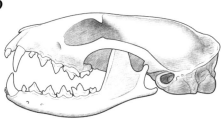

8.6–9.3cm

Cape Genet
Genetta tigrina

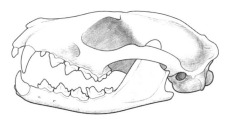

9.2–10.1cm

Pardine Genet
Genetta pardina

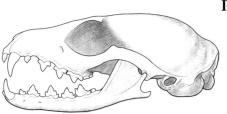

c. 9–10cm

King Genet
Genetta poensis

9.6–10.8cm

Bourlon's Genet
Genetta bourloni

Plate 40

6.4–7.3cm

Central African Oyan
Poiana richardsonii

8.4–11.1cm

African Palm Civet
Nandinia binotata

Plate 41

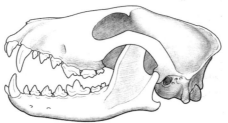

18.5–24cm

African Wild Dog
Lycaon pictus

Plate 42

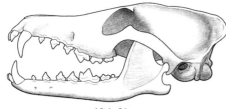

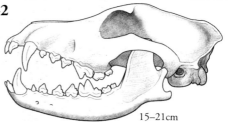

19.1–21cm

Ethiopian Wolf
Canis simensis

15–21cm

Dingo
Canis lupus dingo

Plate 43

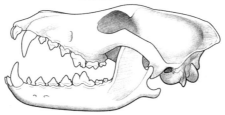

21.1–29.4 cm

Grey Wolf
Canis lupus

Plate 44

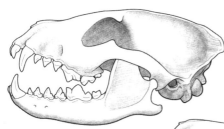

16.6–20cm

Dhole
Cuon alpinus

17–22cm

Coyote
Canis latrans

19.7–26.1cm

Red Wolf
Canis rufus

Plate 45

14.3–17.9cm

Golden Jackal
Canis aureus

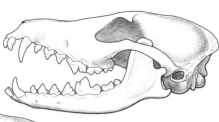

14.3–17cm

Black-backed Jackal
Canis mesomelas

15–17.3cm

Side-striped Jackal
Canis adustus

Plate 46

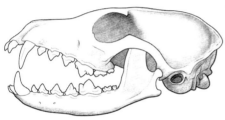

12–13.5cm

Arctic Fox
Alopex lagopus

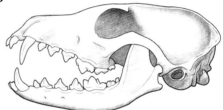

11.7–15.7cm

Red Fox
Vulpes vulpes

Plate 47

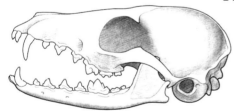

10–12cm

Kit Fox
Vulpes macrotis

10.8–12cm

Swift Fox
Vulpes velox

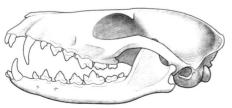

c. 10–12.5cm

Indian Fox
Vulpes bengalensis

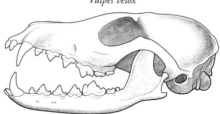

9.8–11.8cm

Corsac Fox
Vulpes corsac

Plate 48

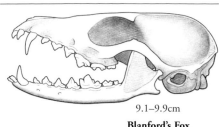

9.1–9.9cm

Blanford's Fox
Vulpes cana

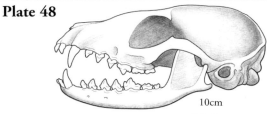

10cm

Pale Fox
Vulpes pallida

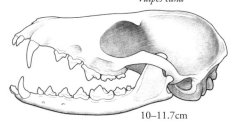

10–11.7cm

Rüppell's Fox
Vulpes ruepellii

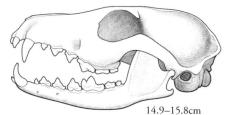

14.9–15.8cm

Tibetan Fox
Vulpes ferrilata

Plate 49

8.3–8.8cm

Fennec Fox
Vulpes zerda

9.3–11.5cm

Cape Fox
Vulpes chama

10.4–12.5cm

Bat-eared Fox
Otocyon megalotis

Plate 50

9.7–11cm

Island Fox
Urocyon littoralis

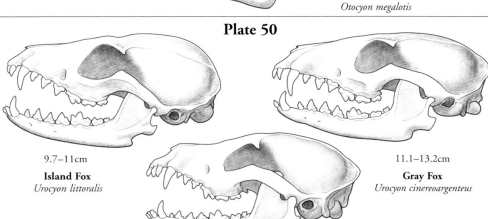

11.1–13.2cm

Gray Fox
Urocyon cinereoargenteus

11–12.7cm

Raccoon Dog
Nyctereutes procyonoides

Plate 51

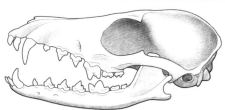

11–14cm

Chilla
Pseudalopex griseus

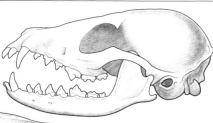

c. 11–13cm

Sechuran Fox
Pseudalopex sechurae

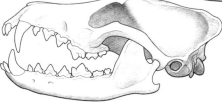

16–18.8cm

Culpeo
Pseudalopex culpaeus

Plate 52

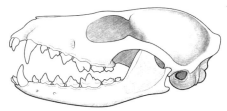

11.1–11.8cm

Hoary Fox
Pseudalopex vetulus

12.9–15.4cm

Pampas Fox
Pseudalopex gymnocercus

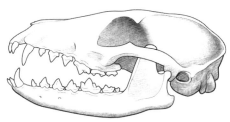

11.2–13.8cm

Crab-eating Fox
Cerdocyon thous

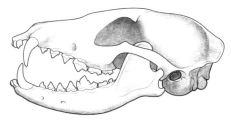

16.1–17.2cm

Short-eared Dog
Atelocynus microtis

Plate 53

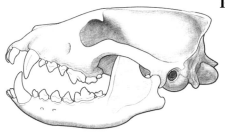

c. 11–13cm

Bush Dog
Speothos venaticus

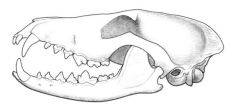

20–23.5cm

Maned Wolf
Chrysocyon brachyurus

Plate 54

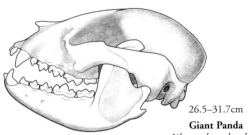

26.5–31.7cm

Giant Panda
Ailuropoda melanoleuca

Plate 55

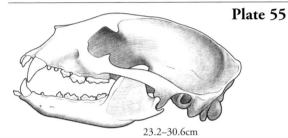

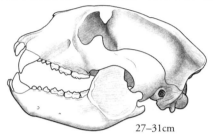

23.2–30.6cm

Asiatic Black Bear
Ursus thibetanus

27–31cm

Andean Bear
Tremarctos ornatus

Plate 56

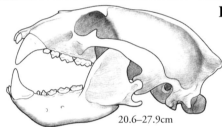

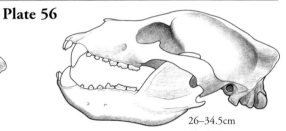

20.6–27.9cm

Sun Bear
Helarctos malayanus

26–34.5cm

Sloth Bear
Melursus ursinus

Plate 57

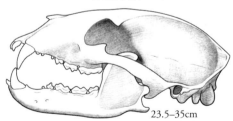

23.5–35cm

American Black Bear
Ursus americanus

Plate 58

26–42.2cm

Brown Bear
Ursus arctos

Plate 59

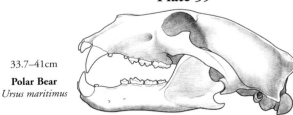

33.7–41cm

Polar Bear
Ursus maritimus

Plate 60

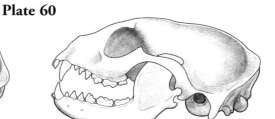

9.3–13.6cm

Northern Raccoon
Procyon lotor

12.9–14.3cm

Crab-eating Raccoon
Procyon cancrivorus

Plate 61

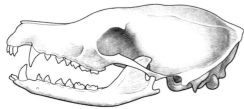

9.7–11.5cm

Mountain Coati
Nasuella olivacea

10.9–12.5cm

South American Coati
Nasua nasua

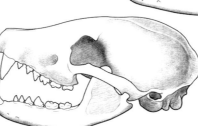

12–13.8cm

White-nosed Coati
Nasua narica

Plate 62

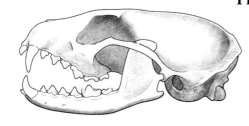

7–8.3cm

Ringtail
Bassariscus astutus

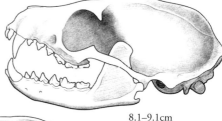

8.1–9.1cm

Cacomistle
Bassariscus sumichrasti

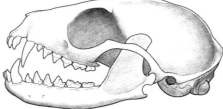

7.5–8.5 cm

Northern Olingo
Bassaricyon gabbi

PROCYONIDAE & AILURIDAE

Plate 63

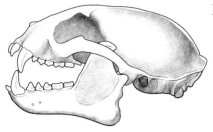

8.4–10cm

Kinkajou
Potos flavus

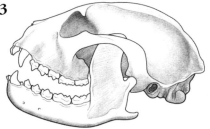

10.5–12.4cm

Red Panda
Ailurus fulgens

MEPHITIDAE

Plate 64

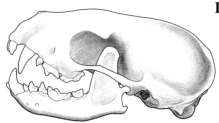

c. 5–7cm

Humboldt's Hog-nosed Skunk
Conepatus humboldtii

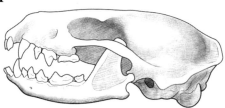

5.5–7.5cm

Molina's Hog-nosed Skunk
Conepatus chinga

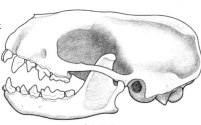

c. 6.5–8.5cm

Striped Hog-nosed Skunk
Conepatus semistriatus

Plate 65

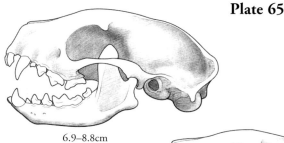

6.9–8.8cm

Striped Skunk
Mephitis mephitis

6.5–8.5cm

American Hog-nosed Skunk
Conepatus leuconotus

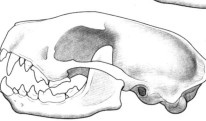

5.5–7.3cm

Hooded Skunk
Mephitis macroura

MEPHITIDAE

Plate 66

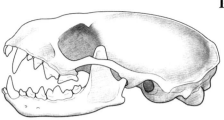

c. 4.9–6.4cm

Eastern Spotted Skunk
Spilogale putorius

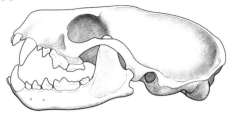

4.9–6.4cm

Western Spotted Skunk
Spilogale gracilis

Plate 67

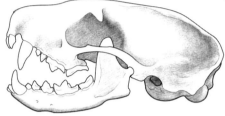

4.2–5cm

Pygmy Spotted Skunk
Spilogale pygmaea

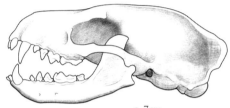

c. 7cm

Palawan Stink Badger
Mydaus marchei

8.4–10.4cm

Sunda Stink Badger
Mydaus javanensis

MUSTELIDAE

Plate 68

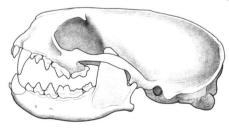

10–13cm

North American Otter
Lontra canadensis

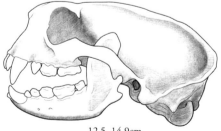

12.5–14.9cm

Sea Otter
Enhydra lutris

Plate 69

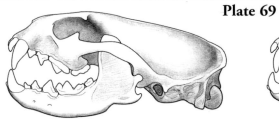

9–11.2cm

Marine Otter
Lontra felina

10.3–11.8cm

Southern River Otter
Lontra provocax

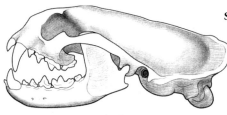

9.4–12cm

Neotropical Otter
Lontra longicaudis

Plate 70

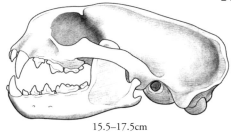

15.5–17.5cm

Giant Otter
Pteronura brasiliensis

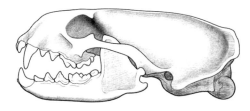

10–13cm

Eurasian Otter
Lutra lutra

Plate 71

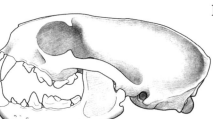

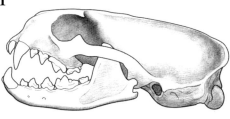

10–12cm

Hairy-nosed Otter
Lutra sumatrana

8.5–9.5cm

Asian Small-clawed Otter
Aonyx cinereus

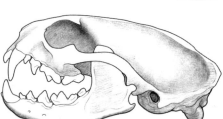

11–13.2cm

Smooth-coated Otter
Lutrogale perspicillata

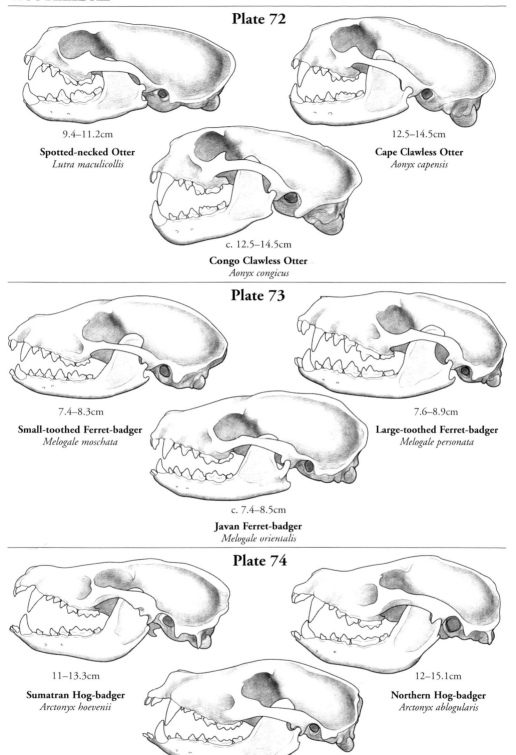

Plate 72

9.4–11.2cm

Spotted-necked Otter
Lutra maculicollis

12.5–14.5cm

Cape Clawless Otter
Aonyx capensis

c. 12.5–14.5cm

Congo Clawless Otter
Aonyx congicus

Plate 73

7.4–8.3cm

Small-toothed Ferret-badger
Melogale moschata

7.6–8.9cm

Large-toothed Ferret-badger
Melogale personata

c. 7.4–8.5cm

Javan Ferret-badger
Melogale orientalis

Plate 74

11–13.3cm

Sumatran Hog-badger
Arctonyx hoevenii

12–15.1cm

Northern Hog-badger
Arctonyx ablogularis

15–17.7cm

Greater Hog-badger
Arctonyx collaris

MUSTELIDAE

Plate 75

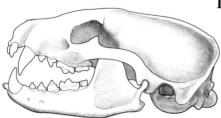

10.5–13.2cm

American Badger
Taxidea taxus

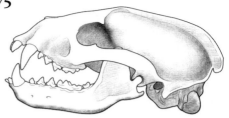

11.6–14.7cm

Eurasian Badger
Meles meles

Plate 76

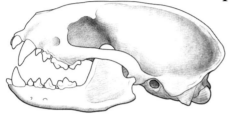

11.2–15.6cm

Honey Badger
Mellivora capensis

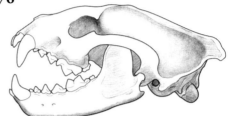

13.6–17.5cm

Wolverine
Gulo gulo

Plate 77

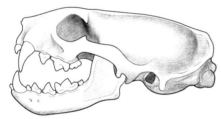

c. 4.8–5.7cm

Patagonian Weasel
Lyncodon patagonicus

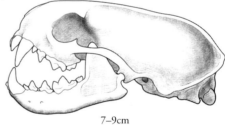

7–9cm

Lesser Grison
Galictis cuja

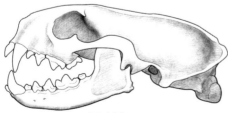

8.9–10.8cm

Greater Grison
Galictis vittata

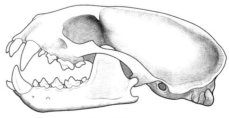

10.3–12.5cm

Tayra
Eira barbara

Plate 78

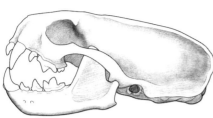

4.8–5.7cm

Striped Weasel
Poecilogale albinucha

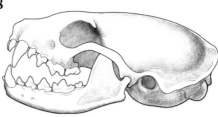

5.2–6cm

Libyan Weasel
Ictonyx libyca

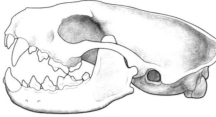

5.9–6.8cm

Zorilla
Ictonyx striatus

Plate 79

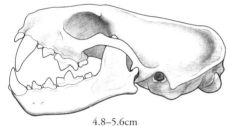

4.8–5.6cm

Marbled Polecat
Vormela peregusna

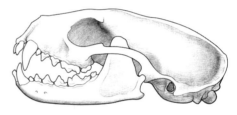

8.9–11.1cm

Yellow-throated Marten
Martes flavigula

Plate 80

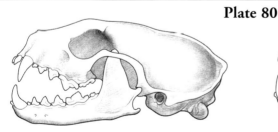

7.8–8.7cm

Stone Marten
Martes foina

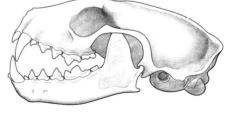

7.5–8.8cm

Pine Marten
Martes martes

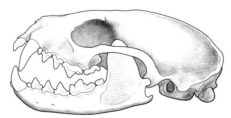

7.5–8.5cm

Japanese Marten
Martes melampus

MUSTELIDAE

Plate 81

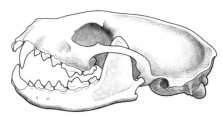

9.9–13.5cm

Fisher
Martes pennanti

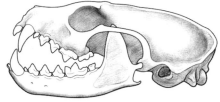

6.8–8.8cm

American Marten
Martes americana

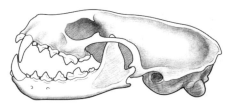

7.9–9.5cm

Sable
Martes zibellina

Plate 82

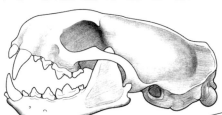

5.7–7cm

Black-footed Ferret
Mustela nigripes

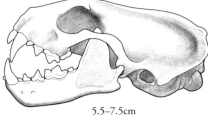

5.5–7.5cm

Steppe Polecat
Mustela eversmanii

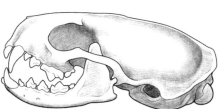

5.6–6.7cm

Siberian Weasel
Mustela sibirica

Plate 83

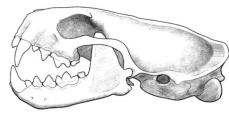

5.4–6.7cm

European Mink
Mustela lutreola

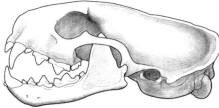

5.1–7cm

European Polecat
Mustela putorius

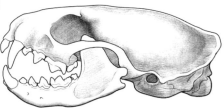

5.3–7.1cm

American Mink
Neovison vison

Plate 84

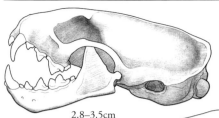

2.8–3.5cm

Least Weasel
Mustela nivalis

3.3–4.6cm

Stoat
Mustela erminea

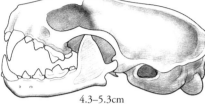

4.3–5.3cm

Altai Weasel
Mustela altaica

Plate 85

c. 3.8–4.5cm

Colombian Weasel
Mustela felipei

3.8–5.6cm

Long-tailed Weasel
Mustela frenata

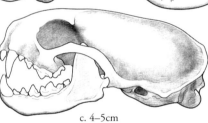

c. 4–5cm

Amazon Weasel
Mustela africana

Plate 86

4.2–5cm

Yellow-bellied Weasel
Mustela kathiah

5.4–6.6cm

Stripe-backed Weasel
Mustela strigidorsa

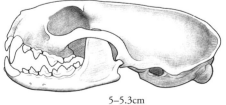

5–5.3cm

Malay Weasel
Mustela nudipes

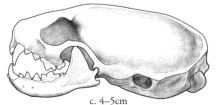

c. 4–5cm

Indonesian Mountain Weasel
Mustela lutreolina

FOOTPRINTS

Far more so than skulls (page 188), carnivores are likely to leave their footprints behind. Indeed, in many cases, it is much more possible to find carnivore tracks than the animal itself. This section includes tracks from a variety of carnivores. For many species, reliable tracks have never been recorded in the field, but all families except the Nandiniidae are represented here.

Carnivores are either digitigrade, in which they stand on the tips of the toes (as in felids, genets and canids) or plantigrade, in which they stand on the entire foot including the heel (as in bears and raccoons).

- Tracks of digitigrade species typically show only the toe pads, including claws when non-protractile, and the plantar pad, which is analogous to the fleshy pads at the bases of the fingers on a human palm. In most feliform carnivores the plantar pad has three lobes, compared to two lobes in caniform species (although there are exceptions in both cases).
- Tracks of plantigrade species show a metacarpal or metatarsal pad underlying the wrist or the heel respectively, which is often entirely or partially fused with the plantar pad, producing one large pad print with the toes, similar to the impression of human feet (compare with bear footprints in the pages that follow).

Most feliform carnivores and canids have five toes on the front foot and four toes on the hind foot. In digitigrade species such as felids, hyaenids, canids, some herpestids and some viverrids, the fifth toe on the front foot (the 'dewclaw') does not touch the ground and is absent from tracks. In some species, this digit has been lost entirely; the African Wild Dog is unique among canids for lacking this toe. Most caniform carnivores and most euplerids have five toes on both the front and hind feet.

Claw impressions typically appear for species lacking the ability to protract the claws – the majority of caniform species, hyaenids and many viverrids, herpestids and euplerids. The tracks of cats (except for the Cheetah), genets, oyans and ringtails (Procyonidae) generally do not show claw marks except in deep substrate such as mud or snow, or when the animal is running.

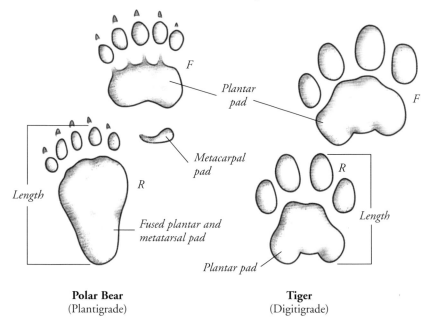

Polar Bear
(Plantigrade)

Tiger
(Digitigrade)

Footprints
Typical plantigrade (Polar Bear) and digitigrade (Tiger) carnivore footprints, showing both fore (F) and rear (R) impressions. The length measurements provided in the following section are taken from the front tips of the middle toe pads to the back of the plantar or metatarsal pad; generally measurements are provided only for the front print (annotated F); where they appear, rear print measurements are annotated R.

FELIDAE

Plate 1

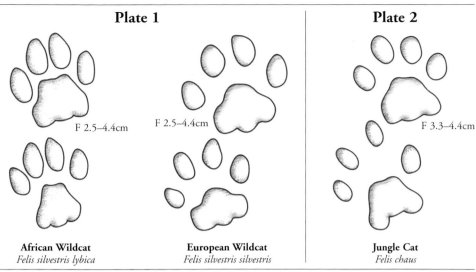

African Wildcat
Felis silvestris lybica

F 2.5–4.4cm

European Wildcat
Felis silvestris silvestris

F 2.5–4.4cm

Plate 2

Jungle Cat
Felis chaus

F 3.3–4.4cm

Plate 3

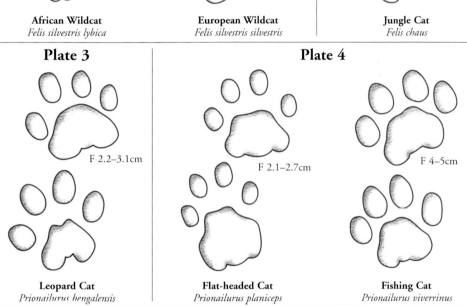

Leopard Cat
Prionailurus bengalensis

F 2.2–3.1cm

Plate 4

Flat-headed Cat
Prionailurus planiceps

F 2.1–2.7cm

Fishing Cat
Prionailurus viverrinus

F 4–5cm

Plate 5

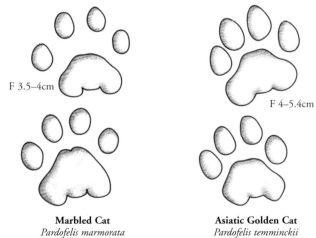

Marbled Cat
Pardofelis marmorata

F 3.5–4cm

Asiatic Golden Cat
Pardofelis temminckii

F 4–5.4cm

Plate 6

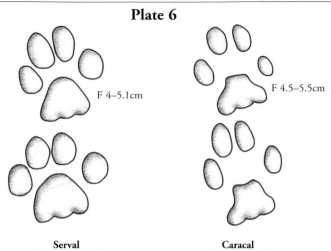

F 4–5.1cm

F 4.5–5.5cm

Serval
Leptailurus serval

Caracal
Caracal caracal

Plate 7

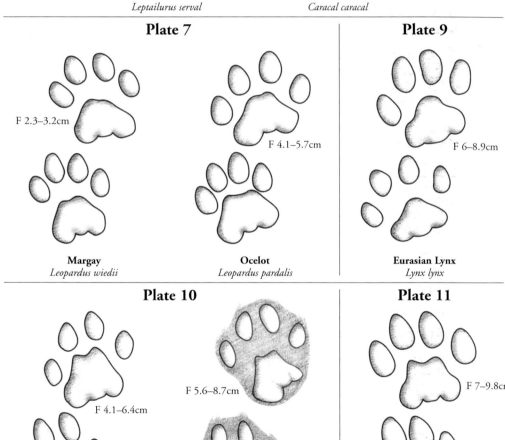

F 2.3–3.2cm

F 4.1–5.7cm

Margay
Leopardus wiedii

Ocelot
Leopardus pardalis

Plate 9

F 6–8.9cm

Eurasian Lynx
Lynx lynx

Plate 10

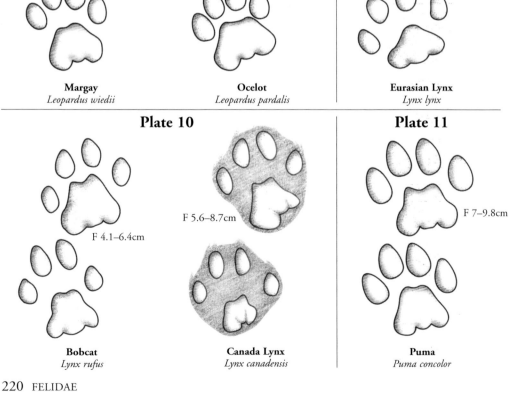

F 4.1–6.4cm

F 5.6–8.7cm

Bobcat
Lynx rufus

Canada Lynx
Lynx canadensis

Plate 11

F 7–9.8cm

Puma
Puma concolor

FELIDAE

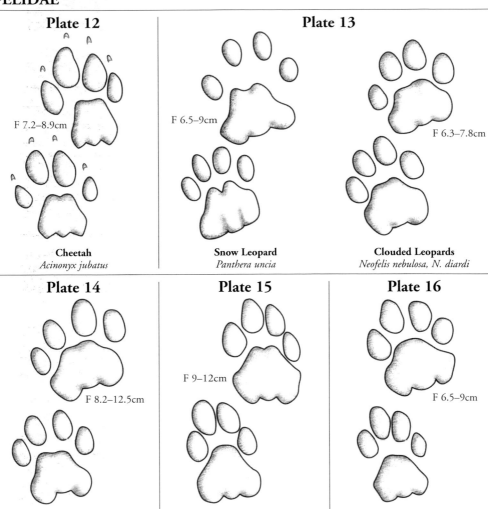

Plate 12

F 7.2–8.9cm

Cheetah
Acinonyx jubatus

Plate 13

F 6.5–9cm

Snow Leopard
Panthera uncia

F 6.3–7.8cm

Clouded Leopards
Neofelis nebulosa, N. diardi

Plate 14

F 8.2–12.5cm

Tiger
Panthera tigris

Plate 15

F 9–12cm

Lion
Panthera leo

Plate 16

F 6.5–9cm

Leopard
Panthera pardus

Plate 17

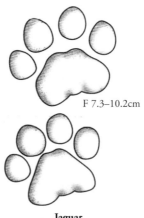

F 7.3–10.2cm

Jaguar
Panthera onca

HYAENIDAE

Plate 18

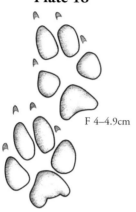

F 4–4.9cm

Aardwolf
Proteles crsistata

Plate 19

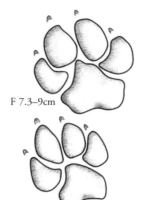

F 7.3–9cm

Brown Hyaena
Parahyaena brunnea

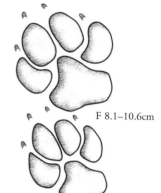

F 8.1–10.6cm

Spotted Hyaena
Crocuta crocuta

HERPESTIDAE

Plate 20

F 2.2–2.6cm

Small Indian Mongoose
Herpestes auropunctatus

Plate 21

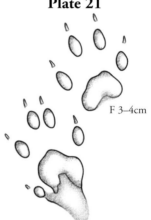

F 3–4cm

Crab-eating Mongoose
Herpestes urva

Plate 22

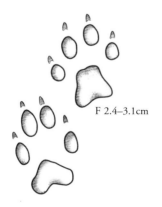

F 2.4–3.1cm

Cape Grey Mongoose
Herpestes pulverulentus

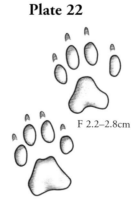

F 2.2–2.8cm

Common Slender Mongoose
Herpestes sanguineus

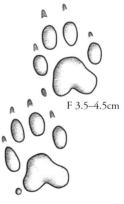

F 3.5–4.5cm

Egyptian Mongoose
Herpestes ichneumon

HERPESTIDAE

Plate 23

Marsh Mongoose
Atilax paludinosus

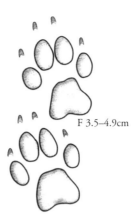

F 3.5–4.9cm

White-tailed Mongoose
Ichneumia albicauda

Plate 25

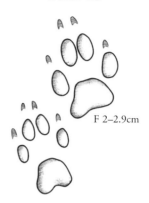

F 2–2.9cm

Yellow Mongoose
Cynictis penicillata

Plate 26

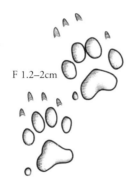

F 1.2–2cm

Common Dwarf Mongoose
Helogale parvula

Plate 27

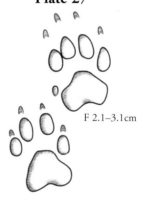

F 2.1–3.1cm

Banded Mongoose
Mungos mungo

EUPLERIDAE

Plate 29

F 4.5–6.2cm

Falanouc
Eupleres goudotii

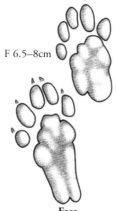

F 6.5–8cm

Fosa
Cryptoprocta ferox

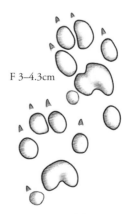

F 3–4.3cm

Fanaloka
Fossa fossana

EUPLERIDAE

Plate 30

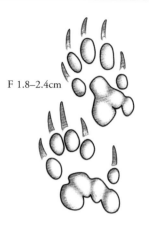

F 1.8–2.4cm

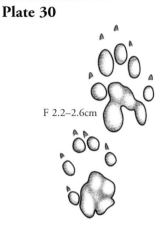

F 2.2–2.6cm

Narrow-striped Boky
Mungotictis decemlineata

Ring-tailed Vontsira
Galidia elegans

PRIONODONTIDAE & VIVERRIDAE

Plate 31

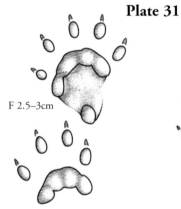

F 2.5–3cm

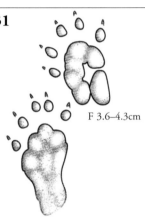

F 3.6–4.3cm

Banded Linsang *Prionodon linsang*
Spotted Linsang *Prionodon pardicolor*

Small-toothed Palm Civet
Arctogalidia trivirgata

Plate 33

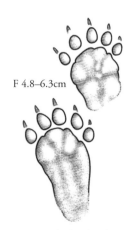

F 4.8–6.3cm

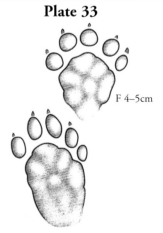

F 4–5cm

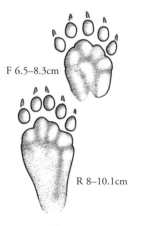

F 6.5–8.3cm

R 8–10.1cm

Masked Palm Civet
Paguma larvata

Common Palm Civet
Paradoxurus hermaphroditus

Binturong
Arctictis binturong

VIVERRIDAE

Plate 34

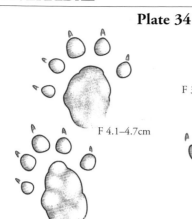

F 4.1–4.7cm

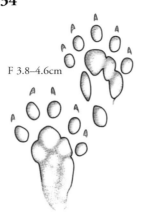

F 3.8–4.6cm

Owston's Civet *Chrotogale owstoni*
Banded Civet *Hemigalus derbyanus*

Otter Civet
Cynogale bennettii

Plate 35

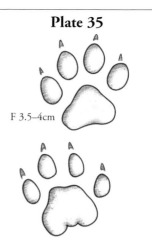

F 3.5–4cm

Large Indian Civet
Viverra zibetha

Plate 36

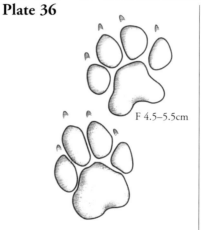

F 2.5–3.2cm

F 4.5–5.5cm

Small Indian Civet
Viverricula indica

African Civet
Civettictis civetta

Plate 39

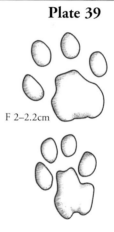

F 2–2.2cm

Cape Genet
Genetta tigrina

Plate 41

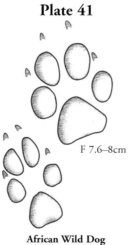

F 7.6–8cm

African Wild Dog
Lycaon pictus

Plate 43

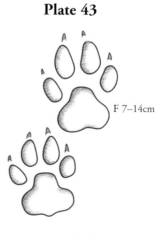

F 7–14cm

Grey Wolf
Canis lupus

CANIDAE

Plate 44

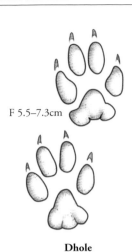

F 5.5–7.3cm

Dhole
Cuon alpinus

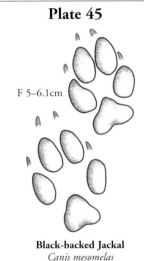

F 5–7.9cm

Coyote
Canis latrans

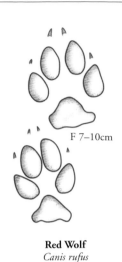

F 7–10cm

Red Wolf
Canis rufus

Plate 45

F 5–6.3cm

Golden Jackal
Canis aureus

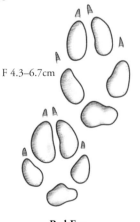

F 5–6.1cm

Black-backed Jackal
Canis mesomelas

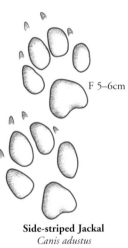

F 5–6cm

Side-striped Jackal
Canis adustus

Plate 46

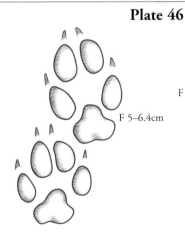

F 5–6.4cm

Arctic Fox
Alopex lagopus

F 4.3–6.7cm

Red Fox
Vulpes vulpes

Plate 47

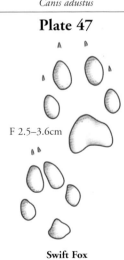

F 2.5–3.6cm

Swift Fox
Vulpes velox

CANIDAE

Plate 49

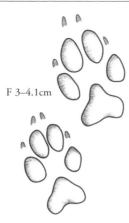

F 3–4.1cm

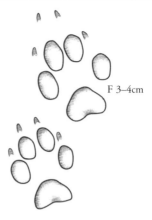

F 3–4cm

Cape Fox
Vulpes chama

Bat-eared Fox
Otocyon megalotis

Plate 50

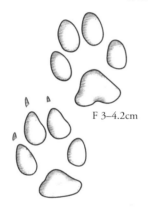

F 3–4.2cm

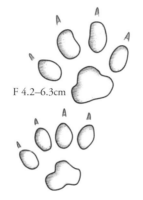

F 4.2–6.3cm

Plate 53

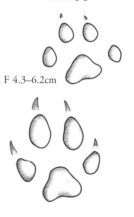

F 4.3–6.2cm

Gray Fox
Urocyon cinereoargenteus

Raccoon Dog
Nyctereutes procyonoides

Bush Dog
Speothos venaticus

URSIDAE (Measurements for bears are to tips of claws.)

Plate 54

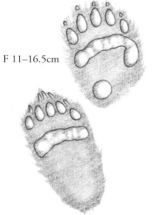

F 11–16.5cm

Plate 55

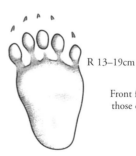

R 13–19cm

Front foot similar to
those of other bears

Asiatic Black Bear
Ursus thibetanus

Giant Panda
Ailuropoda melanoleuca

URSIDAE

Plate 56

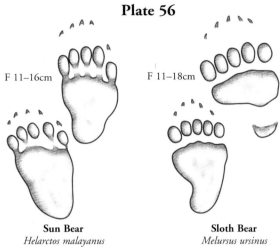

F 11–16cm

F 11–18cm

Sun Bear
Helarctos malayanus

Sloth Bear
Melursus ursinus

Plate 57

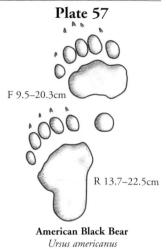

F 9.5–20.3cm

R 13.7–22.5cm

American Black Bear
Ursus americanus

Plate 58

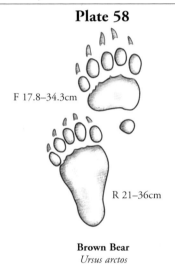

F 17.8–34.3cm

R 21–36cm

Brown Bear
Ursus arctos

Plate 59

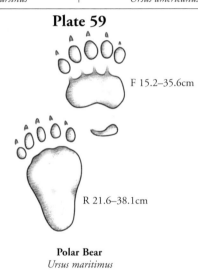

F 15.2–35.6cm

R 21.6–38.1cm

Polar Bear
Ursus maritimus

PROCYONIDAE

Plate 60

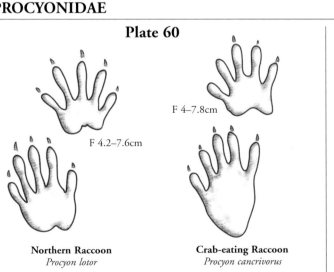

F 4–7.8cm

F 4.2–7.6cm

Northern Raccoon
Procyon lotor

Crab-eating Raccoon
Procyon cancrivorus

Plate 61

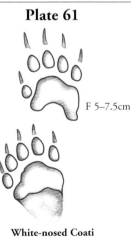

F 5–7.5cm

White-nosed Coati
Nasua narica

PROCYONIDAE & AILURIDAE

Plate 62

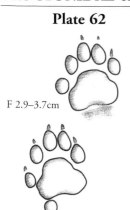

F 2.9–3.7cm

Ringtail
Bassariscus astutus

Plate 63

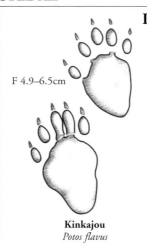

F 4.9–6.5cm

Kinkajou
Potos flavus

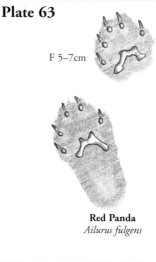

F 5–7cm

Red Panda
Ailurus fulgens

MEPHITIDAE

Plate 65

F 4.3–6.6cm

American Hog-nosed Skunk
Conepatus leuconotus

F 4–5cm

Striped Skunk
Mephitis mephitis

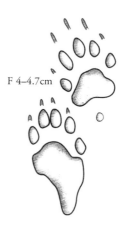

F 4–4.7cm

Hooded Skunk
Mephitis macroura

Plate 66

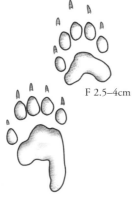

F 2.5–4cm

Eastern Spotted Skunk
Spilogale putorius

F 2.5–4cm

Western Spotted Skunk
Spilogale gracilis

Plate 67

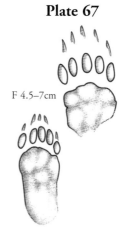

F 4.5–7cm

Sunda Stink Badger
Mydaus javanensis

Plate 68

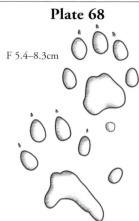

F 5.4–8.3cm

North American Otter
Lontra canadensis

Plate 69

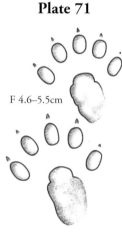

F 4–7.3cm

Neotropical Otter
Lontra longicaudis

Plate 70

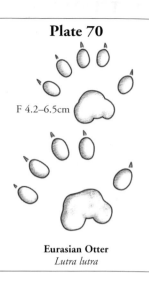

F 4.2–6.5cm

Eurasian Otter
Lutra lutra

Plate 71

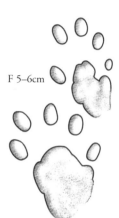

F 5–6cm

Asian Small-clawed Otter
Aonyx cinereus

F 4.6–5.5cm

Hairy-nosed Otter
Lutra sumatrana

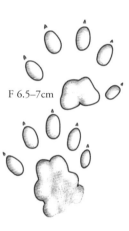

F 6.5–7cm

Smooth-coated Otter
Lutrogale perspicillata

Plate 72

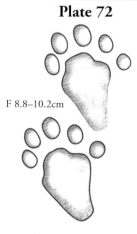

F 8.8–10.2cm

Cape Clawless Otter
Aonyx capensis

Plate 73

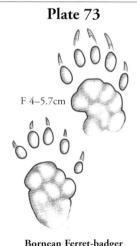

F 4–5.7cm

Bornean Ferret-badger
Melogale everetti

Plate 74

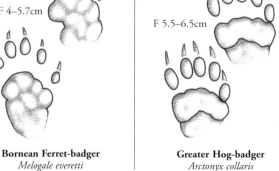

F 5.5–6.5cm

Greater Hog-badger
Arctonyx collaris

Plate 75

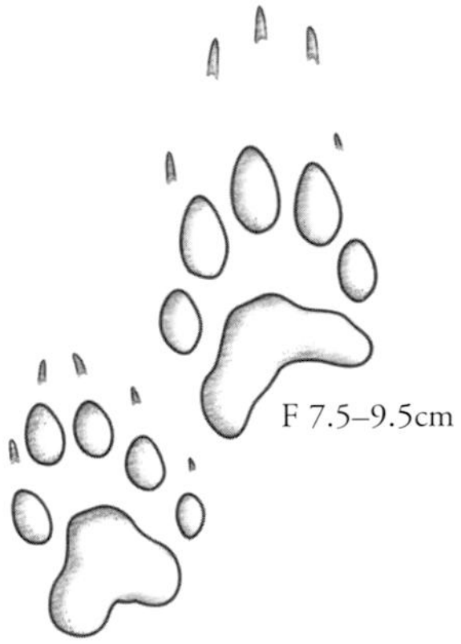

F 8–11.5cm

F 7.5–9.5cm

American Badger
Taxidea taxus

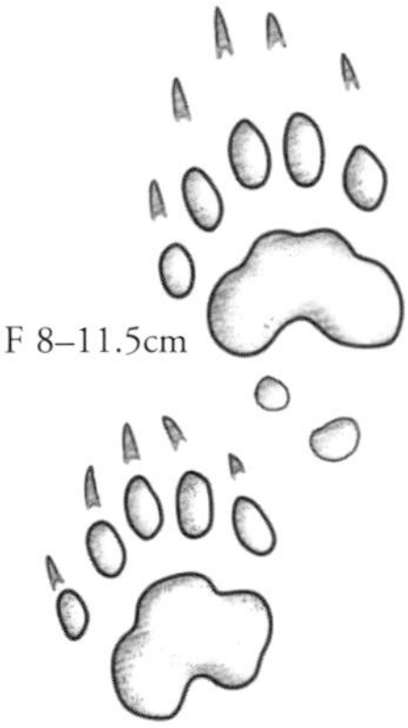

Eurasian Badger
Meles meles

Plate 76

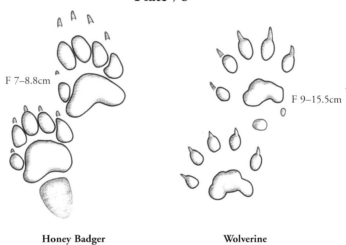

F 7–8.8cm

F 9–15.5cm

Honey Badger
Mellivora capensis

Wolverine
Gulo gulo

Plate 77

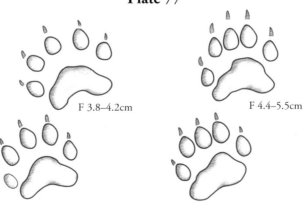

F 3.8–4.2cm

F 4.4–5.5cm

Greater Grison
Galictis vittata

Tayra
Eira barbara

MUSTELIDAE

Plate 79

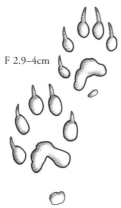

F 2.9–4cm

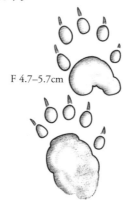

F 4.7–5.7cm

Marbled Polecat
Vormela peregusna

Yellow-throated Marten
Martes flavigula

Plate 80

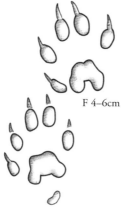

F 4–6cm

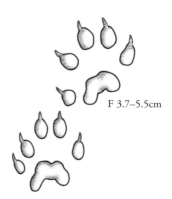

F 3.7–5.5cm

Stone Marten
Martes foina

Pine Marten
Martes martes

Plate 81

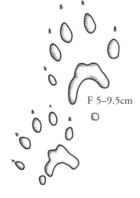

F 5–9.5cm

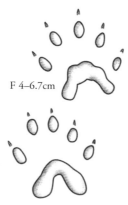

F 4–6.7cm

Fisher
Martes pennanti

American Marten
Martes americana

Plate 82

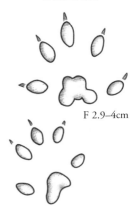

F 2.9–4cm

Black-footed Ferret
Mustela nigripes

MUSTELIDAE

Plate 83

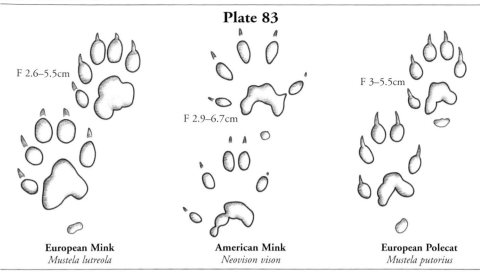

F 2.6–5.5cm

F 2.9–6.7cm

F 3–5.5cm

European Mink
Mustela lutreola

American Mink
Neovison vison

European Polecat
Mustela putorius

Plate 84

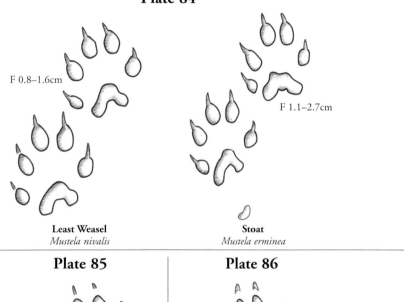

F 0.8–1.6cm

F 1.1–2.7cm

Least Weasel
Mustela nivalis

Stoat
Mustela erminea

Plate 85

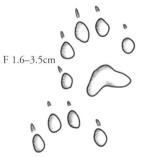

F 1.6–3.5cm

Long-tailed Weasel
Mustela frenata

Plate 86

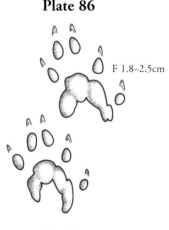

F 1.8–2.5cm

Malay Weasel
Mustela nudipes

albinism Complete or partial absence of pigment in the skin, feathers, hair and eyes due to a reduced ability to produce the pigment melanin, usually inherited as a recessive genetic trait. Albino individuals have white hair with pink eyes and skin.

altiplano High plateau in west-central South America where the Andes are at their widest, dominated by cold, dry desert and grasslands.

anthropogenic Any effect, process or material derived from human activities, e.g. anthropogenic threats are those created by people.

aposematism Referring to adaptations to deter predation, typically in which striking warning colouration signals the unprofitability of a prey item to potential predators, e.g. the black-and-white markings of skunks.

aquatic Living in water for most or all of its life.

arachnid Members of the invertebrate arthropod class Arachnida, chacterized by having eight legs (in some species the front pair converts to a sensory function). Includes spiders, scorpions, mites, ticks and solifuges.

arboreal Living predominantly or entirely in trees.

arthropod Invertebrate animal with an external skeleton (exoskeleton), a segmented body and jointed appendages. Includes insects, spiders, scorpions and crustaceans.

caching (of food) Storing food for future consumption, usually for the lean season, e.g. the northern winter.

camera trap Automated camera used to capture photographs of wild animals, typically for inventory and monitoring efforts.

caniform Referring to the carnivore suborder Caniformia, comprising the 'dog-like' families Canidae, Ursidae, Otariidae, Phocidae, Odobenidae, Procyonidae, Ailuridae, Mephitidae and Mustelidae (and the extinct family Amphicyonidae). Caniform species tend to have non-protractile claws and to be plantigrade (with the exception of the Canidae).

canine (teeth) (also called cuspids, fangs or, in the case of those of the upper jaw, eye teeth) Relatively long, pointed teeth located at the front of the jaw, and often the largest teeth in a mammal's mouth. The canines are used primarily for killing prey, holding or processing food, and occasionally as weapons in fights between conspecifics or in defence against attack.

carnassial In the Carnivora, the modified last (fourth) upper premolar and first lower molar teeth, which are used in concert for shearing tissue with a scissor-like action (the 'carnassial shear').

carnivore Organism that eats mainly animal tissue, whether through predation or scavenging. In scientific terminology, a carnivore is a mammal within the order Carnivora.

cathemeral Active at any time during the day or night without any distinct pattern.

caudal Pertaining to the tail.

cerrado Tropical dry savannah of Brazil, mainly located in the states of Goiás and Minas Gerais.

chaco Semi-arid habitat type dominated by dry savannah and thorn forest found from western Paraguay to south-eastern Bolivia and north-western Argentina.

clade Taxonomic group of organisms classified together on the basis of shared similar features traced to a common ancestor. Not a formal taxonomic classification level (as, e.g., species), but indicates closer relatedness of any two or more taxa compared with other taxa.

conspecific Of the same species.

crepuscular Active primarily during twilight, i.e. at dawn and dusk.

crustacean Large group of arthropod invertebrates usually treated as a subphylum, Crustacea, which includes lobsters, crayfish, crabs, barnacles, shrimp and krill.

cyclical abundance 'Boom-bust' cycles in which some species reproduce rapidly during periods of food abundance, with very rapid population growth that declines equally as rapidly during lean periods, resulting in a population crash. Natural feature of many species' population cycles, most famously the Snowshoe Hare.

deciduous In reference to plants, losing its foliage seasonally.

delayed implantation (embryonic diapause) In which development of the embryo is postponed shortly after conception by delaying implantation in the uterus. The embryo is maintained in a state of dormancy as long as it remains unattached to the uterine lining, extending the normal gestation period for up to a year. Thought to optimize reproductive success by timing mating and birth to avoid severe lean periods such as northern winters and prolonged drought.

dewclaw Fifth innermost digit of the foot of many mammals (and some other vertebrates), equivalent in position to the human thumb. In digitigrade species, it does not make contact with the ground when the animal is standing. Vestigial with no function in many species.

digitigrade In terrestrial mammalian locomotion, an animal that stands or walks on the tips of its digits, technically on its distal and intermediate phalanges ('fingerbones'), e.g. in felids and canids.

disperser Cohort of an animal population that leaves its natal range to seek out unoccupied habitat in which to settle and establish a home range. In carnivores, dispersers tend to be young adults that have been evicted from their natal range or social group.

distal Usually of appendages, meaning the part of an organ furthest from the point of attachment to the body, e.g the distal vertebra of the tail is the tip.

disturbed forest Forest that has significant areas of disturbance by people, including clearing, logging and anthropogenic fires.

diurnal Active primarily during the day.

dorsal Upper side of animals that run, fly or swim in a horizontal position.

erythristism Over-production of red (or orange) pigment, giving animals red-coloured fur or skin.

evergreen In regard to plants, having foliage in all seasons.

feliform Referring to the carnivore suborder Feliformia, comprising the 'cat-like' families Felidae, Hyaenidae, Herpestidae, Eupleridae, Prionodontidae, Viverridae and Nandiniidae (and the extinct family Nimravidae). Protractile claws and digitigrade locomotion are more widespread among feliform species than in caniform species.

feral A feral organism is one that has escaped from domestication and returned, partly or wholly, to a wild state.

form In zoology, term used to describe discrete variation in animals.

frugivore Organism in which fruit is the preferred food, e.g. the Kinkajou.

fynbos ('fine bush' in Afrikaans) Unique, highly diverse heathland vegetation occurring in a small belt of the Western Cape of South Africa.

genus (plural: genera) Low-level taxonomic rank (a taxon) used in the classification of living and fossil organisms.

haul out Behaviour associated mainly with pinnipeds of temporarily leaving the water between periods of foraging for rest sites on land or ice.

herbivore Organism that eats plants.

herptile Collective term for reptiles (class Reptilia) and amphibians (class Amphibia).

hibernation State of inactivity and metabolic depression in animals, characterized by lowered body temperature, reduced breathing rate and reduced metabolic rate. Enables animals to survive lean-season (usually winter) food shortages by slowly metabolizing fat reserves.

home range Area occupied by an individual, pair or group of animals required to satisfy the basic requirements for surviving and reproducing. Home range and territory are essentially equivalent, but the term 'home range' carries no implication of active defence against intruders.

hypercarnivore Usually defined as a species in which more than 70 per cent of the diet is tissue from animals, non-animal foods such as fungi, fruits and other plant material making up the remainder.

hyperphagia Meaning 'excessive eating'. In biology, occurs during the period prior to hibernation in which animals prepare for winter by constant foraging to lay down fat reserves.

incisor (from Latin *incidere*, 'to cut') Frontmost type of tooth in heterodont mammals (those with different types of teeth). Used primarily for processing food, grooming and defensive biting.

invertebrate Animal without a backbone (vertebral column). Ninety-five per cent of all animal species are invertebrates, ranging from simple organisms such as sea sponges and flatworms to complex animals such as arthropods and molluscs.

keratin Key structural material making up the outer layer of vertebrate skin, as well as hair, horn, feathers, hoofs, nails, claws and bills.

kleptoparasitism Form of feeding in which one animal takes prey or other food from another that has caught, collected or otherwise prepared the food, including food that has been stored.

kopje Isolated rock hill, knob, ridge or small mountain that rises abruptly from a level surrounding plain.

lagomorph Members of the order Lagomorpha (meaning 'hare form'), made up of two living families, the Leporidae (hares and rabbits), and the Ochotonidae (pikas).

leucism Genetic condition characterized by reduced pigmentation in animals and humans. Unlike albinism (a defect in pigment production), leucism renders the skin unable to support pigment cells. Leucistic animals are white or pale coloured, and have pigmented eyes (unlike albinos), e.g. so-called white lions.

melanism Elevated dark pigmentation of skin, feathers, hair and eyes due to an excess of the pigment melanin, usually inherited as a recessive genetic trait. Melanistic individuals are black or near-black, and occur in many carnivores, especially felids, viverrids, herpestids and canids.

mesic Characterized by having a moderate water supply, e.g. mesic habitats are moderately well watered.

miombo Mesic to semi-arid woodland characterized by *Brachystegia* trees found across south-central Africa.

molar Rearmost and often most complex tooth type in most mammals. In carnivores, typically employed to grind or crush food items ranging from nuts to bone.

mollusc Phylum of invertebrate animals that includes snails, slugs, squid, octopus, cuttlefish, clams, oysters, scallops and mussels.

morph In biology, occurs when two or more clearly different phenotypes exist in the same population of a species.

myrmecophagy Feeding behaviour defined by consumption of termites and/or ants.

natal range Same as 'birth' range, the area in which an animal is born and raised until it disperses.

nocturnal Active primarily during the night.

nocturno-crepuscular Active primarily during dusk, night and dawn.

obligate carnivore Animal depending solely on the nutrients found in animal flesh for survival. Lacks the physiology required for efficient digestion of plant material; any vegetation consumed is typically an emetic rather than for nutritional gain.

oestrus Stage during a female animal's reproductive cycle in which she is receptive to the male for mating and ovulates (sometimes stimulated by copulation).

paramo Neotropical, largely Andean ecosystem of glacier-formed valleys and wet grasslands interspersed with lakes, peat bogs, shrublands and forest patches.

plantigrade In terrestrial mammalian locomotion, an animal that stands or walks with the metacarpals or metatarsals flat on the ground, so that the entire foot makes contact with the ground, e.g. in humans and ursids.

primary forest Forest of native species with little or no evidence of human activities and where ecological processes are not significantly disturbed.

protractile Of claws, the ability to deliberately extend the claws from a relaxed withdrawn position, as in felids and genets. This is usually termed 'retractile', but in fact retraction is the relaxed state and protractile more accurately describes the ability.

proximal Usually of appendages, meaning the part of an organ closest to the point of attachment to the body, e.g the proximal vertebra of the tail is the base.

riparian Habitat occurring at the interface between land and a freshwater waterway such as a river or stream.

scavenging Behaviour in which a predator consumes carcasses or carrion not killed by itself or members of its social group.

secondary forest Forest or woodland that has regrown or is in the process of regrowing after a major natural or anthropogenic disturbance such as fire or logging. It is younger than primary forest and lacks its old-growth structure, typified by a denser and more 'weedy' subcanopy.

scent-marking Behaviour used by animals to demarcate their territory, usually carried out by depositing strong-smelling secretions such as urine at prominent locations within the range.

species Often defined as a group of organisms capable of interbreeding and producing fertile offspring. While in many cases this definition is adequate, more precise or differing measures are often used, such as similarity of DNA, morphology or ecological niche.

subspecies (commonly abbreviated to subsp. or ssp.) Either a taxonomic rank subordinate to species, or a taxonomic unit in that rank (plural: subspecies). A subspecies cannot be recognized in isolation: a species will either be recognized as having no subspecies at all or two or more, never just one.

sympatric Of two or more species, occurring in the same area or range.

termitaria Termite mound, also often called 'anthill' (though termites are not ants).

terpenes Defensive organic compounds produced by a wide variety of plants and some insects, such as termites and swallowtail butterflies, to deter predation.

terrestrial Living predominantly or entirely on land.

territory Area occupied by an individual, pair or group of animals, which is demarcated and defended against other members of the same species. A territory can also be viewed as an actively defended home range.

ungulate Several groups of mammals traditionally grouped together due to the fact that they use the tips of their toes, usually hoofed, to support their whole body weight. Includes all hoofed mammals such as deer, antelopes, horses, cattle, giraffes, camels, llamas, tapirs, rhinoceroses, hippopotami and elephants.

Valdivian forest Temperate rainforest unique to the west coast of southern South America, lying mostly in Chile and extending into a small part of Argentina.

vertebrate Animals of the subphylum Vertebrata, having a backbone (vertebral column). Includes fish, amphibians, reptiles, birds and mammals.

vocalization Production of sound by the passage of air across the vocal chords, and used by animals to communicate to conspecifics and other species.

BIBLIOGRAPHY

This book draws on thousands of scientific papers, reports and theses; only major sources are listed here.

Bailey, T.N. 2005. *The African Leopard: Ecology and Behaviour of a Solitary Felid.* The Blackburn Press.

Bekoff, M. 2001. *Coyote: Biology, Behaviour and Management.* The Blackburn Press.

Brown, D.E. & Lopez-Gonzales, C.A. 2001. *Borderland Jaguars: Tigres de la Frontera.* University of Utah Press.

Caro, T.M. 1994. *Cheetahs of the Serengeti Plains: Group Living in an Asocial Species.* University of Chicago Press.

Clark, T.P., Curlee, A.P., Minta, S.C. & Kareiva, P.M. 1999. *Carnivores in Ecosystems: The Yellowstone Experience.* Yale University Press.

Craighead, J.J., Sumner, J.S. & Mitchell, J.A. 1995. *The Grizzly Bears of Yellowstone: Their Ecology in the Yellowstone Ecosystem, 1959–1992.* Island Press.

Creel, S. & Creel, N. 2002. *The African Wild Dog: Behaviour, Ecology and Conservation.* Princeton University Press.

Divyabhanusinh, 2002. *The End of a Trail: the Cheetah in India.* Oxford University Press.

Ewer, R.F. 1986. *The Carnivores.* Cornell University Press

Feldhamer, G.A., Thompson, B.C. & Chapman, J.A. *Wild Mammals of North America: Biology, Management and Conservation.* Johns Hopkins University Press.

Gehrt, S.D., Riley, S.P.D. & Cypher, B.L. 2010. *Urban Carnivores Ecology, Conflict, and Conservation.* Johns Hopkins University Press.

Gittleman, J.L. 1989–1996. *Carnivore Behavior, Ecology and Evolution.* Vol. 1 (1989), Vol. 2 (1996). Cornell University Press.

Gittleman, J.L., Fun, S.M., MacDonald, D.W. & Wayne, R.K. 2001. *Carnivore Conservation.* Cambridge University Press.

Glatston, A.R. 2010. *Red Panda Biology and Conservation of the First Panda.* William Andrew Publishing.

Glatston, A.R. 1994. *The Red Panda, Olingos, Coatis, Raccoons and their Relatives: An Action Plan for the Conservation of Procyonids and Ailurids.* IUCN.

Goswami, A. & Friscia, A. 2010. *Carnivoran Evolution: New Views on Phylogeny, Form and Function.* Cambridge University Press.

Griffiths, H.I. 2000. *Mustelids in a Modern World: Management and Conservation Aspects of Small Carnivore and Human Interactions.* Backhuys.

Hamilton, G. 2008. *Arctic Fox Life at the Top of the World.* A & C Black.

Hatler, D.F. 2008. *Carnivores of British Columbia.* Royal British Columbia Museum.

Hayward, M.W. & Somers, M.J. *The Reintroduction of Top-order Predators.* Wiley-Blackwell Publishing.

Heptner, V.G. & Sludskii, A.A. 1992. *Mammals of the Soviet Union, Carnivora, Vol. II, Part 1a: Sirenia and Carnivora Sea Cows, Wolves and Bears, Vol. II, Part 1b: Weasels; Additional Species, Vol. II, Part 2: Hyaenas and Cats.* E.J. Brill.

Hoogesteijn, R. & Mondolfi, E. 1992. *The Jaguar.* Armitano Editores C.A.

Hornocker, M. & Negri, S. 2009. *Cougar Ecology and Conservation.* University Of Chicago Press.

Jackson, P. & Nowell, K. 1996. *Wild Cats Status Survey and Conservation Action Plan.* IUCN.

Kingdon, J. 1977. *East African Mammals: An Atlas of Evolution in Africa. Volume IIIA Carnivores.* University of Chicago Press.

Kingdon, J.S. & Hoffmann, M. In press. *The Mammals of Africa 5: Carnivora, Pholidota, Perissodactyla.*

Academic Press.

King, C. & Powell, R.A. 2007. *The Natural History of Weasels and Stoats: Ecology, Behaviour and Management.* Oxford University Press.

Kruuk, H. 2002. *Hunter and Hunted: Relationships between Carnivores and People.* Cambridge University Press.

Kruuk, H. 2006. *Otters: Ecology, Behaviour and Conservation.* Oxford University Press.

Krystufek, B., Flajsman, B. & Griffiths, H.I. 2003. *Living with Bears: a Large European Carnivore in a Shrinking World.* MK Trgovina d.d.

Leydet, F. 1988. *The Coyote: Defiant Songdog of the West.* University of Oklahoma Press.

Lindburg, D. & Baragona, K. 2004. *Giant Pandas: Biology and Conservation.* University of California Press.

Logan, K.A. & Sweanor, L.L. 2001. *Desert Puma: Evolutionary Ecology and Conservation of an Enduring Carnivore.* Island Press.

Macdonald, D.W. 1992. *The Velvet Claw.* BBC Books.

Macdonald, D.W. & Loveridge, A.J. 2010. *The Biology and Conservation of Wild Felids.* Oxford University Press.

Macdonald, D.W. & Sillero-Zubiri, C. 2004. *The Biology and Conservation of Wild Canids.* Oxford University Press.

Mech, L.D. 1981. *The Wolf: The Ecology and Behavior of an Endangered Species.* University of Minnesota Press.

Mech, L.D. & Boitani, L. 2003. *Wolves: Behavior, Ecology, and Conservation.* Chicago University Press.

Mills, M.G.L. 1990. *Kalahari Hyaenas: Comparative Behavioural Ecology of Two Species.* The Blackburn Press.

Mills, M.G.L. & Hofer, H. 1998. *Hyaenas: Status Survey and Conservation Action Plan.* IUCN.

Nowak, R.M. 2005. *Walker's Carnivores of the World.* Johns Hopkins University Press.

Nowell, K. & Jackson P. 1996. *Wild Cats: Status Survey and Conservation Action Plan.* IUCN.

Ray, J.C., Hunter, L. & Zigouris, J. 2005. *Setting Conservation and Research Priorities for Larger African Carnivores.* Wildlife Conservation Society.

Ray, J.C., Redford, K.H., Steneck, R.S. & Berger, J. 2005. *Large Carnivores and the Conservation of Biodiversity.* Island Press.

Rosevear, D.R. 1974. *Carnivores of West Africa.* The British Museum.

Ruggiero, L.F., Aubry, K.B., Buskirk, S.W., Koehler, G.M., Krebs, C., McKelvey, K.S. & Squires, J.R. 2006. *Ecology and Conservation of Lynx in the United States.* University Press of Colorado.

Schaller, G.B. 1972. *The Serengeti Lion: A Study of Predator-Prey Relations.* University of Chicago Press.

Schaller, G.B., Jinchu, H., Wenshi, P. & Jing, Z. 1985. *Giant Pandas of Wolong.* University of Chicago Press.

Schreiber, A., Wirth, R., Riffel, M. & van Rompaey, H. 1989. *Weasels, Civets, Mongooses and their Relatives: An Action Plan for the Conservation of Mustelids and Viverrids.* IUCN.

Sillero-Zubiri, C., Hoffmann, M. & Macdonald, D.W. 2004. *Canids: Foxes, Wolves, Jackals and Dogs Status Survey and Conservation Action Plan.* IUCN.

Seidensticker, J., Christie, S. & Jackson, P. *Riding the Tiger: Tiger Conservation in Human-dominated Landscapes.* Cambridge University Press.

Seidensticker, J. & Lumpkin, S. 2004. *Smithsonian Answer Book: Cats.* Smithsonian Books.

Servheen, C., Herrero, S. & Peyton, B. 1999. *Bears: Status Survey and Conservation Action Plan.* IUCN.

Skinner, J.D. & Chimimba, C.T. 2006. *The Mammals of the Southern African Subregion.* Cambridge University Press.

Sovada, M.A., Ludwig, N. & Carbyn, L.N. 2003. *The Swift Fox: Ecology and Conservation of Swift Foxes in a Changing World.* Canadian Plains Research Center.

Sunquist, M. & Sunquist, F. 2002. *Wild Cats of the World.* University of Chicago Press.

Tilson, R. & Nyhus, P. 2010. *Tigers of the World: The Science, Politics and Conservation of* Panthera tigris. Academic Press.

Turner, A. & Anton, M. 1997. *The Big Cats and Their Fossil Relatives: an Illustrated Guide to Their Evolution and Natural History.* Columbia University Press.

Wang, X. & Tedford, R.H. 2008. *Dogs: Their Fossil Relatives and Evolutionary History.* Columbia University Press.

Wilson, D.E. & Mittermeier, R.A. 2009. *Handbook of Mammals of the World, Vol. 1: Carnivores.* Lynx Edicions.

Zeveloff, S.I. 2002. *Raccoons: A Natural History.* Smithsonian Institution Press.

Additional Resources

IUCN/SSC Specialist Groups
All carnivore families are covered by Specialist Groups that produce excellent Action Plans, newsletters and journals. www.iucn.org/about/work/programmes/species/about_ssc/specialist_groups/

Carnivore Ecology & Conservation
An excellent compendium of carnivore knowledge compiled by carnivore biologist Guillaume Chapron. www.carnivoreconservation.org

Mammalian Species
A long-running series that summarizes the biology of a single species in a dedicated account; many carnivores species are covered. www.science.smith.edu/departments/Biology/VHAYSSEN/msi/

INDEX

ACKNOWLEDGEMENTS

Covering such a broad scope, the world's 245 species of terrestrial carnivore, this book relied heavily on a large number of people for comments, data, publications and reference material. Many people provided images or assisted in locating them for the preparation of the plates, in some cases to illustrate species or forms that had never been accurately depicted. For this, particular thanks go to: Francesco Angelici, Sixto Angulo, Rosario Arispe, Jane Ashley-Edmonds, Christos Astaras, Guy Balme, Eyal Bartov, Jerry Belant, Abelardo Rodriguez Bolaños, Jeffrey P. Bonner, Adam Britt, Milo Burcham, Duncan Butchart, Marcelo Carrera, Erika Cuellar, Rogerio Cunha de Paula, Daniela DeLuca, James Eaton, Mohammad Farhadinia, Charles Foley, Nick Garbutt, Oscar E. Murillo García, Arash Ghoddousi, Varad B. Giri, Helle Goldman and Jon Winther-Hansen, Lon Grassman, Andy Hearn and Jo Ross, Philipp Henschel, Bob and Kris Inman, Andy Jennings, Jaime Jimenez, Arlyne Johnson, Calvin Jones, Roland Kays, Marcella Kelly, Barney Long, Leo Maffei, Sean Matthews, Divya Mudappa, Andy Noss, Andres Novaro, Stephane Ostrowski, Antonio Rossano Mendes Pontes, Ingrid Porton, Shankar Raman, Justina Ray, Scott Roberton, SOMASPA (Panama), Chris Roche, Brian Rode, Carlos A. Saavedra Rodríguez, Fabio Rohe, Steve Ross, Ricardo Sampaio, Jim Sanderson, Kevin Schafer, Alex Sliwa, Rob Steinmetz, Simon Stobbs, Chris and Tilde Stuart, Narong Suannarong, Tim Tetzlaff, Sara Tromp and Joe Walston.

I am grateful to Darrin Lunde and Eileen Westwig (American Museum of Natural History, New York) and to Bill Stanley (Field Museum, Chicago) for making available their carnivore skull specimens, and to Sarah Arnoff, Joanna Cagan and Graeme Patterson for photographing them. Sarah also compiled additional reference material for carnivore skulls and tracks. A special thanks to Sally McClarty for executing the wonderful illustrations for these sections. I am indebted to the following colleagues who reviewed sections of the text and commented on plates: Arturo Caso, Andrew Derocher, Philippe Gaubert, John Goodrich, Andy Hearn and Jo Ross, Kris Helgen, Marna Herbst, Jan Kamler, David MacDonald, Tom McCarthy, Dale Miquelle, Divya Mudappa, Tadeu de Oliveira, Alan Rabinowitz, Alan Root and Howard Quigley. For addressing specific queries and providing publications, unpublished data and other comments, I thank Nick Brickle, Vincent Burke, Sasha Carvaja, Armando X. Castellanos, Jerry Dragoo, Colin Groves, Frank Hawkins, Erin McCloskey, Kate McFadden, Javier Pereira, Miguel Pinto, Pamela Racobs and Steve Ross.

I am especially grateful to Philippe Gaubert, who provided photographs, publications, unpublished data and invaluable comments on little-known genets, and Geraldine Veron who did likewise for viverrids and herpestids. Kris Helgen very kindly made available his team's revised classification of hog-badgers before it was published and, with Roland Kays, he assisted similarly with his ongoing effort to revise olingo taxonomy. Will Duckworth commented on the nomenclature and taxonomy of all species. Mike Hoffmann provided a constant stream of new information, publications and contact details from his extensive network.

At Panthera, I am grateful to David Katz for compiling measural data and the glossary, and especially to Erin Archuleta, who double-checked many details during the book's preparation.

Finally and most importantly, this book would never have been published without the support and encouragement of Thomas Kaplan and Daphne Recanati Kaplan. It is a small part of their extraordinary commitment to the conservation of the world's carnivores.